Universitext

Alexander Isaev

Introduction to Mathematical Methods in Bioinformatics

With 76 Figures and 3 Tables

 Springer

Alexander Isaev
Australian National University
Department of Mathematics
Canberra, ACT 0200
Australia
e-mail: alexander.isaev@maths.anu.edu.au

Corrected Second Printing 2006

Mathematics Subject Classification (2000): 91-01 (Primary)
91D20 (Secondary)

Library of Congress Control Number: 2006930998

ISBN: 3-540-21973-0
ISBN: 9783540219736

Springer is a part of Springer Science+Business Media
springer.com
Springer-Verlag Berlin Heidelberg 2006

Cover design: Erich Kirchner, Heidelberg
Typesetting by the author and SPi using a Springer LaTeX macro package

Printed on acid-free paper: SPIN:11809142   41/2141/SPi - 5 4 3 2 1 0

To Esya and Masha

Preface

Broadly speaking, Bioinformatics can be defined as a collection of mathematical, statistical and computational methods for analyzing biological sequences, that is, DNA, RNA and amino acid (protein) sequences. Numerous projects for sequencing the DNA of particular organisms constantly supply new amounts of data on an astronomical scale, and it would not be realistic to expect that biologists will ever be able to make sense of it without resorting to help from more quantitative disciplines.

Many studies in molecular biology require performing specific computational procedures on given sequence data, for example, simply organizing the data conveniently, and therefore analysis of biological sequences is often viewed as part of computational science. As a result, bioinformatics is frequently confused with *computational sequence analysis*, which is somewhat narrower. However, understanding biological sequences now increasingly requires profound ideas beyond computational science, specifically, mathematical and statistical ideas. For example, the protein folding problem incorporates serious differential geometry and topology; understanding the evolution of sequences is dependent upon developing better probabilistic evolutionary models; analysis of microarray data requires new statistical approaches. Generally, when one goes beyond algorithms and starts either looking for *first principles* behind certain biological processes (which is an extremely difficult task) or paying more attention to *modeling* (which is a more standard approach), one crosses the boundary between computational science and mathematics/statistics. These days many mathematicians are becoming interested in bioinformatics and beginning to contribute to research in biology. This is also my personal story: I am a pure mathematician specializing in several complex variables, but now I work in bioinformatics as well.

Despite this mathematical trend, bioinformatics is still largely taught by computational groups and computer science departments, and partly by engineering and biology (in particular, genetics) departments. Naturally, in courses developed by these departments emphasis is placed on algorithms and their implementation in software. Although it is useful to know how a particular piece of software works, this software-oriented education does not always reveal the mathematical principles on which the algorithms are based. Such incompleteness may lead to certain problems for the graduates. Suppose, for example, that a commonly used model is implemented in software and the students are taught how to use it. Of course, the model makes some simplifying *assumptions* about the biological processes it attempts to describe, and these assumptions are buried in the mathematical core of the model. If the students are taught only how to use the software and are not taught the mathematical foundations of the model, they will know nothing about the assumptions and therefore *limitations* of the model; this in turn means that they will not be able to interpret correctly the results of applying the software to biological data.

This situation with education in bioinformatics is now beginning to change as mathematics departments around the world are starting to teach this subject. I have been teaching two bioinformatics courses at the Department of Mathematics of the Australian National University (ANU) in Canberra, for two years now. When I started teaching them I quickly realized that all the textbooks that I found on the subject were skewed towards computational issues, reflecting, of course, the dominant teaching culture at the time. Those textbooks were not very satisfying from a mathematician's point of view and were unacceptable for my purposes. What I needed was a clear and mathematically rigorous exposition of procedures, algorithms and models commonly used in bioinformatics. As a result, I began writing my own lecture notes, and eventually they formed the basis for this book.

The book has two parts corresponding to the two courses. The first course is for second-year students and requires two medium-level first year mathematics courses as prerequisites. It concerns four important topics in bioinformatics (sequence alignment, profile hidden Markov models, protein folding and phylogenetic reconstruction) and covers them in considerable detail. Many mathematical issues related to these topics are discussed, but their probabilistic and statistical aspects are not covered in much depth there, as the students are not required to have a background in these areas. The second course (intended for third-year students) includes elements of probability and statistics; this allows one both to explore additional topics in sequence alignment, and to go back to some of the issues left unexplained in the first course, treating them from the general probabilistic and statistical point of view.

The second course is much more demanding mathematically because of its probabilistic and statistical component. At the same time, the chapters on probability and statistics (Chaps. 6 and 8) contain very few proofs. The path taken in these chapters is to give the reader all the main constructions (for instance, the construction of probability measure) and to illustrate them by many examples. Such a style is more gentle on students who only have taken a couple of mathematical courses and do not possess the mathematical maturity of a student majoring in mathematics. In fact, this is the general approach taken in the book: I give very few proofs, but a lot of discussions and examples.

Nevertheless, the book is quite mathematical in its logical approach, rigor and paying attention to subtle details.

Thus, for someone who wants to get a mathematical overview of some of the important topics in bioinformatics but does not want to go too deeply into the associated probabilistic and statistical issues, Part I of the book is quite sufficient. But it should be stressed that without reading Part II, one's understanding of various procedures from Part I will be incomplete.

Although Parts I and II together cover a substantial amount of material, none of the topics discussed in the book is treated comprehensively. For example, the chapter on protein folding (Chap. 4) and the one on phylogenetic reconstruction (Chap. 5) could each easily be expanded into a separate book. The amount of material included in the book is what realistically can be taught as two one-semester courses. Certainly, if the probability and statistics components of the book are taught separately in a different course, one can fit in more genuine bioinformatics topics, for example, the analysis of microarray data, currently not represented in the book at all.

The book concentrates on the mathematical basics of bioinformatics rather than on recent progress in the area. Even the material included in the book is found quite demanding by many students, and this is why I decided to select for it only a few topics in bioinformatics. Thus, this book is by no means a comprehensive guide to bioinformatics.

This is primarily a textbook for students with some mathematical background. At the same time, it is suitable for any mathematician, or, indeed, anyone who appreciates quantitative thinking and mathematical rigor, and who wants to learn about bioinformatics. It took me a substantial effort to explain various bioinformatics procedures in a way suitable for a general mathematical audience, and hence this book can be thought of, at least to some extent, as a translation and adaptation of some topics in bioinformatics for mathematicians. On top of this, the book contains a mathematical introduction to statistics that I have tried to keep as rigorous as possible.

I would like to thank my colleagues Prof. Sue Wilson and Prof. Simon Easteal of the Mathematical Sciences Institute (MSI) and the Centre for Bioinformation Science (CBiS) at the ANU who first suggested that I should turn my lecture notes into a book and encouraged me during the course of writing. I would like to thank Prof. Peter Hall of the MSI for patiently answering my many questions on the theory of statistics. Finally, I am grateful to Prof. John Hutchinson of the Department of Mathematics for encouragement and general discussions.

Canberra,
March 2004   *Alexander Isaev*

Contents

Part I Sequence Analysis

1   **Introduction: Biological Sequences** 3

2   **Sequence Alignment** 7
    2.1   Sequence Similarity 7
    2.2   Dynamic Programming: Global Alignment 9
    2.3   Dynamic Programming: Local Alignment 10
    2.4   Alignment with Affine Gap Model 12
    2.5   Heuristic Alignment Algorithms 14
        2.5.1   FASTA 14
        2.5.2   BLAST 16
    2.6   Significance of Scores 16
    2.7   Multiple Alignment 16
        2.7.1   MSA 19
        2.7.2   Progressive Alignment 20
    Exercises 23

3   **Markov Chains and Hidden Markov Models** 25
    3.1   Markov Chains 25
    3.2   Hidden Markov Models.................................... 33
    3.3   The Viterbi Algorithm 37
    3.4   The Forward Algorithm 40
    3.5   The Backward Algorithm and Posterior Decoding 41
    3.6   Parameter Estimation for HMMs 45
        3.6.1   Estimation when Paths are Known 46
        3.6.2   Estimation when Paths are Unknown 47
    3.7   HMMs with Silent States 56
    3.8   Profile HMMs.. 63
    3.9   Multiple Sequence Alignment by Profile HMMs 66
    Exercises 68

4   Protein Folding 75
    4.1   Levels of Protein Structure 75
    4.2   Prediction by Profile HMMs 79
    4.3   Threading 79
    4.4   Molecular Modeling 82
    4.5   Lattice HP-Model 84
    Exercises 88

5   Phylogenetic Reconstruction 89
    5.1   Phylogenetic Trees 89
    5.2   Parsimony Methods 94
    5.3   Distance Methods 96
    5.4   Evolutionary Models 122
        5.4.1   The Jukes-Cantor Model 127
        5.4.2   The Kimura Model 128
        5.4.3   The Felsenstein Model 129
        5.4.4   The Hasegawa-Kishino-Yano (HKY) Model 130
    5.5   Maximum Likelihood Method 130
    5.6   Model Comparison 137
    Exercises 139

Part II Mathematical Background for Sequence Analysis

6   Elements of Probability Theory 145
    6.1   Sample Spaces and Events 145
    6.2   Probability Measure 151
    6.3   Conditional Probability 159
    6.4   Random Variables 162
    6.5   Integration of Random Variables 163
    6.6   Monotone Functions on the Real Line 172
    6.7   Distribution Functions 176
    6.8   Common Types of Random Variables 181
        6.8.1   The Discrete Type 181
        6.8.2   The Continuous Type 182
    6.9   Common Discrete and Continuous Distributions 184
        6.9.1   The Discrete Case 184
        6.9.2   The Continuous Case 188
    6.10 Vector-Valued Random Variables 191
    6.11 Sequences of Random Variables 196
    Exercises 205

7   Significance of Sequence Alignment Scores 209
   7.1   The Problem 209
   7.2   Random Walks 211
   7.3   Significance of Scores 220
   Exercises 228

8   Elements of Statistics 231
   8.1   Statistical Modeling 231
   8.2   Parameter Estimation 235
   8.3   Hypothesis Testing 256
   8.4   Significance of Scores for Global Alignments 266
   Exercises 269

9   Substitution Matrices 271
   9.1   The General Form of a Substitution Matrix 271
   9.2   PAM Substitution Matrices 273
   9.3   BLOSUM Substitution Matrices 279
   Exercises 283

References 285

Index 289

Part I

Sequence Analysis

1

Introduction: Biological Sequences

This book is about analyzing sequences of letters from a finite alphabet $\mathcal{Q}$. Although most of what follows can be applied to sequences derived from arbitrary alphabets, our primary interest will be in *biological sequences*, that is, DNA, RNA and protein sequences.

DNA (deoxyribonucleic acid) sequences are associated with the four-letter *DNA alphabet* $\{A, C, G, T\}$, where A, C, G and T stand for the *nucleic acids* or *nucleotides* adenine, cytosine, guanine and thymine respectively. Most DNA sequences currently being studied come from DNA molecules found in chromosomes that are located in the nuclei of the cells of living organisms. In fact, a DNA molecule consists of *two strands* of nucleotides (attached to a sugar-phosphate backbone) twisted into the well-known double-helical arrangement. The two-strand structure is important for the replication of DNA molecules. There is a pairing (called *hybridization*) of nucleotides across the two strands: A is bonded to T, and C is bonded to G. Therefore, if one knows the sequence of one strand of a DNA molecule, that of the other strand can easily be reconstructed, and DNA sequences are always given as sequences of single, not paired nucleotides. The chemistry of the backbone of each strand of a DNA molecule determines a particular *orientation* of the strand, the so-called 5′ *to* 3′ *orientation*. This is the orientation in which DNA sequences are written. It should be noted that the orientations of the two strands in a DNA molecule are opposite (for this reason the strands are said to be *antiparallel*), and therefore, although the sequences of the strands determine one another, they are read in opposite directions.

Traditionally, DNA research has been focused on special stretches of the strands of DNA molecules called *protein-coding genes*; they are found on both strands, and rarely overlap across the strands. Protein-coding genes are used to produce proteins which are linear polymers of 20 different *amino acids* linked by *peptide bonds*. The single-letter amino acid notation is given in Table 1.1.

Table 1.1.

Single letter code	Amino acid
A	Alanine
R	Arginine
N	Asparagine
D	Aspartic acid
C	Cysteine
Q	Glutamine
E	Glutamic acid
G	Glycine
H	Histidine
I	Isoleucine
L	Leucine
K	Lysine
M	Methionine
F	Phenylalanine
P	Proline
S	Serine
T	Threonine
W	Tryptophan
Y	Tyrosine
V	Valine

Thus, protein sequences are associated with the 20-letter *amino-acid alphabet* $\{A, C, D, E, F, G, H, I, K, L, M, N, P, Q, R, S, T, V, W, Y\}$. The three-dimensional structure of a protein molecule results from the folding of the polypeptide chain and is much more complicated than that of a DNA molecule. A protein can only function properly if it is correctly folded. Deriving the correct three-dimensional structure from a given protein sequence is the famous *protein folding problem* that is still largely unsolved (see Chap. 4).

The main components of a protein-coding gene are *codons*. Each codon is a triplet of nucleotides coding for a single amino acid. The process of producing proteins from genes is quite complex. Every gene begins with one of the standard *start codons* indicating the beginning of the process and ends with one of the standard *stop codons* indicating the end of it. A particular way codons code for amino acids is called a *genetic code*. Several genetic codes are known, and different ones apply to different DNA molecules depending on their origin (see, e.g., [Kan]). When the sequence of a gene is read from the start to the stop codon, a growing chain of amino acids is made which, once the stop codon has been reached, becomes a complete protein molecule. It has a natural orientation inherited from that of the gene used to produce it, and this is the orientation in which protein sequences are written. The start of a protein chain is called the *amino end* and the end of it the *carboxy end*.

In fact, proteins are derived from genes in two steps. Firstly, RNA (ribonucleic acid) is made (this step is called *transcription*) and, secondly, the RNA is used to produce a protein (this step is called *translation*). RNA is another linear macromolecule, it is closely related to DNA. RNA is single-stranded, its backbone is slightly different from that of DNA, and instead of the nucleic acid thymine the nucleic acid uracil denoted by U is used. Thus, RNA sequences are associated with the four-letter *RNA alphabet* $\{A, C, G, U\}$. An RNA molecule inherits its orientation from that of the DNA strand used to produce it, and this is the orientation in which RNA sequences are written. Since RNA is single-stranded, parts of it can hybridize with its other parts which gives rise to non-trivial three-dimensional structures essential for the normal functioning of the RNA. There are in fact many types of RNA produced from not necessarily protein-coding genes, but from *RNA-coding genes*. The RNA derived from a protein-coding gene is called *messenger RNA* or *mRNA*. As an example of RNA of another type we mention *transfer RNA* or *tRNA* that takes part in translating mRNA into protein.

In this book we concentrate on DNA and protein sequences although everything that follows can be applied, at least in principle, to RNA sequences as well (subject to the availability of RNA sequence data). In fact, most procedures are so general that they work for sequences of letters from any finite alphabet, and for illustration purposes we often use the artificial two- and three-letter alphabets $\{A, B\}$ and $\{A, B, C\}$.

Table 1.2.

Database	Principal function	Organization	Address
MEDLINE	Bibliographic	National Library of Medicine	www.nlm.nih.gov
GenBank	Nucleotide sequences	National Center for Biotechnology Information	www.ncbi.nlm.nih.gov
EMBL	Nucleotide sequences	European Bioinformatics Institute	www.ebi.ac.uk
DDBJ	Nucleotide sequences	National Institute of Genetics, Japan	www.ddbj.nig.ac.jp
SWISS-PROT	Amino acid sequences	Swiss Institute of Bioinformatics	www.expasy.ch
PIR	Amino acid sequences	National Biomedical Research Foundation	www-nbrf.georgetown.edu
PRF	Amino acid sequences	Protein Research Foundation, Japan	www.prf.or.jp
PDB	Protein structures	Research Collaboratory for Structural Bioinformatics	www.rcsb.org
CSD	Protein structures	Cambridge Crystallographic Data Centre	www.ccdc.cam.ac.uk

Biological sequences are organized in databases, many of which are public. In Table 1.2 we list all major public molecular biology databases. Detailed information on them can be found in [Kan].

2

Sequence Alignment

2.1 Sequence Similarity

New DNA, RNA and protein sequences develop from pre-existing sequences rather than get invented by nature from scratch. This fact is the foundation of any sequence analysis. If we manage to relate a newly discovered sequence to a sequence about which something (e.g., structure or function) is already known, then chances are that the known information applies, at least to some extent, to the new sequence as well. We will think of any two related sequences as sequences that arose from a common ancestral sequence during the course of evolution and say that they are *homologous*. It is *sequence homology* that will be of interest to us during much of the book. Of course, if we believe that all life forms on earth came from the same origin and apply the above definition directly, then all sequences are ultimately homologous. In practice, two sequences are called homologous, if one can establish their relatedness by *currently available methods*, and it is the sensitivity of the methods that produces a borderline between sequences called homologous and ones that are not called homologous. For example, two protein sequences can be called homologous, if one can show experimentally that their functions in the respective organisms are related. Thus, sequence homology is a dynamic concept, and families of homologous sequences known at the moment may change as the sensitivity of the methods improves.

The first step towards inferring homology is to look for *sequence similarity*. If two given sequences are very long, it is not easy to decide whether or not they are similar. To see if they are similar, one has to properly *align* them. When sequences evolve starting from a common ancestor, their residues can undergo *substitutions* (when residues are replaced by some other residues). Apart from substitutions, during the course of evolution sequences can accumulate a number of events of two more types: *insertions* (when new residues appear in a sequence in addition to the existing ones) and *deletions* (when some residues disappear). Therefore, when one is trying to produce the best possible *alignment* between two sequences, residues must be allowed to be aligned not only to other residues but also to *gaps*. The presence of a gap in an alignment represents either an insertion or deletion event. Consider, for example, the following two very short nucleotide sequences, each consisting of only seven residues

$$x : TACCAGT$$
$$y : CCCGTAA.$$

The sequences are of the same length, and there is only one way to align them, if one does not allow gaps in alignments

$$x : T\ A\ C\ C\ A\ G\ T$$
$$y : C\ C\ C\ G\ T\ A\ A.$$

However, if we allow gaps, there are many possible alignments. In particular, the following alignment seems to be much more informative than the preceding one

$$x : T\ A\ C\ C\ A\ G\ T - -$$
$$y : C - C\ C - G\ T\ A\ A. \tag{2.1}$$

Alignment (2.1) indicates that the subsequence $CCGT$ may be an evolutionarily conserved region, which means that both x and y may have evolved from a common ancestral sequence containing the subsequence $CCGT$ in appropriate positions. Another possible alignment that also looks reasonable is

$$x : T\ A\ C\ C\ A\ G\ T - -$$
$$y : - - C\ C\ C\ G\ T\ A\ A. \tag{2.2}$$

How can one choose between alignments (2.1) and (2.2)? Are there any better alignments? To answer these questions we need to be able to *score* any possible alignment. Then the alignments that have the highest score are by definition the best or *optimal* ones (there may be more than one such alignments).

The most popular scoring schemes assume independence among the columns in an alignment and set the total score of the alignment to be equal to the sum of the scores of each column. Therefore, for such schemes one only needs to specify the scores $s(a, b) = s(b, a)$ and the *gap penalty* $s(-, a) = s(a, -)$, with $a, b \in \mathcal{Q}$, where $\mathcal{Q}$ is either the 4-letter DNA or RNA alphabet, or the 20-letter amino acid alphabet, depending on the kind of sequences that we are interested in aligning. The resulting best alignments between two sequences depend, of course, on the scoring scheme. It is possible that for two different scoring schemes the best alignments will be entirely different. As an example of a scoring scheme one can set $s(a, a) = 1$ (the score of a match), $s(a, b) = -1$, if $a \neq b$ (the score of a mismatch), and $s(-, a) = s(a, -) = -2$ (the gap penalty) – see Exercise 2.1. However, it is important to keep in mind that a scoring scheme must be *biologically relevant* in order to produce a sensible alignment. Such a scheme must take into account factors that constrain sequence evolution. For example, many popular scoring schemes introduce some degree of dependence among the columns in an alignment by making the score of a continuous gap region an *affine* function of its length (note that in the example above the score of a gap region is *linear* in its length).

The numbers $s(a,b)$ form a *scoring* or *substitution matrix*. Substitution matrices are always symmetric and must possess some specific properties in order to be successfully used for sequence comparison. These properties will be discussed in detail in Chaps. 7 and 9. The most popular substitution matrices are so-called PAM and BLOSUM matrices used to align amino acid sequences. These 20×20-matrices are derived by statistically analyzing known amino acid sequences. The derivation of PAM and BLOSUM matrices will be explained in Chap. 9. For the purposes of the first part of the book we assume that the substitution matrix is given. For simplicity, we will initially restrict our considerations only to DNA sequences, thus assuming that $Q = \{A, C, G, T\}$. This setup will be sufficient to demonstrate the main principles of alignment procedures.

2.2 Dynamic Programming: Global Alignment

In this section we assume a *linear gap model* (that is, $s(-, a) = s(a, -) = -d$ for $a \in Q$, with $d > 0$, so that the score of a gap region of length L is equal to $-dL$) and present an algorithm, the *Needleman-Wunsch algorithm* [NW], [G], that always finds *all* optimal *global* alignments (there are frequently more than one such alignments).

The idea is to produce an optimal alignment from optimal alignments of subsequences. Algorithms that achieve optimization by means of performing optimization for smaller amounts of data (in this case subsequences) are generally called *dynamic programming algorithms*. Suppose we are given two sequences $x = x_1 x_2 \ldots x_i \ldots x_n$ and $y = y_1 y_2 \ldots y_j \ldots y_m$. We construct an $(n + 1) \times (m + 1)$-matrix F. Its (i, j)th element $F(i, j)$ for $i = 1, \ldots, n$, $j = 1, \ldots, m$ is equal to the score of an optimal alignment between $x_1 \ldots x_i$ and $y_1 \ldots y_j$. The element $F(i, 0)$ for $i = 1, \ldots, n$ is the score of aligning $x_1 \ldots x_i$ to a gap region of length i. Similarly, the element $F(0, j)$ for $j = 1, \ldots, m$ is the score of aligning $y_1 \ldots y_j$ to a gap region of length j. We build F recursively initializing it by the condition $F(0, 0) = 0$ and then proceeding to fill the matrix from the top left corner to the bottom right corner. If $F(i - 1, j - 1)$, $F(i - 1, j)$ and $F(i, j - 1)$ are known, $F(i, j)$ is clearly calculated as follows

$$F(i, j) = \max \begin{cases} F(i - 1, j - 1) + s(x_i, y_j), \\ F(i - 1, j) - d, \\ F(i, j - 1) - d. \end{cases}$$

Indeed, there are three possible ways to obtain the best score $F(i, j)$: x_i can be aligned to y_j (see the first option in the formula above), or x_i is aligned to a gap (the second option), or y_j is aligned to a gap (the third option). Calculating

$F(i, j)$ we keep a pointer to the option from which $F(i, j)$ was produced. When we reach $F(n, m)$ we trace back the pointers to recover optimal alignments. The value $F(n, m)$ is exactly their score. Note that more than one pointers may come out of a particular cell of the matrix which results in several optimal alignments.

Example 2.1. Let $x = CTTAGA$, $y = GTAA$, and suppose that we are using the scoring scheme: $s(a, a) = 1$, $s(a, b) = -1$, if $a \neq b$, and $s(-, a) = s(a, -) = -2$. The corresponding matrix F with pointers is shown in Fig. 2.1. Tracing

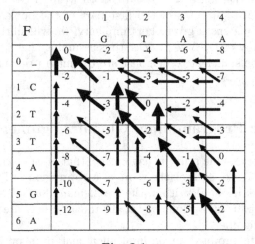

Fig. 2.1.

back the pointers gives the following three optimal alignments

$$x : C\ T\ T\ A\ G\ A$$
$$y : G - T\ A - A,$$

$$x : C\ T\ T\ A\ G\ A$$
$$y : G\ T - A - A,$$

$$x : C\ T\ T\ A\ G\ A$$
$$y : - G\ T\ A - A$$

with score -2. The corresponding paths through the matrix F are shown with thick arrows.

2.3 Dynamic Programming: Local Alignment

A more biologically interesting alignment problem is to find all pairs of subsequences of two given sequences that have the highest-scoring alignments. We will only be interested in subsequences of consecutive elements or *segments*. Any such subsequence of a sequence $x_1 x_2 \dots x_n$ has the form $x_i x_{i+1} \dots x_{i+k}$ for some $1 \leq i \leq n$ and $k \leq n - i$. This alignment problem is called the *local alignment problem*. Here we present the *Smith-Waterman algorithm* [SW] that solves the problem for a linear gap model.

We construct an $(n + 1) \times (m + 1)$-matrix as in the previous section, but the formula for its entries is slightly different

$$F(i, j) = \max \begin{cases} 0, \\ F(i - 1, j - 1) + s(x_i, y_j), \\ F(i - 1, j) - d, \\ F(i, j - 1) - d. \end{cases} \tag{2.3}$$

Taking the first option in the above formula is equivalent to starting a new alignment: if an optimal alignment up to some point has a negative score, it is better to start a new alignment, rather than to extend the current one.

Another difference is that now an alignment can end anywhere in the matrix, so instead of taking the value $F(n, m)$ in the bottom right corner of the matrix for the best score, we look for a maximal elements in the matrix F and start traceback from there. The traceback ends when we get to a cell with value 0, which corresponds to the start of the alignment.

Example 2.2. For the sequences from Example 2.1 the only best local alignment is

$$x : T \ A$$
$$y : T \ A,$$

and its score is equal to 2. The corresponding dynamic programming matrix F is shown in Fig. 2.2, where the thick arrows represent traceback. Note that if an element of F is equal to 0 and no arrows come out of the cell containing this element, then the element is obtained as the first option in formula (2.3).

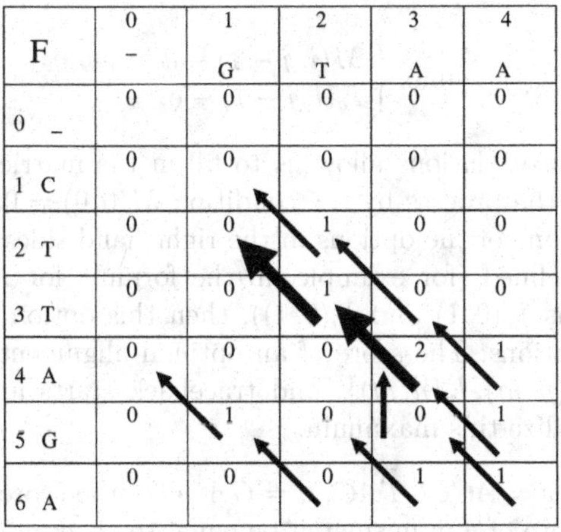

Fig. 2.2.

2.4 Alignment with Affine Gap Model

In this section we will assume an *affine gap model*, that is, we set the score of any gap region of length L to be equal to $-d - e(L-1)$ for some $d > 0$ and $e > 0$. In this case $-d$ is called the *gap opening penalty* and $-e$ the *gap extension penalty*. Usually, e is set to be smaller than d, which reflects the fact known from biology that starting a gap region is harder than extending it. We will only discuss a global alignment algorithm here, its local version can be readily obtained as in the preceding section (see also [G]). The algorithm requires three matrices: one $(n+1) \times (m+1)$-matrix and two $n \times m$-matrices. Let $M(i,j)$ for $i = 1,\ldots,n$ and $j = 1,\ldots,m$ be the score of an optimal alignment between $x_1 \ldots x_i$ and $y_1 \ldots y_j$, given that the alignment ends with x_i aligned to y_j. The element $M(i,0)$ for $i = 1,\ldots,n$ is the score of aligning $x_1 \ldots x_i$ to a gap region of length i. Similarly, the element $M(0,j)$ for $j = 1,\ldots,m$ is the score of aligning $y_1 \ldots y_j$ to a gap region of length j. Further, let $I_x(i,j)$ for $i = 1,\ldots,n$ and $j = 1,\ldots,m$ be the score of an optimal alignment between $x_1 \ldots x_i$ and $y_1 \ldots y_j$ given that the alignment ends with x_i aligned to a gap. Finally, let $I_y(i,j)$ for $i = 1,\ldots,n$ and $j = 1,\ldots,m$ be the score of an optimal alignment between $x_1 \ldots x_i$ and $y_1 \ldots y_j$ given that the alignment ends with y_j aligned to a gap. Then, if we assume that a deletion is never followed directly by an insertion (unless the deletion starts at the beginning of an alignment), we have

$$M(i,j) = \max \begin{cases} M(i-1, j-1) + s(x_i, y_j), \\ I_x(i-1, j-1) + s(x_i, y_j), \\ I_y(i-1, j-1) + s(x_i, y_j), \end{cases}$$

$$I_x(i,j) = \max \begin{cases} M(i-1, j) - d, \\ I_x(i-1, j) - e, \end{cases}$$

$$I_y(i,j) = \max \begin{cases} M(i, j-1) - d, \\ I_y(i, j-1) - e. \end{cases}$$

These recurrence relations allow us to fill in the matrices M, I_x and I_y, once we initialize the process by the condition $M(0,0) = 0$. We note that if, for some i and j, one of the options in the right-hand sides of the recurrence relations is not defined (for example, in the formula for $M(1,2)$ the right-hand side contains $I_x(0,1)$ and $I_y(0,1)$), then this option is not taken into account in calculations. The score of an optimal alignment is then given by $\max\{M(n,m), I_x(n,m), I_y(n,m)\}$, and traceback starts at the element (or elements) that realize this maximum.

Example 2.3. Let $x = ACGGTAC$, $y = GAGGT$, the score of any match be equal to 1, the score of any mismatch be equal to -1, $d = 3$ and $e = 2$. Then we obtain the dynamic programming matrices shown in Fig. 2.3.

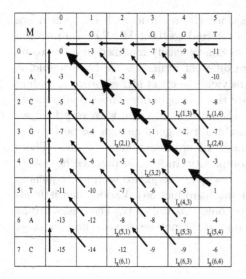

I_x	1	2	3	4	5
	G	A	G	G	T
1 A	-6 M(0,1)	-8 M(0,2)	-10 M(0,3)	-12 M(0,4)	-14 M(0,5)
2 C	-4 M(1,1)	-5 M(1,2)	-9 M(1,3)	-11 M(1,4)	-13 M(1,5)
3 G	-6	-5 M(2,2)	-6 M(2,3)	-9 M(2,4)	-11 M(2,5)
4 G	-7 M(3,1)	-7	-4 M(3,3)	-5 M(3,4)	-10 M(3,5)
5 T	-9 M(4,1)	-8 M(4,2)	-6	-3 M(4,4)	-6 M(4,5)
6 A	-11	-10 M(5,2)	-8	-5	-2 M(5,5)
7 C	-13	-11 M(6,2)	-10	-7	-4

I_y	1	2	3	4	5
	G	A	G	G	T
1 A	-6 M(1,0)	-4 M(1,1)	-5 M(1,2)	-7	-9
2 C	-8 M(2,0)	-7 M(2,1)	-5 M(2,2)	-6 M(2,3)	-8
3 G	-10 M(3,0)	-7 M(3,1)	-8 M(3,2)	-4 M(3,3)	-5 M(3,4)
4 G	-12 M(4,0)	-9 M(4,1)	-8 M(4,2)	-7 M(4,3)	-3 M(4,4)
5 T	-14 M(5,0)	-13 M(5,1)	-10 M(5,2)	-9 M(5,3)	-8 M(5,4)
6 A	-16 M(6,0)	-15 M(6,1)	-11 M(6,2)	-11 M(6,3)	-10 M(6,4)
7 C	-18 M(7,0)	-17 M(7,1)	-15 M(7,2)	-12 M(7,3)	-12 M(7,4)

Fig. 2.3.

The arrows and labels indicate from which elements of the three matrices each number was produced (in addition, we draw the vertical and horizontal arrows in the 0th column and 0th row of the matrix M). The thick arrows show traceback; it starts at $I_x(7,5) = -4$. The corresponding optimal alignment with score -4 is

$$x : A\,C\,G\,G\,T\,A\,C$$
$$y : G\,A\,G\,G\,T - -.$$

There are many other variants of the basic dynamic programming algorithm: for overlap matches, repeated matches, more complex gap models, etc.

We do not discuss all of them here since the main ideas behind them are very similar. The interested reader is referred to [DEKM]. We also mention that dynamic programming algorithms can be easily modified in order to be applicable to aligning sequences from different alphabets as, for example, is required in Sect. 4.3. All the algorithms yield the exact highest score according to a given scoring scheme and generate the corresponding optimal alignments by a traceback procedure. However, when one has to deal with very long sequences, a time complexity $O(nm)$ might not be good enough for performing the required search on a computer in an acceptable amount of time. There is a similar problem with memory complexity. Indeed, in practice homology search takes the form of comparing a new query sequence with *all* sequences in a large database (see Table 1.2 in the Introduction). Various heuristic algorithms have been developed to overcome this difficulty. These algorithms are faster, but the trade-off is that one might not necessarily find the best possible alignments.

2.5 Heuristic Alignment Algorithms

In this section we will briefly discuss two alignment algorithms that are not guaranteed to find an optimal alignment, but are substantially faster than the dynamic programming algorithms introduced above.

2.5.1 FASTA

Calculation of the dynamic programming matrices for a pair of sequences takes a substantial amount of time and memory. However, as we have seen, the areas of interest in these matrices (that is, the areas formed by the cells that take part in a traceback procedure) are minuscule in comparison to the entire areas of the matrices. The FASTA algorithm [PL] is designed to limit the areas of the matrices that the dynamic programming examines. First, FASTA determines candidate areas (that is, the areas that are likely to produce optimal alignments) and then the relevant dynamic programming algorithm is applied with the additional condition that all the elements of the matrices that lie outside the candidate areas are equal to $-\infty$. With these condition satisfied, the corresponding traceback procedure never leaves the candidate areas. Since the sizes of the candidate areas are usually very small compared to the sizes of the full matrices, this reduction gives a significant increase in computational speed.

Candidate areas can be identified, for example, by considering the *dot matrix* for two given sequences. It is formed by dots that are entered for all positions corresponding to matching pairs of letters in the two sequences. Local similarities can then be detected as diagonal stretches of consecutive dots. An example of a dot matrix is shown in Fig. 2.4, where diagonal stretches of length at least 3 are shaded.

Dot Matrix	G	T	C	A	G	A	C	G	C	T	C	A
C			*				*		*		*	
A				*		*						*
G	*				*			*				
A				*		*						*
G	*				*			*				
T		*								*		
T		*								*		
A				*		*						*
C			*				*		*		*	
G	*				*			*				
T		*								*		
C			*				*		*		*	
A				*		*						*

Fig. 2.4.

Once diagonal stretches of length at least k have been identified, we extend them to full diagonals, and candidate areas are constructed as bands of width b around the extensions, for some preset values of k and b. Figure 2.5 shows the candidate areas for the sequences from Fig. 2.4 with $k = 3$ and $b = 1$ in the case of alignments with linear gap model (we omitted the 0th row and column that remain unchanged). Such areas can also be defined for each of the three matrices required in the case of alignments with affine gap model.

F	G	T	C	A	G	A	C	G	C	T	C	A
C					$-\infty$	$-\infty$	$-\infty$	$-\infty$	$-\infty$	$-\infty$	$-\infty$	$-\infty$
A						$-\infty$	$-\infty$	$-\infty$	$-\infty$	$-\infty$	$-\infty$	$-\infty$
G							$-\infty$	$-\infty$	$-\infty$	$-\infty$	$-\infty$	$-\infty$
A								$-\infty$	$-\infty$	$-\infty$	$-\infty$	$-\infty$
G	$-\infty$								$-\infty$	$-\infty$	$-\infty$	$-\infty$
T	$-\infty$	$-\infty$								$-\infty$	$-\infty$	$-\infty$
T	$-\infty$	$-\infty$	$-\infty$								$-\infty$	$-\infty$
A	$-\infty$	$-\infty$	$-\infty$	$-\infty$								$-\infty$
C		$-\infty$	$-\infty$	$-\infty$	$-\infty$							
G			$-\infty$	$-\infty$	$-\infty$	$-\infty$						
T				$-\infty$	$-\infty$	$-\infty$	$-\infty$					
C	$-\infty$				$-\infty$	$-\infty$	$-\infty$	$-\infty$				
A	$-\infty$	$-\infty$				$-\infty$	$-\infty$	$-\infty$	$-\infty$			

Fig. 2.5.

Of course, FASTA searches can miss optimal alignments. The values of k and b are trade-offs between time and optimality. An implementation of FASTA with many user-adjustable parameters (including k and b) can be found on http://bioweb.pasteur.fr/seqanal/interfaces/fasta.html.

2.5.2 BLAST

Unlike FASTA, the BLAST (Basic Local Alignment Search Tool) [AL] algorithms is not based on dynamic programming. Initially, BLAST looks for short stretches of identities (just as FASTA does) and then attempts to extend them in either direction in search of a good longer alignment. This strategy is reasonable biologically, since two related sequences tend to share well-conserved segments. Although the search algorithm implemented in BLAST is entirely heuristic, it is the most successful search tool up to date. One of many implementations of BLAST can be found on http://bioweb.pasteur.fr/seqanal/interfaces/blast2.html.

2.6 Significance of Scores

A major concern when interpreting database search results is whether a similarity found between two sequences is biologically significant, especially when the similarity is only marginal. Since good alignments can occur by chance alone, one needs to perform some further statistical analysis to assess their significance. This issue will be discussed in considerable detail in Chap. 7 (see also Sect. 8.4). Assessment of the significance of sequence alignment scores is part of both FASTA and BLAST algorithms.

2.7 Multiple Alignment

In sequence analysis one is often interested in determining common features among a collection of sequences. To identify such a feature, one needs to determine an optimal *multiple alignment* for the whole collection. As in the case of two sequences, in order to produce an optimal multiple alignment, one needs to be able to score any multiple alignment by using some scoring scheme. Analogously to the case of two sequences, most alignment methods assume that the individual columns of an alignment not containing gaps are independent and so use a scoring function of the form

$$S(M) = G(M) + \sum_i s(M_i),$$

where M denotes a multiple alignment, M_i is the ith column not containing gaps, $s(M_i)$ is the score of M_i, and G is a function for scoring columns with gaps.

In the standard (but not very satisfactory) methods for scoring multiple alignments, columns not containing gaps are scored by the "sum of pairs" (*SP*) function. The *SP-score* for a column M_i not containing gaps is defined as

$$s(M_i) = \sum_{k<l} s(M_i^k, M_i^l),$$

where the sum is taken over all pairs (M_i^k, M_i^l), $k < l$, of elements of M_i, and the scores $s(a, b)$, for $a, b \in \mathcal{Q}$, come from a substitution matrix used for scoring pairwise sequence alignments (such as PAM and BLOSUM matrices discussed in Chap. 9). Often gaps are scored by defining $s(-, a) = s(a, -)$, setting $s(-, -) = 0$ and introducing the corresponding *SP*-score for columns containing gaps. We call any such way of scoring gap regions a *linear gap model for multiple alignments*. Summing all pairwise substitution scores may seem to be a natural thing to do, but in fact there is no statistical justification for the *SP*-score.

Once a scheme for scoring multiple alignments has been fixed, it is possible to generalize pairwise dynamic programming algorithms to aligning $n \geq 3$ sequences. In problems involving multiple alignments one is usually interested in global alignments, and what follows is a generalization of the Needleman-Winsch algorithm. Here we assume a scoring scheme for which

$$\mathcal{S}(M) = \sum_i s(M_i), \tag{2.4}$$

with the sum taken over *all* columns of the alignment (including the ones containing gaps). We note that a multidimensional dynamic programming algorithm with affine gap model exists as well.

Suppose we have n sequences $x^1 = x_1^1 \ldots x_{m_1}^1$, $x^2 = x_1^2 \ldots x_{m_2}^2$, ..., $x^n = x_1^n \ldots x_{m_n}^n$. Let $i_1, \ldots, i_n$ be integers with $0 \leq i_j \leq m_j$, $j = 1, \ldots, n$, where at least one number is non-zero. Define $F(i_1, \ldots, i_n)$ as the maximal score of an alignment of the subsequences ending with $x_{i_1}^1 \ldots x_{i_n}^n$ (if for some j we have $i_j = 0$, then the other subsequences are aligned to a gap region). The recursion step of the dynamic programming algorithm is given by

$$F(i_1, \ldots, i_n) = \max \begin{cases} F(i_1 - 1, \ldots, i_n - 1) + s(x_{i_1}^1, \ldots, x_{i_n}^n), \\ F(i_1, i_2 - 1 \ldots, i_n - 1) + s(-, x_{i_2}^2, \ldots, x_{i_n}^n), \\ F(i_1 - 1, i_2, i_3 - 1 \ldots, i_n - 1) + s(x_{i_1}^1, -, x_{i_3}^3, \ldots, x_{i_n}^n), \\ \vdots \\ F(i_1 - 1, \ldots, i_{n-1} - 1, i_n) + s(x_{i_1}^1, \ldots, x_{i_{n-1}}^{n-1}, -), \\ F(i_1, i_2, i_3 - 1 \ldots, i_n - 1) + s(-, -, x_{i_3}^3, \ldots, x_{i_n}^n), \\ \vdots \end{cases}$$

where all combinations of gaps occur except the one where all residues are replaced by gaps. The algorithm is initialized by setting $F(0, \ldots, 0) = 0$. Traceback starts at $F(m_1, \ldots, m_n)$ and is analogous to that for pairwise alignments. The matrix $\left(F(i_1, \ldots, i_n) \right)$ with $0 \leq i_j \leq m_j$, $j = 1, \ldots, n$, is an $(m_1 + 1) \times \ldots \times (m_n + 1)$-matrix, and it is convenient to visualize it by considering its two-dimensional sections.

Example 2.4. We will find all optimal alignments of the three sequences $x = AATC$, $y = GTC$, $z = AAG$ using the following scoring scheme: the score of an alignment is calculated from the scores of its columns M_i's from formula (2.4); if M_i contains three identical symbols, set $s(M_i) = 2$; if it contains exactly two identical symbols, but no gaps, set $s(M_i) = 1$; if it contains three distinct symbols, but no gaps, set $s(M_i) = -1$; if it contains exactly one gap, set $s(M_i) = -2$; if it contains two gaps, set $s(M_i) = -4$. In this case the indices i_1, i_2 and i_3 correspond to sequences x, y and z respectively. Figure 2.6 shows the four sections of the matrix F in the direction of i_3.

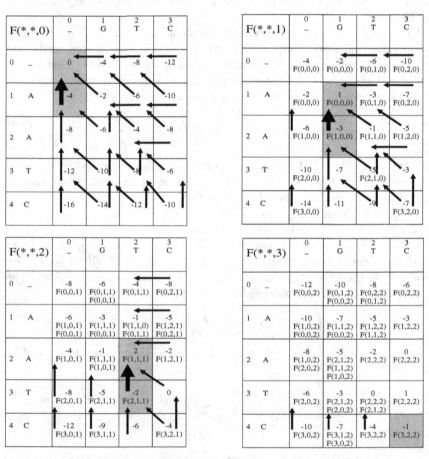

Fig. 2.6.

As before, the arrows and labels indicate from which elements each number was derived. The shaded cells and thick arrows correspond to traceback; it starts at $F(4, 3, 3) = -1$ and goes through the shaded cells until we reach $F(0, 0, 0)$. The traceback produces the following three paths

$$F(4, 3, 3) \rightarrow F(3, 2, 2) \rightarrow F(2, 1, 1) \rightarrow F(1, 0, 0) \rightarrow F(0, 0, 0),$$
$$F(4, 3, 3) \rightarrow F(3, 2, 2) \rightarrow F(2, 1, 1) \rightarrow F(1, 1, 1) \rightarrow F(0, 0, 0),$$
$$F(4, 3, 3) \rightarrow F(3, 2, 2) \rightarrow F(2, 2, 2) \rightarrow F(1, 1, 1) \rightarrow F(0, 0, 0).$$

They respectively give rise to the following three optimal alignments with score -1

$$
\begin{array}{lll}
x : A\ A\ T\ C & x : A\ A\ T\ C & x : A\ A\ T\ C \\
y : -\ G\ T\ C, & y : G\ -\ T\ C, & y : G\ T\ -\ C. \\
z : -\ A\ A\ G & z : A\ -\ A\ G & z : A\ A\ -\ G
\end{array}
$$

Because of the memory and time complexity, the above algorithm in practice cannot be applied to align a large number of sequences. Therefore alternative algorithms (mainly heuristic) have been developed. Below we briefly mention some of them.

2.7.1 MSA

MSA always finds all optimal alignments and is based on the multi-dimensional dynamic programming algorithm. The idea is to reduce the number of elements of the dynamic programming matrix that need to be examined to find an optimal multiple alignment [CL], [LAK]. MSA can successfully align up to seven protein sequences of length up to 300 residues.

Here we assume an SP-scoring system for both residues and gaps (that is, an SP-scoring scheme with linear gap model), hence the score of a multiple alignment is the sum of the scores of all pairwise alignments induced by the multiple alignment. Let M denote a multiple alignment, and M^{kl} the induced pairwise alignment between the kth and lth sequences. Then we have

$$\mathcal{S}(M) = \sum_{k<l} \mathcal{S}(M^{kl}),$$

where $\mathcal{S}(M^{kl})$ is the score of M^{kl}. Clearly, if s^{kl} is the score of an optimal global alignment between the kth and lth sequences, then $\mathcal{S}(M^{kl}) \leq s^{kl}$.

Assume that we have a lower bound τ on the score of an optimal multiple alignment. Such a bound can be found by any fast (and not necessarily very precise) heuristic multiple alignment algorithm such as the *Star Alignment algorithm* discussed below. Then, for an optimal multiple alignment M_0 we have

$$\tau \leq \mathcal{S}(M_0) = \sum_{k'<l'} \mathcal{S}(M_0^{k'l'}) \leq \mathcal{S}(M_0^{kl}) - s^{kl} + \sum_{k'<l'} s^{k'l'},$$

for all k and l. Hence

$$\mathcal{S}(M_0^{kl}) \geq t^{kl},$$

where

$$t^{kl} = \tau + s^{kl} - \sum_{k' < l'} s^{k'l'}. \tag{2.5}$$

The scores of the pairwise optimal alignments in the right-hand side of the above formula can be calculated as discussed in the preceding sections and hence t^{kl} can be determined (at least approximately).

Thus, we only need to look for such multiple alignments that induce pairwise alignments with scores no less than the numbers t^{kl}. This observation substantially reduces the number of elements in the multi-dimensional dynamic programming matrix that need to be examined and hence increases computational speed.

2.7.2 Progressive Alignment

Progressive alignment methods are heuristic in nature. They produce multiple alignments from a number of pairwise alignments. A common scheme for such methods is as follows: first, two sequences are chosen and aligned, then a third sequence is chosen and aligned (in some specific way) to the alignment of the first two sequences, and the process continues until all the sequences have been used. Perhaps the most widely used algorithms of this type is CLUSTALW [THG]. Despite being heuristic, this method uses evolutionary relationships among the sequences of interest. Specifically, it produces a phylogenetic tree (see Sect. 5.1) by the neighbor-joining method (see Sect. 5.3) with pairwise "distances" calculated from the Kimura model (see Sects. 5.4 and 5.5).

Multiple alignments obtained by methods of this kind are almost always adjusted by the eye. Below we describe one simple progressive alignment algorithm. For more information on progressive alignment methods the reader is referred to [DEKM].

Star Alignment

The Star Alignment algorithm is a fast heuristic method for producing multiple alignments. Of course, as any other heuristic algorithm, it does not guarantee to find an optimal alignment. The general idea is to first select a sequence which is most similar to all the other sequences, and then to use it as the center of a "star" aligning all the other sequences to it. We will describe this algorithm by means of considering the following example.

Example 2.5. Suppose we are given the following five DNA sequences

$$
\begin{aligned}
x^1 &: ATTGCCATT \\
x^2 &: ATGGCCATT \\
x^3 &: ATCCAATTTT \\
x^4 &: ATCTTCTT \\
x^5 &: ACTGACC.
\end{aligned}
$$

We assume the same scoring scheme for pairwise alignments as in Example 2.1 and consider the corresponding SP-scoring scheme with linear gap model, calculate all pairwise optimal scores (that is, the scores found by the global pairwise alignment algorithm described in Sect. 2.2), write them in the matrix below and find the sum in each row

	x^1	x^2	x^3	x^4	x^5	Total Score
x^1		7	-2	0	-3	2
x^2	7		-2	0	-4	1
x^3	-2	-2		0	-7	-11
x^4	0	0	0		-3	-3
x^5	-3	-4	-7	-3		-17

Of all the sequences, x^1 has the best total score (equal to 2), and is selected to be at the center of the future star. Optimal alignments between x^1 and each of the other sequences found by the global alignment algorithm from Sect. 2.2 are as follows

$$x^1 : A\ T\ T\ G\ C\ C\ A\ T\ T$$
$$x^2 : A\ T\ G\ G\ C\ C\ A\ T\ T,$$

$$x^1 : A\ T\ T\ G\ C\ C\ A\ T\ T\ -\ -$$
$$x^3 : A\ T\ C\ -\ C\ A\ A\ T\ T\ T\ T,$$

$$x^1 : A\ T\ T\ G\ C\ C\ A\ T\ T$$
$$x^4 : A\ T\ C\ T\ T\ C\ -\ T\ T,$$

$$x^1 : A\ T\ T\ G\ C\ C\ A\ T\ T$$
$$x^5 : A\ C\ T\ G\ A\ C\ C\ -\ -.$$

We now merge the above alignments using the "once a gap – always a gap" principle. We start with x^1 and x^2

$$x^1 : A\ T\ T\ G\ C\ C\ A\ T\ T$$
$$x^2 : A\ T\ G\ G\ C\ C\ A\ T\ T$$

and add x^3, but, since x^3 is longer than x^1 and x^2, we add gaps at the ends of x^1 and x^2

$$x^1 : A\ T\ T\ G\ C\ C\ A\ T\ T\ -\ -$$
$$x^2 : A\ T\ G\ G\ C\ C\ A\ T\ T\ -\ -$$
$$x^3 : A\ T\ C\ -\ C\ A\ A\ T\ T\ T\ T.$$

These gaps are never removed. If we introduce a gap somewhere, it will be carried through.

Finally, we add x^4 and x^5. They too must have gaps added to their ends. The resulting multiple alignment is as follows

$$x^1 : A\,T\,T\,G\,C\,C\,A\,T\,T - -$$
$$x^2 : A\,T\,G\,G\,C\,C\,A\,T\,T - -$$
$$x^3 : A\,T\,C - C\,A\,A\,T\,T\,T\,T$$
$$x^4 : A\,T\,C\,T\,T\,C - T\,T - -$$
$$x^5 : A\,C\,T\,G\,A\,C\,C - - - -.$$

The Star Alignment algorithm may not find an optimal multiple alignment as shown in the example below.

Example 2.6. Consider the following three sequences

$$x^1 : GGCAA$$
$$x^2 : GCACA$$
$$x^3 : GGACA,$$

and assume the same scoring scheme as in Example 2.5. Here x^1 is again the center of the star, and the optimal alignments between x^1 and each of the other two sequences are

$$x^1 : G\,G\,C\,A - A$$
$$x^2 : G - C\,A\,C\,A,$$

$$x^1 : G\,G - C\,A\,A$$
$$x^3 : G\,G\,A\,C\,A -.$$

Merging these alignments, we obtain

$$x^1 : G\,G - C\,A - A$$
$$x^2 : G - - C\,A\,C\,A$$
$$x^3 : G\,G\,A\,C\,A - -.$$

The above alignment is not optimal, its score is equal to -5. The best score is in fact equal to 0. One optimal alignment is shown below.

$$x^1 : G\,G\,C\,A - A$$
$$x^2 : G - C\,A\,C\,A$$
$$x^3 : G\,G - A\,C\,A.$$

Machine Learning Approach

Another way to produce pairwise and multiple alignments is by utilizing the *machine learning approach*. This means that one builds a kind of a "learning machine" that can be trained from initially unaligned sequences. After that the trained machine can attempt to align the sequences from the training data.

There are many types of learning machines, and, apart from sequence alignment, they have other numerous applications in bioinformatics and other areas. In the following chapter we will deal with one type of learning machines, *Hidden Markov Models* or *HMMs*, and discuss some of their applications in bioinformatics, including multiple sequence alignment (see Sect. 3.9).

Exercises

2.1. Consider the following scoring scheme for DNA sequences: $s(a, a) = 1$ (the score of any match), $s(a, b) = 1$ if $a \neq b$ (the score of any mismatch), $s(-, a) = s(a, -) = -2$ (the gap penalty), for all $a, b \in \mathcal{Q} = \{A, C, G, T\}$. Let

$$x : ACTGTCCA$$
$$y : CTGAATCAGA.$$

Find the score of the following alignment

$$x : A\ C\ T\ G - - - T - C\ C\ A$$
$$y : C - T\ G\ A\ A\ T - C\ A\ G\ A.$$

Is there a higher-scoring alignment between x and y?

2.2. Consider the following scoring scheme with linear gap model for DNA sequences: the score of any match is equal to 5, the score of any mismatch is equal to -4, the gap penalty is equal to -2. Using the relevant dynamic programming algorithms with this scoring scheme find

(i) all optimal global alignments of the sequences

$$x : AAGTTCGT$$
$$y : CAGTAAT,$$

(ii) all optimal local alignments of the sequences

$$x : AGTGGCATT$$
$$y : TGTCGCAT.$$

2.3. Consider the following scoring scheme with affine gap model for DNA sequences: the score of any match is equal to 2, the score of any mismatch is equal to -3, the gap opening penalty is -4, the gap extension penalty is -1. Using the relevant dynamic programming algorithm with this scoring scheme find all optimal global alignments of the following sequences

$$x : TGGCAAC$$
$$y : CTGGA.$$

2.4. Consider the following scoring scheme with linear gap model for DNA sequences: the score of any match is equal to 1, the score of any mismatch is equal to -2, the gap penalty is equal to -1. Suppose that we are using the FASTA algorithm to look for optimal local alignment between the following sequences

$$x : CGACTCCGAT$$
$$y : CGACTAAAACCGAT.$$

Set $k = 3$ and $b = 1$. Will the algorithm find an optimal alignment? What will change if we set $k = 4$ and $b = 2$?

2.5. For an alignment M of three sequences define the score $\mathcal{S}(M)$ by formula (2.4). The score $s(M_i)$ of the column M_i is defined as follows: if the column contains three identical symbols, set $s(M_i) = 3$; if it contains two identical symbols, but no gaps, set $s(M_i) = 2$; if it contains three distinct symbols, but no gaps, set $s(M_i) = 1$; if it contains exactly one gap, set $s(M_i) = -1$; if it contains two gaps, set $s(M_i) = -2$. Applying the relevant dynamic programming algorithm find all optimal alignments of the three sequences

$$x = CAGC$$
$$y = CTG$$
$$z = TAC.$$

2.6. Using the Star Alignment algorithm with the scoring scheme for pairwise alignments from Exercise 2.1, construct a multiple alignment for the following sequences

$$x^1 : AGTCCT$$
$$x^2 : ACTGTTC$$
$$x^3 : CCGCGTT$$
$$x^4 : GGGTCCT.$$

Using the SP-score of this alignment find the lower bounds t^{kl} for MSA for the sequences x^1, x^2, x^3, x^4, as defined in (2.5).

3

Markov Chains and Hidden Markov Models

3.1 Markov Chains

We introduce *discrete-time finite Markov chains* or *discrete-time finite Markov models* (or simply *Markov chains* or *Markov models*) as follows. Consider some finite set X of all possible *states* $G_1, \ldots, G_N$. At each of the time points $t = 1, 2, 3, \ldots$ a Markov chain occupies one of these states. In each time step t to $t + 1$ the process either stays in the same state or moves to some other state in X. It does so in a probabilistic way, rather than in a deterministic way, that is, if at time t the process is in state G_i, then at time $t + 1$ the process moves to any possible state G_j with a certain probability. This probability is assumed to depend *only* on i and j, *not* on t, or the states that the process occupied before state G_i. The probabilities are then denoted by $p_{G_i G_j}$, or simply p_{ij}, to indicate their dependence on i and j only. The term "probability" here is understood as follows: if we let the Markov chain run freely, then, for every pair i, j, the proportion of observed transitions from G_i to G_j among all observed transitions from G_i tends to p_{ij}, as the number of model runs tends to infinity (this will be explained in more detail below). A rigorous introduction to the probability theory and statistics (in particular, a rigorous treatment of Markov chains) will be given in Chaps. 6 and 8.

The probabilities p_{ij}, $i, j = 1, \ldots N$, are called the *transition probabilities* of the Markov chain and are arranged in the matrix

$$P = \begin{pmatrix} p_{11} & p_{12} & \cdots & p_{1N} \\ p_{21} & p_{22} & \cdots & p_{2N} \\ \vdots & \vdots & \vdots & \vdots \\ p_{N1} & p_{N2} & \cdots & p_{NN} \end{pmatrix},$$

called the *matrix of transition probabilities*. Any row of the matrix corresponds to the state *from* which transitions are made, and any column to the state *to* which transitions are made. Thus the probabilities in any particular row in the matrix P sum up to 1.

A Markov chain is often represented graphically as a collection of circles (states), from which arrows representing transitions come out, as in the example shown in Fig. 3.1. To each arrow a label (the corresponding transition probability) is often assigned. If for some i and j we have $p_{ij} = 0$, then instead of drawing an arrow going from state G_i to state G_j with label 0, we do not draw an arrow at all. Hence the notion of the *connectivity* of a Markov chain arises: it is the graph made of the states connected with arrows, where the corresponding labels are ignored. Markov chains for which all the transition probabilities are non-zero are called *fully connected*. A fixed connectivity defines a family of Markov chains parametrized by the transition probabilities corresponding to the arrows present in the graph. In some applications we will be required to choose particular values of the transition probabilities in accordance with given biological data. In this case we say that the transition probabilities are *estimated* from the data. In such situations the connectivity of the chain will be fixed in advance, and all transition probabilities corresponding to the arrows present in the graph will be allowed to take all possible values (modulo an additional constraint that will be introduced later). As a result of the estimation process, some of these probabilities may be set to 0, and thus the connectivity of the resulting chain may be more constrained than the originally assumed connectivity (that we will call the *a priori connectivity*). In such cases we will sometimes draw all the *a priori* allowed arrows, even if some of the corresponding transition probabilities are eventually set to 0.

One can think of a Markov chain as a process of generating all possible sequences of any fixed finite length $L \geq 2$, that is, sequences of the form $x_1 x_2 \ldots x_L$, where $x_j \in X$ (in fact, the chain generates all possible infinite sequences, but we will restrict our attention only to their first L elements). In order to initiate the process, we need to fix in advance the probabilities $P(G_j)$ for all $j = 1, \ldots, N$ (called the *initialization probabilities*), that is, N non-negative numbers that sum up to 1. We will arrange the initialization probabilities in a vector of length N and call it the *vector of initialization probabilities*. For each j, $P(G_j)$ is the probability of the sequence generation process starting at state G_j. Choosing these values means that we impose the following condition on the process of sequence generation: as the number of runs of the process goes to infinity, the proportion of sequences that start with G_j among all generated sequences tends to $P(G_j)$ for all $j = 1, \ldots, N$. Next, if we generate a collection of sequences of length L and for every pair i, j determine the ratio of the number of times for which G_i in the generated sequences is immediately followed by G_j and the number of times for which G_i is immediately followed by any element from the set X, the resulting ratio is required to tend to p_{ij}, as the number of model runs tends to infinity. This statement is the exact meaning of the phrase "the process moves from state G_i to state G_j with probability p_{ij}". Since we are given that the generation of every element (apart from the first one) of any sequence depends only on the preceding element, it can be shown that the frequency of a particular sequence

$x = x_1 \ldots x_L$ tends to

$$P(x) = P(x_1)p_{x_1 x_2} p_{x_2 x_3} \times \ldots \times p_{x_{L-1} x_L}, \tag{3.1}$$

as the number of runs increases. We will call the number in the right-hand side of formula (3.1) the *probability of the sequence* x. It is easy to prove that the sum of probabilities of all sequences of length L is equal to 1 (see Exercise 3.1). Thus, in accordance with the above discussion, we can define a Markov chain as a pair consisting of a vector of initialization probabilities and a matrix of transition probabilities.

The term "time" used above is appropriate if, for example, we consider the evolution through time of a given nucleotide site and try to model it by means of a Markov chain. For such purposes, however, continuous-time Markov chains are more appropriate; they will be discussed in Sect. 5.4. The role of time can be also played by other characteristics of the data in question. As an example, consider a DNA sequence read from left to right. We can use a Markov chain to model dependence of a nucleotide site on its neighbor on the left, which can be useful if we want to model a specific type of sequence composition.

Consider for instance protein-coding genes in prokaryotic organisms (for example, in bacteria). Typically, such genes occupy as much as 90% of bacterial DNA. A prokaryotic gene consists of a coding region flanked by a start codon (usually ATG, but sometimes GTG or TTG) at the beginning (on the left) and a stop codon (one of TAA, TAG, TGA) at the end (on the right). The length of the region between the start and stop codons is divisible by 3, and each triplet of nucleotides (codon) codes for an amino acid. Stretches of DNA that possess this property are called *open reading frames* or *ORFs*, hence every gene is an ORF to which a start codon at the beginning and a stop codon at the end are added. We note in passing that the structure of a protein-coding gene in eukaryotes is more complicated, in particular, coding regions (called *exons*) are interrupted by non-coding ones (called *introns*).

What sort of probabilistic model might we use for modeling prokaryotic genes? As one possibility, we can consider a Markov chain for which the *a priori* connectivity is shown in Fig. 3.1. Residues in prokaryotic genes are not independent, but interact with each other and modeling such genes by means of a Markov chain takes into account at least the dependence of each residue on the residue that immediately precedes it. For a particular choice of transition and initialization probabilities, if we let the chain run freely, it will generate all possible sequences of letters from the DNA alphabet $\{A, C, G, T\}$. If we want the composition of sequences with high probabilities to resemble that of prokaryotic genes, the transition and initialization probabilities must be selected in a certain way. In order to achieve this goal we can use real DNA data.

Suppose that in a set of prokaryotic DNA sequences n genes were experimentally identified. Then, it is natural to set the transition probabilities as follows

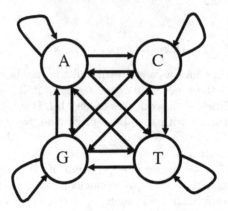

Fig. 3.1.

$$p_{ab} = \frac{H_{ab}}{\displaystyle\sum_{c \in \mathcal{Q}} H_{ac}}, \tag{3.2}$$

where H_{ab} is the number of times nucleotide b follows nucleotide a in the data (if a particular nucleotide a_0 does not occur in the data at all, then the data is not sufficient for estimating $p_{a_0 b}$, and these transition probabilities can be chosen arbitrarily). The initialization probabilities $P(a)$ for each $a \in \mathcal{Q}$ are also calculated in the spirit of formula (3.2), that is, $P(a)$ is the number of times nucleotide a appears at the beginning of the sequences, divided by n. For this Markov chain we hope that sequences with high probabilities resemble those of real prokaryotic genes. We say that the resulting Markov chain is a *model* for the n sequences that were used to set the parameters, or that the n sequences are *modeled* by it. Often we will also say that these sequences are the *training data* for the resulting Markov chain.

Note that for the Markov chain constructed in this way we did not constrain its connectivity in advance, that is, we did not require *a priori* that any particular transition probabilities must be set to 0. The resulting Markov chain may or may not be fully connected depending on the transition probabilities obtained from the data.

Suppose now that we want to do a *search* with the Markov model obtained above. This means that we want to use the model to determine whether or not a given unannotated DNA sequence is the sequence of a prokaryotic gene. Search for prokaryotic genes may seem easy at first sight: indeed, since every gene is an ORF, one can look for a start codon and the first stop codon that follows it, and hope that the stretch of DNA so obtained is a gene. However, it may happen (and it does in fact happen in 50% of cases) that two start codons are encountered without an intervening stop codon (see Exercise 3.4). Which of the corresponding ORFs is then a gene? To answer this question one has to somehow assess the composition of the ORFs and to compare it with that of known gene sequences. This can be attempted by using the simple DNA model constructed above.

Let the unannotated DNA sequence be $x = x_1 \ldots x_L$. We then treat x as if it was generated by the Markov chain, that is, assess the probability $P(x)$ of x found from formula (3.1), which is often called the *probability with which x arises from the model*. If $P(x)$ is in some sense large (say, greater than some preset threshold), then we say that there is a good chance that x came from a prokaryotic gene. If, on the other hand, the probability $P(x)$ is small, we say that x probably has no relation to the training data, and hence no relation to prokaryotic genes (if we believe that the training data is sufficiently representative). Of course, one should bear in mind that the above model is very simple, and its search power is therefore limited. It can produce low probabilities for real prokaryotic genes (false negatives) and high probabilities for sequences unrelated to the training data (false positives).

The procedures described above can be applied to any Markov chain. Indeed, suppose we wish to estimate the transition probabilities of a Markov chain with state set $X = \{G_1, \ldots, G_N\}$ from some training data. The training data consists of a finite number of finite sequences of elements from X. We assume that some *a priori* connectivity is fixed, that is, some of the transition probabilities are set to 0 from the start and that the *training data agrees with the connectivity*, that is, if, for some i and j, the connectivity requires that $p_{ij} = 0$, then the training sequences do not contain G_i followed immediately by G_j anywhere. The same restriction applies to the initialization probabilities: if it is *a priori* required that the chain starts at state G_j with probability 0, then none of the training sequences begins with G_j. If these conditions are satisfied, we can estimate the transition probabilities of the Markov chain from the training data by using formula (3.2), and the initialization probabilities as the frequencies of the occurrences of G_j at the beginning of the sequences for all j. The connectivity of the resulting Markov chain may be more constrained than the *a priori* assumed connectivity (see Exercise 3.3). One should also remember that if a certain state G_i does not occur in any of the sequences, then the data is not sufficient for estimating p_{ij}, and these transition probabilities can be set arbitrarily (we will encounter such a situation in a more general setting in Example 3.13). Once the transition and initialization probabilities have been fixed, search can be done in exactly the same way as described for the DNA model above. The purpose of searches of this type is to detect potential membership of an unannotated sequence in the sequence family used as training data.

To use formula (3.1) we must separately specify the initialization probability $P(x_1)$. To avoid this inhomogeneity, it is possible to add an extra *begin state* to the Markov chain; we denote it by $\mathcal{B}$. Then for the initialization probabilities we have $P(G_j) = p_{\mathcal{B} G_j}$ (we will sometimes also use the notation $p_{\mathcal{B} j}$ for them), $j = 1, \ldots, N$. In the future we will always consider Markov chains with begin states. Similarly, unless explicitly stated otherwise, we will always add an *end state* $\mathcal{E}$ to ensure that the end of the sequence is modeled (see Part 4 of Example 6.11 for an explanation), and we also assume that there is no transition from $\mathcal{B}$ to $\mathcal{E}$. The end state is an *absorbing state*, that is, a state, for which the transition probability into itself is equal to 1. We, however, will ignore this transition and think of a Markov chain with an end state as a process that generates either infinite sequences $\mathcal{B} x_1 x_2 \ldots$, or sequences of finite length of the form $\mathcal{B} x_1 x_2 \ldots x_L \mathcal{E}$, where $x_j \in X$ (we will often omit $\mathcal{B}$ at the beginning and $\mathcal{E}$ at the end of generated sequences). We will only be interested in finite sequences, but we no longer fix their length and consider sequences of all possible lengths. For such Markov chains formula (3.1) changes to

$$P(x) = p_{\mathcal{B} x_1} p_{x_1 x_2} p_{x_2 x_3} \times \ldots \times p_{x_{L-1} x_L} p_{x_L \mathcal{E}}, \tag{3.3}$$

where $p_{x_L \mathcal{E}}$ denotes the probability of transition from x_L to $\mathcal{E}$. The probabilities $p_{G_j \mathcal{E}}$, $j = 1, \ldots, N$, are called the *termination probabilities* (we will sometimes also use the notation $p_{j \mathcal{E}}$ for them). In the future we will apply the term "transition probabilities" to all of $p_{\mathcal{B} j}$, p_{ij} and $p_{j \mathcal{E}}$. We note that a Markov chain with an end state may have other absorbing states, hence the end state is just a distinguished absorbing state for which the transition into itself is ignored.

Note that no arrows go into $\mathcal{B}$ and out of $\mathcal{E}$. Therefore, to simplify notation we will often denote $\mathcal{B}$ and $\mathcal{E}$ by 0 and, instead of writing $p_{\mathcal{B} x_1}$ and $p_{x_L \mathcal{E}}$ we will frequently write $p_{0 x_1}$ and $p_{x_L 0}$ respectively (to simplify notation even further we will also write $p_{0 G_j} = p_{0j}$ and $p_{G_j 0} = p_{j0}$ for $j = 1, \ldots, N$). Then formula (3.3) takes the form

$$P(x) = p_{0 x_1} p_{x_1 x_2} p_{x_2 x_3} \times \ldots \times p_{x_{L-1} x_L} p_{x_L 0}. \tag{3.4}$$

In the future we will only consider Markov models whose connectivity satisfies the following assumption: for every state G_j for which there exists a *path* from $\mathcal{B}$ to G_j, there also exists a path from G_j to $\mathcal{E}$. We will call such models *non-trivially connected*. Not every Markov chain is non-trivially connected. Note that if the transition probabilities are obtained from training data by means of formula (3.2) naturally extended to transitions from $\mathcal{B}$ and into $\mathcal{E}$, the resulting model is non-trivially connected (whereas the *a priori* assumed connectivity may not be that of a non-trivially connected model, as in Exercise 3.3). Of course, for a Markov chain with an end state none of the training sequences is allowed to end with G_j, if the *a priori* connectivity requires that the termination probability p_{j0} is equal to 0; this condition is part of the requirement that the training data agrees with the *a priori* connectivity for such a chain. It can be shown that for non-trivially connected models the sum of probabilities $P(x)$ over all sequences x of all finite lengths is equal to 1 (see Part 5 of Example 6.11 and Exercise 3.2). For more discussion on the justification of the definition of probabilities for a Markov chain with an end state also see Part 5 of Example 6.11.

Figure 3.2 shows the connectivity from Fig. 3.1 modified by adding the begin and end states.

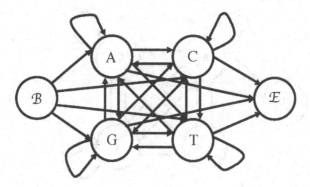

Fig. 3.2.

We will now give an example of parameter estimation for this connectivity taken as an *a priori* connectivity.

Example 3.1. Let $n = 4$ and suppose that we are given the following sequences that are known to be the sequences of some prokaryotic genes

$$
\begin{aligned}
x^1 &: \textbf{ATG}CTATTGATT\textbf{TAA} \\
x^2 &: \textbf{GTG}AAAGACTTC\textbf{TAA} \\
x^3 &: \textbf{ATG}CCCGATGAACGC\textbf{TAG} \\
x^4 &: \textbf{ATG}AAGCATGAT\textbf{TAA}.
\end{aligned}
\tag{3.5}
$$

Then, ignoring the start and stop codons (shown in bold), that is, taking into account only the corresponding ORFs, from formula (3.2) we obtain

$$
p_{0A} = \frac{1}{2}, \quad p_{0C} = \frac{1}{2}, \quad p_{0G} = 0, \quad p_{0T} = 0,
$$

$$
p_{AA} = \frac{4}{13}, p_{AC} = \frac{2}{13}, p_{AG} = \frac{2}{13}, p_{AT} = \frac{5}{13}, p_{A0} = 0,
$$

$$
p_{CA} = \frac{1}{9}, \quad p_{CC} = \frac{2}{9}, \quad p_{CG} = \frac{2}{9}, \quad p_{CT} = \frac{2}{9}, \quad p_{C0} = \frac{2}{9},
$$

$$
p_{GA} = \frac{5}{7}, \quad p_{GC} = \frac{2}{7}, \quad p_{GG} = 0, \quad p_{GT} = 0, \quad p_{G0} = 0,
$$

$$
p_{TA} = \frac{1}{10}, p_{TC} = \frac{1}{10}, p_{TG} = \frac{3}{10}, p_{TT} = \frac{3}{10}, p_{T0} = \frac{1}{5}.
$$

Hence, the resulting model is not fully connected and in fact has the connectivity shown in Fig. 3.3. Note, in particular, that this model is indeed non-trivially connected.

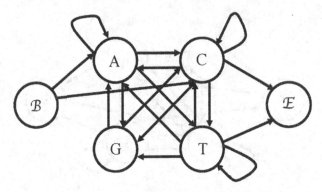

Fig. 3.3.

For an example of searching with this model see Exercise 3.4.

Software package called Glimmer is a primary microbial gene finder. It is based on the ideas described above. However, instead of Markov chains discussed in this section it uses *higher order Markov chains* that model dependence of every nucleotide site on a number of preceding sites, and their generalizations (see [DS]).

In Example 3.1 the *a priori* connectivity did not have any constraints. We will now give an example of parameter estimation with *a priori* constraints.

Example 3.2. Consider the *a priori* connectivity shown in Fig. 3.4. Suppose

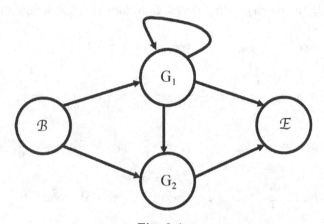

Fig. 3.4.

that the training data consists of the following three sequences

$$x^1 : G_1G_1G_1G_1$$
$$x^2 : G_1G_1G_2$$
$$x^3 : G_1G_2.$$

We see that the training data agrees with the connectivity, and from formula (3.2) obtain

$$p_{01} = 1, \quad p_{02} = 0,$$

$$p_{11} = \frac{4}{7}, p_{12} = \frac{2}{7}, p_{10} = \frac{1}{7}.$$

It is also obvious that $p_{22} = 0, p_{20} = 1$ (this is clear from the given connectivity, but also follows from the estimation procedure).

Observe that the connectivity of the resulting Markov chain is more constrained than the original connectivity, namely, it has no arrow from $\mathcal{B}$ to G_2.

It will be explained in Chap. 8 (see Part 5 of Example 8.12 and Exercise 8.4) that formula (3.2) does not just seem intuitively right, but gives values of the transition probabilities that guarantee that the *likelihood of the training data* $P(x^1) \times \ldots \times P(x^n)$ is maximal possible. As we will be see in Chap. 8, the likelihood can be thought of as the joint probability of the occurrence of the sequences $x^1, \ldots, x^n$ in n "independent runs" of the model. Therefore, it is often said that with this estimation procedure the sequences in the training data are assumed to be *independent*.

3.2 Hidden Markov Models

The approach to look for new prokaryotic genes described in the previous section required extracting all ORFs from the DNA sequence in question and analyzing each ORF separately. It would be useful, however, if we could design an algorithm that could analyze unannotated DNA sequences directly, without making the preprocessing step of extracting all possible ORFs. The need for such an algorithm becomes even more clear, if instead of the problem of searching for prokaryotic genes we consider the problem of searching for some other DNA features that do not have such well-defined boundaries as genes (start and stop codons). One example of a search of this kind is the determination of *CpG* islands discussed in [DEKM].

To construct such an algorithm, we have to model both the sequence composition of genes and that of intergenic regions. One possibility that can be considered is to use a model with connectivity shown in Fig. 3.1 for genes, another model with the same connectivity for intergenic regions, allow all possible transitions between the states of the two models, and add the begin and end states. Suppose we have somehow set the transition probabilities in a reasonable way. We will call the resulting model the *two-block model*. Let us denote the states of the first model by A_g, C_g, G_g, T_g and the states of the second model by $A_{ig}, C_{ig}, G_{ig}, T_{ig}$. The complication with this new model is that there is no longer a one-to-one correspondence between the letters of the DNA alphabet $\{A, C, G, T\}$ and the non-zero states of the model. Therefore, if we are given an unannotated DNA sequence, say $x^0 = ACCTG$, and we wish to treat x^0 as a sequence generated by the model, it is not clear how one should calculate the probability with which x^0 arises from it. Indeed, x^0 could be produced, for example, by the sequence of states $A_g C_{ig} C_g T_g G_{ig}$, in which case according to formula (3.4) the probability of x^0 should be computed as

$$p_0 A_g p_{A_g C_{ig}} p_{C_{ig} C_g} p_{C_g T_g} p_{T_g G_{ig}} p_{G_{ig} 0}. \tag{3.6}$$

Alternatively, x^0 could be produced in several other ways, for example, by the sequence of states $A_{ig} C_{ig} C_g T_g G_g$, in which case the probability of x should be computed as

$$p_0 A_{ig} p_{A_{ig} C_{ig}} p_{C_{ig} C_g} p_{C_g T_g} p_{T_g G_g} p_{G_g 0}. \tag{3.7}$$

This indeterminacy arises because we do not *a priori* know the way x^0 was generated by the model, that is, the corresponding sequence of states. In this situation it is common to say that the state sequence is *hidden*. The two-block DNA model described above is an example of a *hidden Markov model* or an *HMM*.

We will now give a general definition of HMM. An HMM is an ordinary discrete-time finite Markov chain (that we always assume to be non-trivially connected) with states $G_1, \ldots, G_N$, transition probabilities $p_{0j}, p_{ij}, p_{j0}, i, j = 1, \ldots, N$, that, in addition, at each state emits symbols from an alphabet $\mathcal{Q}$ (the DNA alphabet in our examples). For each state G_k and each symbol $a \in \mathcal{Q}$ an *emission probability* $q_{G_k}(a) = q_k(a)$ is specified. For every $k = 1, \ldots, N$, the probabilities $q_k(a)$ sum up to 1 over all $a \in \mathcal{Q}$. The Markov chain will be referred to as the *underlying Markov chain of the HMM*. The two-block DNA model can be thought of as an HMM where at each state one letter (the letter used in the name of the state) is emitted with probability 1 and the rest of the letters are emitted with probability 0.

One can think of an HMM as a process of sequence generation. There are two possible interpretations of this process, each interpretation being suitable for particular types of problems. One way is to think of an HMM as a process that generates *pairs of sequences* (x, π), where x is a sequence of letters from $\mathcal{Q}$ and π is a sequence of non-zero states of the Markov chain (we will generally call sequences of states *paths through the underlying Markov chain* or simply *paths*). The lengths of x and π are equal and can be either finite or infinite. We will only be interested in finite sequences and write x and π respectively as $x_1 x_2 \ldots x_L$ and $\pi = \mathcal{B} \pi_1 \pi_2 \ldots \pi_L \mathcal{E} = 0 \pi_1 \pi_2 \ldots \pi_L 0 = \pi_1 \pi_2 \ldots \pi_L$, where x_j is the element of $\mathcal{Q}$ emitted at the state π_j, for $j = 1, \ldots, L$. As the number of runs of such a model increases, the frequency of an element $a \in \mathcal{Q}$ emitted at state G_k among all pairs of sequences whose length does not exceed any given number, tends to $q_k(a)$. This is the exact meaning of the phrase "the state G_k emits the symbol a with probability $q_k(a)$". This interpretation of an HMM is useful, when one applies it to analyzing biological sequences for which paths through the underlying Markov chain can be assumed known (for example, when the two-block DNA model is used to model DNA sequences where prokaryotic genes have been determined experimentally). Another way is to think of an HMM as a process that generates sequences of letters from $\mathcal{Q}$, that is, in this interpretation we ignore paths along which sequences are generated. This interpretation of an HMM is useful, when one applies it to analyzing biological sequences for which paths through the underlying Markov chain are unknown, as in the example of the sequence x^0 above. A rigorous treatment of HMMs incorporating both approaches will be given in Chaps. 6 and 8.

As in the case of Markov chains, any *a priori* connectivity defines a family of HMMs parametrized by the transition probabilities corresponding to the arrows present in the connectivity graph (subject to constraints dictated by the condition of non-trivial connectedness) and the emission probabilities. Analogously to Markov chains, HMMs are often derived as *models* for particular *training data*, from which the transition and emission probabilities can be *estimated*. This process of parameter estimation is more complicated than that for Markov chains (based on formula (3.2)) and will be discussed in Sect. 3.6. We emphasize that the underlying Markov chain obtained as a result of the estimation procedures described in Sect. 3.6 is always non-trivially connected, even if the *a priori* assumed connectivity is not that of a non-trivially connected Markov chain.

Let $x = x_1 \ldots x_L$ be a sequence of letters from $\mathcal{Q}$ and $\pi = \pi_1 \ldots \pi_L$ be a path of the same length. We will now define the probability $P(x, \pi)$ of the pair (x, π) as follows

$$P(x, \pi) = p_{0\,\pi_1} q_{\pi_1}(x_1) p_{\pi_1 \pi_2} q_{\pi_2}(x_2) \times \ldots \times p_{\pi_{L-1} \pi_L} q_{\pi_L}(x_L) p_{\pi_L\,0}. \qquad (3.8)$$

It is natural to think of this probability as the probability of the sequence x being generated along the path π. The definition is reasonable since it can be shown that the frequency of x being generated by the model along π among all truncations of pairs of sequences at length greater than L tends to the number in the right-hand side of formula (3.8) – see Parts 5 and 6 of Example 6.11 for a precise statement. Formula (3.8) is the HMM analogue of formula (3.4), and (3.6), (3.7) are in fact calculated for the two-block DNA model in accordance with (3.8) and equal respectively to $P(x^0, \pi^1)$ and $P(x^0, \pi^2)$ with $\pi^1 = A_g C_{ig} C_g T_g G_{ig}$ and $\pi^2 = A_{ig} C_{ig} C_g T_g G_g$. It can be shown that for a general HMM we have $\sum_{(x,\pi)} P(x, \pi) = 1$, where the summation is taken over all pairs of sequences of all finite lengths (see Exercise 3.5).

However, formula (3.8) is not so useful in practice because in general for a biological sequence we do not know the path. In the next section we will describe an algorithm that for a given sequence x finds all the *most probable paths*, that is, paths that maximize $P(x, \pi)$ over all paths π of length equal to the length of x. One of the most probable paths is often regarded as the path along which x is generated by the model.

Further, for the purposes of analyzing sequences for which paths are not known, it is natural to define "the total probability" $P(x)$ of x as

$$P(x) = \sum_{\text{all } \pi \text{ of length } L} P(x, \pi). \tag{3.9}$$

It follows from Exercise 3.5 that $\sum_x P(x) = 1$, where the summation is taken over all sequences of all finite lengths. In Sects. 3.4 and 3.5 we will describe algorithms that for a given sequence x calculate $P(x)$.

HMMs can be used for *searching* in a way similar to that for Markov models, with the purpose of detecting potential membership of an unannotated sequence in a sequence family used as training data (see Exercise 3.6). For example, suppose we are given an unannotated DNA sequence $x = x_1 \ldots x_L$, and we wish to determine whether or not the sequence x has something in common with the training DNA data used to estimate the parameters of the two-block DNA model. One way to do it is by assessing the probability $P(x)$ of x found from formula (3.9), which is often called the *probability with which x arises from the model*. Alternatively, one can assess $P(x, \pi^*)$, the *probability with which x arises from the model along the path π^**, calculated from formula (3.8) for one of the most probable paths π^*. If either of these quantities is in some sense large (say, greater than some preset threshold), then we say that there is a good chance that x is related to the training data. If, on the other hand, the probability $P(x)$ is small, we say that x probably has no relation to the training data. Often instead of $P(x)$ and $P(x, \pi^*)$ different (but related) probabilistic quantities are used for assessing unannotated sequences. Of course, one must bear in mind that searches of this kind may lead to false negatives and false positives.

Another kind of search that one can perform using HMMs is, for a query sequence x, determine a likely path through the Markov model. For example, if we search with the two-block DNA model, we are interested in knowing which elements of x are likely to have come from the states labeled using the subscript "g" since continuous path stretches consisting of such states may correspond to prokaryotic genes. Recovering likely paths for sequences is called *decoding* and will be discussed in Sects. 3.3 and 3.5. Of course, searches of this kind can lead to errors as well.

The two-block DNA model has a very large number of parameters, and one needs a very large training dataset to reasonably estimate all of them. We will now give a more appropriate example of an HMM for modeling prokaryotic DNA. Consider the *a priori* connectivity shown in Fig. 3.5.

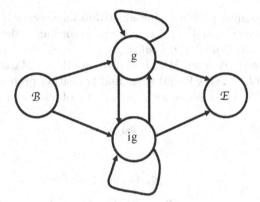

Fig. 3.5.

The state "g" is associated with genes and the state "ig" with intergenic regions. Each of the states is allowed to emit letters from the DNA alphabet $\{A, C, G, T\}$ with certain probabilities (note, however, that a more realistic model would single out the start codons on the way from "ig" to "g" and stop codons on the way from "g" to "ig"). In Sect. 3.6 we will give an example of parameter estimation for this model.

While search for prokaryotic genes can in fact be done avoiding HMMs (see the previous section), generalizations of the HMM arising from the connectivity shown in Fig. 3.5 are useful for finding genes in eukaryotes. The structure of eukaryotic genes is much more complicated than that of prokaryotic ones, and a separate state is required to model every particular gene feature, as well as the intergenic region. Also, at each state not single letters, but sequences of letters are emitted. Such a model is implemented in GENSCAN which is a popular successful gene finder for human DNA. Details on GENSCAN can be found in [EG] and the original paper [BK].

3.3 The Viterbi Algorithm

In the previous section we mentioned a particular kind of search performed with HMMs called decoding. With a search of this type, if x is a sequence of letters from the alphabet $\mathcal{Q}$, one wishes to know a likely path for x through the underlying Markov chain. Of course, we first have to specify what "likely" means. One reasonable approach is to attempt to find a path π^* that maximizes the probability $P(x, \pi)$ over all paths π with length equal to the length of x (another approach will be discussed in Sect. 3.5). Such a most probable path may not be unique. One can try to find all the most probable paths directly by listing all possible paths and calculating $P(x, \pi)$ for each of them. However, the number of all possible paths increases exponentially with the length of the sequence, and therefore such a strategy is not practical. There is a much faster dynamic programming algorithm for determining all the most probable paths, called the *Viterbi algorithm*. Accordingly, the most probable paths are often called the *Viterbi paths*.

Suppose we are given an HMM whose underlying Markov chain has a state set $X = \{G_1, \ldots, G_N\}$, end state and transition probabilities p_{0j}, p_{ij}, p_{j0}, $i, j = 1, \ldots, N$. Fix a sequence $x = x_1 \ldots x_L$ of letters from the alphabet Q and define

$$v_k(1) = p_{0k} q_k(x_1) \tag{3.10}$$

for $k = 1, \ldots, N$, and

$$v_k(i) = \max_{\pi_1, \ldots, \pi_{i-1} \in X} p_{0\,\pi_1} q_{\pi_1}(x_1) p_{\pi_1\,\pi_2} q_{\pi_2}(x_2) \times \ldots$$
$$\times p_{\pi_{i-2}\,\pi_{i-1}} q_{\pi_{i-1}}(x_{i-1}) p_{\pi_{i-1}\,G_k} q_k(x_i),$$

for $i = 2, \ldots, L$ and $k = 1, \ldots, N$. Clearly, we have

$$v_k(i+1) = q_k(x_{i+1}) \max_{l=1,\ldots,N} (v_l(i) p_{lk}) \tag{3.11}$$

for all $i = 1, \ldots, L-1$ and $k = 1, \ldots, N$. Therefore, $v_k(i)$ for $i = 2, \ldots, L$ and $k = 1, \ldots, N$ can be calculated from initial conditions (3.10) and recursion relations (3.11). At each recursion step we form a set $\mathcal{V}_k(i)$ that consists of all integers m for which $v_m(i) p_{mk} = \max_{l=1,\ldots,N} (v_l(i) p_{lk})$, $i = 1, \ldots, L-1$, $k = 1, \ldots, N$.

Further, we have

$$\max_{\text{all } \pi \text{ of length } L} P(x, \pi) = \max_{l=1,\ldots,N} (v_l(L) p_{l0}),$$

and define $\mathcal{V}(L)$ to be the set that consists of all integers m for which $v_m(L) p_{m0} = \max_{l=1,\ldots,N} (v_l(L) p_{l0})$. Every Viterbi path can be recovered by the following traceback procedure: choose $m_L \in \mathcal{V}(L)$, then choose $m_{L-1} \in \mathcal{V}_{m_L}(L-1)$, then choose $m_{L-2} \in \mathcal{V}_{m_{L-1}}(L-2)$, and proceed until we choose $m_1 \in \mathcal{V}_{m_2}(1)$. For the resulting path $\pi^* = 0 G_{m_1} G_{m_2} \ldots G_{m_L} 0 = G_{m_1} G_{m_2} \ldots G_{m_L}$ we have

$$P(x, \pi^*) = \max_{\text{all } \pi \text{ of length } L} P(x, \pi).$$

The Viterbi algorithm is implemented in GENSCAN, where decoding corresponds to human gene finding.

We will now give an example of applying the Viterbi algorithm.

Example 3.3. Consider the HMM for which Q is the two-letter alphabet $\{A, B\}$, whose underlying Markov chain is shown in Fig. 3.6 and whose emission probabilities are as follows

$$q_1(A) = 0.5, \ q_1(B) = 0.5,$$
$$q_2(A) = 0.1, \ q_2(B) = 0.9,$$
$$q_3(A) = 0.9, \ q_3(B) = 0.1.$$

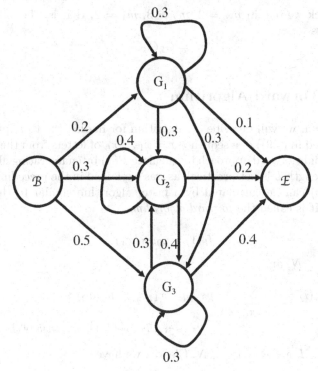

Fig. 3.6.

We let $x = BAB$ and find the (unique) Viterbi path π^* for it. We have

$$v_1(1) = 0.5 \times 0.2 = 0.1,$$
$$v_2(1) = 0.9 \times 0.3 = 0.27,$$
$$v_3(1) = 0.1 \times 0.5 = 0.05,$$

$$
\begin{aligned}
v_1(2) &= 0.5 \times 0.1 \times 0.3 = 0.015, & \mathcal{V}_1(1) &= \{1\}, \\
v_2(2) &= 0.1 \times 0.27 \times 0.4 = 0.0108, & \mathcal{V}_2(1) &= \{2\}, \\
v_3(2) &= 0.9 \times 0.27 \times 0.4 = 0.0972, & \mathcal{V}_3(1) &= \{2\}, \\
v_1(3) &= 0.5 \times 0.015 \times 0.3 = 0.00225, & \mathcal{V}_1(2) &= \{1\}, \\
v_2(3) &= 0.9 \times 0.0972 \times 0.3 = 0.026244, & \mathcal{V}_2(2) &= \{3\}, \\
v_3(3) &= 0.1 \times 0.0972 \times 0.3 = 0.002916, & \mathcal{V}_3(2) &= \{3\}.
\end{aligned}
$$

This gives

$$\max_{\text{all } \pi \text{ of length 3}} P(x, \pi) = 0.026244 \times 0.2 = 0.0052488,$$

and

$$\mathcal{V}(3) = \{2\}.$$

Tracing back we obtain $m_3 = 2$, $m_2 = 3$, $m_1 = 2$, that is, the only Viterbi path for x is

$$\pi^* = G_2 G_3 G_2.$$

3.4 The Forward Algorithm

In this section we will describe an algorithm for finding the total probability $P(x)$ (defined in (3.9)) of a sequence $x = x_1 \ldots x_L$ of letters from the alphabet $\mathcal{Q}$. The straightforward approach to calculate $P(x)$ by enumerating all possible paths π of length L is not practical, as was explained in the preceding section. In fact $P(x)$ can be calculated by a faster algorithm similar to the Viterbi algorithm. It is called the *forward algorithm*.

Define

$$f_k(1) = p_{0k} q_k(x_1) \tag{3.12}$$

for $k = 1, \ldots, N$, and

$$f_k(i) = \sum_{\pi_1, \ldots, \pi_{i-1} \in X} p_{0\,\pi_1} q_{\pi_1}(x_1) p_{\pi_1\,\pi_2} q_{\pi_2}(x_2) \times \ldots$$
$$\times p_{\pi_{i-2}\,\pi_{i-1}} q_{\pi_{i-1}}(x_{i-1}) p_{\pi_{i-1}\,G_k} q_k(x_i),$$

for $i = 2, \ldots, L$ and $k = 1, \ldots, N$. Clearly, we have

$$f_k(i+1) = q_k(x_{i+1}) \sum_{l=1}^{N} f_l(i) p_{lk} \tag{3.13}$$

for all $i = 1, \ldots, L-1$ and $k = 1, \ldots, N$. Therefore, $f_k(i)$ for $i = 2, \ldots, L$ and $k = 1, \ldots, N$ can be calculated from initial conditions (3.12) and recursion relations (3.13). Further, we have

$$P(x) = \sum_{k=1}^{N} f_k(L) p_{k0}.$$

The name of this algorithm is related to the fact that it reads the sequence x forward. In the next section we will see another algorithm for calculating $P(x)$ that reads the sequence backward.

We will now give an example of applying the forward algorithm.

Example 3.4. Consider the HMM from Example 3.3 and find $P(x)$, where $x = AAB$. We have

$f_1(1) = 0.5 \times 0.2 = 0.1,$

$f_2(1) = 0.1 \times 0.3 = 0.03,$

$f_3(1) = 0.9 \times 0.5 = 0.45,$

$f_1(2) = 0.5 \times 0.1 \times 0.3 = 0.015,$

$f_2(2) = 0.1 \times (0.1 \times 0.3 + 0.03 \times 0.4 + 0.45 \times 0.3) = 0.0177,$

$f_3(2) = 0.9 \times (0.1 \times 0.3 + 0.03 \times 0.4 + 0.45 \times 0.3) = 0.1593,$

$f_1(3) = 0.5 \times 0.015 \times 0.3 = 0.00225,$

$f_2(3) = 0.9 \times (0.015 \times 0.3 + 0.0177 \times 0.4 + 0.1593 \times 0.3) = 0.053433,$

$f_3(3) = 0.1 \times (0.015 \times 0.3 + 0.0177 \times 0.4 + 0.1593 \times 0.3) = 0.005937,$

which gives

$$P(x) = 0.00225 \times 0.1 + 0.053433 \times 0.2 + 0.005937 \times 0.4 = 0.0132864.$$

3.5 The Backward Algorithm and Posterior Decoding

In this section we will describe another algorithm for calculating $P(x)$, the *backward algorithm*. The name of this algorithm is related to the fact that, unlike the forward algorithm, it reads the sequence x backward. However, as we will explain later in the section, this algorithm also has other uses. In particular, there is a decoding procedure (alternative to the one based on the Viterbi algorithm) that requires quantities calculated by both the forward and backward algorithms.

For a fixed sequence $x = x_1 \ldots x_L$ of letters from the alphabet $\mathcal{Q}$ define

$$b_k(i) = \sum_{\pi_{i+1}, \ldots, \pi_L \in X} p_{G_k \, \pi_{i+1}} q_{\pi_{i+1}}(x_{i+1}) p_{\pi_{i+1} \, \pi_{i+2}} q_{\pi_{i+2}}(x_{i+2}) \times \cdots$$
$$\times p_{\pi_{L-1} \, \pi_L} q_{\pi_L}(x_L) p_{\pi_L \, 0},$$

for $i = 1, \ldots, L - 1$, $k = 1, \ldots, N$ and

$$b_k(L) = p_{k0}, \tag{3.14}$$

for $k = 1, \ldots, N$. Clearly, we have

$$b_k(i) = \sum_{l=1}^{N} p_{kl} q_l(x_{i+1}) b_l(i+1) \tag{3.15}$$

for $i = 1, \ldots, L - 1$, $k = 1, \ldots, N$. Therefore $b_k(i)$ for $i = 1, \ldots, L - 1$, $k = 1, \ldots, N$ can be found from initial conditions (3.14) and recursion relations (3.15). Further, we have

$$P(x) = \sum_{k=1}^{N} p_{0k} q_k(x_1) b_k(1).$$

We will now give an example of applying the backward algorithm.

Example 3.5. Consider the HMM and the sequence $x = AAB$ from Example 3.4 and find $P(x)$ using the backward algorithm. We have

$b_1(3) = 0.1,$
$b_2(3) = 0.2,$
$b_3(3) = 0.4,$
$b_1(2) = 0.3 \times 0.5 \times 0.1 + 0.3 \times 0.9 \times 0.2 + 0.3 \times 0.1 \times 0.4 = 0.081,$
$b_2(2) = 0.4 \times 0.9 \times 0.2 + 0.4 \times 0.1 \times 0.4 = 0.088,$
$b_3(2) = 0.3 \times 0.9 \times 0.2 + 0.3 \times 0.1 \times 0.4 = 0.066,$
$b_1(1) = 0.3 \times 0.5 \times 0.081 + 0.3 \times 0.1 \times 0.088 + 0.3 \times 0.9 \times 0.066 = 0.03261,$
$b_2(1) = 0.4 \times 0.1 \times 0.088 + 0.4 \times 0.9 \times 0.066 = 0.02728,$
$b_3(1) = 0.3 \times 0.1 \times 0.088 + 0.3 \times 0.9 \times 0.066 = 0.02046.$

This gives

$$P(x) = 0.2 \times 0.5 \times 0.03261 + 0.3 \times 0.1 \times 0.02728 + 0.5 \times 0.9 \times 0.02046 = 0.0132864,$$

which agrees with the result in Example 3.4.

We will now introduce some concepts from the probability theory. We will only do it for the special case associated with HMMs. A general treatment of these concepts is postponed until Chap. 6.

Suppose we are given an HMM for which we assume, as usual, that the underlying Markov chain is non-trivially connected. Define the *sample space* as follows

$$S = \Big\{ (y, \pi) : y \text{ is a sequence of letters from } Q \text{ of finite length and}$$

$$\pi \text{ is a path of the same length through the underlying Markov chain} \Big\}.$$

The sample space is the collection of all outcomes of the HMM, if we think of it as a process that generates pairs of finite sequences as was explained in Sect. 3.2. Subsets of S are called *events*. We say that an event E *occurs* at a particular time, if the element of the sample space generated by the model at that time belongs to E. For an arbitrary event E we define the *probability of the occurrence of E* or simply the *probability of E* as follows

$$P(E) = \sum_{(y,\pi) \in E} P(y, \pi).$$

Next, for two events E_1 and E_2 with $P(E_2) > 0$ define the *conditional probability* $P(E_1|E_2)$ as follows

$$P(E_1|E_2) = \frac{P(E_1 \cap E_2)}{P(E_2)}.$$

The conditional probability $P(E_1|E_2)$ is often called the *probability of the event E_1 given the event E_2* and has the meaning of the probability of the event E_1 calculated with the prior knowledge that the event E_2 occurs.

For a fixed sequence $x = x_1 \ldots x_L$ consider the event

$$E(x) = \{(y, \pi) \in S : y = x\}.$$

Clearly, $P(E(x)) = P(x)$, where $P(x)$ was defined in (3.9). Next, fix two integers $1 \leq i \leq L$ and $1 \leq k \leq N$, and consider the event

$$E_{i,k} = \{(y, \pi) \in S : \text{the length of } y \text{ and } \pi \text{ is } \geq i, \text{ and } \pi_i = G_k\}.$$

We suppose that $P(x) > 0$ (otherwise x is not an interesting sequence) and calculate the conditional probability $P(E_{i,k}|E(x))$. Assuming that $2 \leq i \leq L - 1$ we obtain

$$P(E_{i,k}|E(x)) = \frac{P(E_{i,k} \cap E(x))}{P(x)} = \frac{1}{P(x)} \sum_{\pi_1, \ldots, \pi_{i-1}, \pi_{i+1}, \ldots, \pi_L \in X} p_{0 \, \pi_1} q_{\pi_1}(x_1)$$

$$\times p_{\pi_1 \, \pi_2} q_{\pi_2}(x_2) \ldots p_{\pi_{i-2} \, \pi_{i-1}} q_{\pi_{i-1}}(x_{i-1}) p_{\pi_{i-1} \, G_k} q_k(x_i) p_{G_k \, \pi_{i+1}} q_{\pi_{i+1}}(x_{i+1})$$

$$\times p_{\pi_{i+1} \, \pi_{i+2}} q_{\pi_{i+2}}(x_{i+2}) \ldots p_{\pi_{L-1} \, \pi_L} q_{\pi_L}(x_L) p_{\pi_L \, 0} = \frac{1}{P(x)} \sum_{\pi_1, \ldots, \pi_{i-1} \in X} p_{0 \, \pi_1}$$

$$\times q_{\pi_1}(x_1) p_{\pi_1 \, \pi_2} q_{\pi_2}(x_2) \ldots p_{\pi_{i-2} \, \pi_{i-1}} q_{\pi_{i-1}}(x_{i-1}) p_{\pi_{i-1} \, G_k} q_k(x_i)$$

$$\times \sum_{\pi_{i+1}, \ldots, \pi_L \in X} p_{G_k \, \pi_{i+1}} q_{\pi_{i+1}}(x_{i+1}) p_{\pi_{i+1} \, \pi_{i+2}} q_{\pi_{i+2}}(x_{i+2}) \ldots p_{\pi_{L-1} \, \pi_L}$$

$$\times q_{\pi_L}(x_L) p_{\pi_L \, 0} = \frac{f_k(i) b_k(i)}{P(x)},$$

where $f_k(i)$ are the quantities defined in the previous section and found from the forward algorithm. Analogous calculations for $i = 1$ and $i = L$ show that the same result holds in these cases as well and thus we have

$$P(E_{i,k}|E(x)) = \frac{f_k(i) b_k(i)}{P(x)} \tag{3.16}$$

for $i = 1, \ldots, L$ and $k = 1, \ldots, N$. Hence the probabilities $P(E_{i,k}|E(x))$ can be efficiently calculated by applying both the forward and backward algorithms at the same time (the probability $P(x)$ can be found from either of them).

For every $i = 1, \ldots, L$ and $k = 1, \ldots, N$ the number $P(E_{i,k}|E(x))$ is the probability with which the ith element x_i of x is emitted at state G_k and is also called the *posterior probability of state G_k at observation i given x*. Posterior probabilities can be used for decoding in the following way. For every $i = 1, \ldots, L$ consider the set $\mathcal{B}(i)$ that consists of all states $G_m \in X$ for which $P(E_{i,m}|E(x)) = \max_{k=1,\ldots,N} P(E_{i,k}|E(x))$. In other words, elements of $\mathcal{B}(i)$ are the *most probable states for x_i*, $i = 1,\ldots,L$. The determination of all such states for every x_i is called *posterior decoding*.

For decoding it is also desirable to know complete paths for x rather than the most probable states for every x_i, and we can consider all paths $\hat{\pi}$ of length L such that $\hat{\pi}_i \in \mathcal{B}(i)$ for $i = 1,\ldots,L$. Note, however, that for such a path $\hat{\pi}$ the probability $P(x,\hat{\pi})$ may be very low (and, in fact, is often equal to 0).

There are also other uses of posterior probabilities. Consider, for example, the two-block model from Sect. 3.2. For a DNA sequence x we are interested in determining which of the elements of x are likely to have been generated at one of the states labeled using the subscript "g". The *posterior probability of these states at observation i given x* is equal to the sum of the posterior probabilities for each of the states. If this probability is large (say, greater than some preset threshold) for every element in a long continuous subsequence of x, then the subsequence is accepted as part of a prokaryotic gene. This procedure of gene finding is alternative to the one used in GENSCAN.

Another application of posterior probabilities is parameter estimation that will be discussed in the next section.

We will now give an example of posterior decoding.

Example 3.6. Consider the HMM and the sequence $x = AAB$ from Examples 3.4 and 3.5 and find the posterior probabilities $P(E_{i,k}|E(x))$ for $i,k = 1,2,3$. From the calculations in Examples 3.4 and 3.5 and formula (3.16) we obtain

$$P(E_{1,1}|E(x)) = 0.2454389, \quad P(E_{1,2}|E(x)) = 0.06159682,$$
$$P(E_{1,3}|E(x)) = 0.6929642, \quad P(E_{2,1}|E(x)) = 0.0914469,$$
$$P(E_{2,2}|E(x)) = 0.1172327, \quad P(E_{2,3}|E(x)) = 0.7913204,$$
$$P(E_{3,1}|E(x)) = 0.01693461, \quad P(E_{3,2}|E(x)) = 0.8043262,$$
$$P(E_{3,3}|E(x)) = 0.1787392,$$

and therefore posterior decoding is

$$\mathcal{B}(1) = \{G_3\}, \quad \mathcal{B}(2) = \{G_3\}, \quad \mathcal{B}(3) = \{G_2\}.$$

This decoding produces a single path

$$\hat{\pi} = G_3 G_3 G_2,$$

for which

$$P(x,\hat{\pi}) = 0.5 \times 0.9 \times 0.3 \times 0.9 \times 0.3 \times 0.9 \times 0.2 = 0.006561.$$

Next, for each x_i, $i = 1,2,3$, we will calculate the probability with which it is generated by either G_1 or G_2. The result for $x_1 = A$ is $P(E_{1,1}|E(x)) + P(E_{1,2}|E(x)) = 0.3070357$, for $x_2 = A$ is $P(E_{2,1}|E(x)) + P(E_{2,2}|E(x)) = 0.2086796$, and for $x_3 = B$ is $P(E_{3,1}|E(x)) + P(E_{3,2}|E(x)) = 0.8212608$.

For comparison, we will now find the (unique) Viterbi path for x. We have

$$v_1(1) = 0.5 \times 0.2 = 0.1,$$
$$v_2(1) = 0.1 \times 0.3 = 0.03,$$
$$v_3(1) = 0.9 \times 0.5 = 0.45,$$

$v_1(2) = 0.5 \times 0.1 \times 0.3 = 0.015,$	$\mathcal{V}_1(1) = \{1\},$
$v_2(2) = 0.1 \times 0.45 \times 0.3 = 0.0135,$	$\mathcal{V}_2(1) = \{3\},$
$v_3(2) = 0.9 \times 0.45 \times 0.3 = 0.1215,$	$\mathcal{V}_3(1) = \{3\},$
$v_1(3) = 0.5 \times 0.015 \times 0.3 = 0.00225,$	$\mathcal{V}_1(2) = \{1\},$
$v_2(3) = 0.9 \times 0.1215 \times 0.3 = 0.032805,$	$\mathcal{V}_2(2) = \{3\},$
$v_3(3) = 0.1 \times 0.1215 \times 0.3 = 0.003645,$	$\mathcal{V}_3(2) = \{3\}.$

This gives

$$\max_{\text{all } \pi \text{ of length } 3} P(x, \pi) = 0.032805 \times 0.2 = 0.006561 = P(x, \hat{\pi}),$$

and

$$\mathcal{V}(3) = \{2\}.$$

Tracing back we obtain $m_3 = 2$, $m_2 = 3$, $m_1 = 3$, and the only Viterbi path for x is

$$\pi^* = G_3 G_3 G_2 = \hat{\pi}.$$

Thus in this example the posterior decoding procedure gives the same result as the one based on the Viterbi algorithm.

3.6 Parameter Estimation for HMMs

As we have already mentioned in Sect. 3.2, a fixed *a priori* Markov chain connectivity defines a family of HMMs parametrized by the corresponding transition and emission probabilities, where the transition probabilities are required to satisfy the conditions arising from non-trivial connectedness. In this section we will explain how for a particular *training dataset* one can choose values of the parameters in a reasonable way. A training dataset consists either of a number of pairs of sequences $(x^1, \pi^1), \ldots, (x^n, \pi^n)$, or a number of sequences $x^1, \ldots, x^n$, where x^j is a finite sequence of letters from the alphabet $\mathcal{Q}$ and π^j is a path of the same length through the graph representing the *a priori* connectivity, that starts at $\mathcal{B}$ and ends at $\mathcal{E}$, $j = 1, \ldots, n$. We will attempt to select parameter values in such a way that, for datasets of the first type, $P(x^1, \pi^1) \times \ldots \times P(x^n, \pi^n)$ is maximal possible and, for datasets of the second type, $P(x^1) \times \ldots \times P(x^n)$ is maximal possible. The resulting HMM is said to *model* the training data; we also say that the training data is *modeled* by the HMM.

The quantities $P(x^1, \pi^1) \times \ldots \times P(x^n, \pi^n)$ and $P(x^1) \times \ldots \times P(x^n)$ are called the *likelihoods of the training datasets*. As we will explain in detail in Chap. 8, for a fixed set of parameters the first one can be thought of as the joint probability of the occurrence of the pairs $(x^1, \pi^1), \ldots, (x^n, \pi^n)$ in n "independent runs" of the corresponding HMM, if we treat it as a process that generates pairs of sequences; the second one can be thought of as the probability of the occurrence of $x^1, \ldots, x^n$ in n "independent runs" of the HMM if we treat it as a process that generates single sequences. Therefore, it is often said that parameter estimation procedures that attempt to maximize the likelihoods assume that the components of the datasets are *independent*.

3.6.1 Estimation when Paths are Known

We will now explain the estimation procedure for datasets of the first type. Suppose we are given n pairs of sequences $(x^1, \pi^1), \ldots, (x^n, \pi^n)$, where for each j, x^j is a finite sequence of letters from the alphabet $\mathcal{Q}$ and π^j is a path of the same length through the graph representing the *a priori* connectivity, that starts at $\mathcal{B}$ and ends at $\mathcal{E}$ (in particular, the path component of the dataset agrees with the *a priori* connectivity). Then we can count the number of times each particular transition or emission is used in the set of training sequences. Let these be $H_{\alpha\beta}$ and $J_l(a)$ with $\alpha = \mathcal{B}, 1, \ldots, N$, $\beta = 1, \ldots, N, \mathcal{E}$, $l = 1, \ldots, N$, $a \in \mathcal{Q}$. Then we set

$$p_{\alpha\beta} = \frac{H_{\alpha\beta}}{\displaystyle\sum_{\gamma=1,\ldots,N,\mathcal{E}} H_{\alpha\gamma}}, \qquad q_l(a) = \frac{J_l(a)}{\displaystyle\sum_{b\in\mathcal{Q}} J_l(b)}, \qquad (3.17)$$

which is the HMM analogue of formula (3.2). Note that with this estimation process the Markov chain underlying the resulting HMM is automatically non-trivially connected. As in the case of Markov chains, one should bear in mind that if in the training data a particular state G_i does not occur at all, then the transition and emission probabilities p_{ij}, p_{i0}, $q_i(a)$ cannot be estimated from the data and can be set arbitrarily (see Example 3.13).

It will be explained in Chap. 8 (see Part 6.a of Example 8.12 and Exercise 8.5) that this estimation procedure does not simply look reasonable, but in fact gives values of transition and emission probabilities that guarantee that the likelihood $P(x^1, \pi^1) \times \ldots \times P(x^n, \pi^n)$ is maximal possible.

Example 3.7. Consider the *a priori* connectivity shown in Fig. 3.5. We wish to use it for modeling prokaryotic genes, and suppose that we are given four DNA sequences that are extensions of sequences (3.5)

$$
\begin{aligned}
&x^1 : ATGACT\,\mathbf{ATGCTATTGATTTAA}\,CGC \\
&x^2 : CCC\,\mathbf{ATGGTGAAAGACTTCTAA}\,GAT \\
&x^3 : AAAGTGACT\,\mathbf{ATGCCCGATGAACGCTAGG}\,AA \\
&x^4 : ATGGAT\,\mathbf{ATGAAGCATGATTAA}\,CAT.
\end{aligned}
\qquad (3.18)
$$

Assume that in this data all gene sequences have been determined experimentally (shown in bold). Thus, the experimental information provides us with the following four paths through the graph in Fig. 3.5, corresponding to the four sequences

π^1 : 0 ig ig ig ig ig ig g g g g g g g g g g g g g g g g ig ig ig 0
π^2 : 0 ig ig ig ig ig ig g g g g g g g g g g g g g g g g g ig ig ig 0
π^3 : 0 ig ig ig ig ig ig ig ig ig g g g g g g g g g g g g g g g g g g ig ig ig 0
π^4 : 0 ig ig ig ig ig ig g g g g g g g g g g g g g g g g ig ig ig 0.

Then from formulas (3.17) we obtain

$$p_{0\,g} = 0, \qquad p_{0\,\mathrm{ig}} = 1,$$

$$p_{g\,g} = \frac{59}{63}, \quad p_{g\,\mathrm{ig}} = \frac{4}{63}, \quad p_{g\,0} = 0,$$

$$p_{\mathrm{ig}\,g} = \frac{4}{39}, \quad p_{\mathrm{ig}\,\mathrm{ig}} = \frac{31}{39}, \quad p_{\mathrm{ig}\,0} = \frac{4}{39},$$

$$q_g(A) = \frac{23}{63}, \, q_g(C) = \frac{1}{7}, \quad q_g(G) = \frac{13}{63}, \, q_g(T) = \frac{2}{7},$$

$$q_{\mathrm{ig}}(A) = \frac{1}{3}, \, q_{\mathrm{ig}}(C) = \frac{8}{39}, \, q_{\mathrm{ig}}(G) = \frac{3}{13}, \, q_{\mathrm{ig}}(T) = \frac{3}{13}.$$

3.6.2 Estimation when Paths are Unknown

We will now discuss estimation procedures for datasets of the second type. Suppose we are given n sequences $x^1, \ldots, x^n$ of letters from the alphabet $\mathcal{Q}$. In this case we would like to choose parameter values that maximize the likelihood $P(x^1) \times \ldots \times P(x^n)$ (assuming that such optimal values exist). In this situation there is no explicit formula that would produce optimal values from the training data $x^1, \ldots, x^n$. One approach to determining optimal parameter values in practical applications is to apply standard optimization algorithms to $P(x^1) \times \ldots \times P(x^n)$ treated as a function of the parameters. All such general algorithms are iterative; they start with some initial parameter values (of course, the initial values of the transition probabilities must agree with the condition of non-trivial connectedness) and have certain stopping criteria (for example, the algorithm stops if the change in values of $P(x^1) \times \ldots \times P(x^n)$ from iteration to iteration becomes smaller than some predetermined threshold). A common problem with optimization algorithms is that they may converge to a point close to a point of local, not global maximum of $P(x^1) \times \ldots \times P(x^n)$ in the parameter space. Since the result depends on the initial parameter values, one way to improve search for a global maximum is to let the optimizer have several runs starting with different sets of initial parameter values.

For the special case of maximizing the function $P(x^1) \times \ldots \times P(x^n)$, there exists a particular iteration method that is commonly used, known as the *Baum-Welch training algorithm* [B]. It is possible to show that when the algorithm is applied, the likelihood $P(x^1) \times \ldots \times P(x^n)$ does not decrease from iteration to iteration (in fact, it increases if the parameter values change) – see, e.g., [DEKM]. As all general optimization algorithms, the Baum-Welch algorithm may converge to a point close to a point of local, not global maximum of $P(x^1) \times \ldots \times P(x^n)$. The algorithm first estimates $H_{\alpha\beta}$ and $J_l(a)$ using the current values of $p_{\alpha\beta}$ and $q_l(a)$; then (3.17) are used to produce new values of $p_{\alpha\beta}$ and $q_l(a)$, $\alpha = \mathcal{B}, 1, \ldots, N$, $\beta = 1, \ldots, N, \mathcal{E}$, $l = 1, \ldots, N$, $a \in \mathcal{Q}$.

We will now explain the algorithm in detail. As in Sect. 3.5, for a fixed sequence $x = x_1 \ldots x_L$ consider the event $E(x)$ and recall that $P(E(x)) = P(x)$. Next, fix three integers $1 \le i \le L - 1$, $1 \le k, l \le N$, and consider the event

$$E_{i,(k,l)} = \{(y, \pi) \in S : \text{the length of } y \text{ and } \pi \text{ is } \ge i + 1,$$
$$\text{and } \pi_i = G_k, \pi_{i+1} = G_l\}.$$

We suppose that $P(x) > 0$ (otherwise x is not an interesting sequence) and calculate the conditional probability $P(E_{i,(k,l)}|E(x))$. Assuming that $2 \le i \le L - 2$ we obtain

$$P(E_{i,(k,l)}|E(x)) = \frac{P(E_{i,(k,l)} \cap E(x))}{P(x)} = \frac{1}{P(x)} \sum_{\pi_1,\ldots,\pi_{i-1},\pi_{i+2},\ldots,\pi_L \in X} p_{0\,\pi_1}$$

$$\times q_{\pi_1}(x_1) p_{\pi_1\,\pi_2} q_{\pi_2}(x_2) \ldots p_{\pi_{i-2}\,\pi_{i-1}} q_{\pi_{i-1}}(x_{i-1}) p_{\pi_{i-1}\,G_k} q_k(x_i) p_{kl} q_l(x_{i+1}) p_{G_l\,\pi_{i+2}}$$

$$\times q_{\pi_{i+2}}(x_{i+2}) \ldots p_{\pi_{L-1}\,\pi_L} q_{\pi_L}(x_L) p_{\pi_L\,0} = \frac{1}{P(x)} \sum_{\pi_1,\ldots,\pi_{i-1} \in X} p_{0\,\pi_1} q_{\pi_1}(x_1) p_{\pi_1\,\pi_2}$$

$$\times q_{\pi_2}(x_2) \ldots p_{\pi_{i-2}\,\pi_{i-1}} q_{\pi_{i-1}}(x_{i-1}) p_{\pi_{i-1}\,G_k} q_k(x_i) p_{kl} q_l(x_{i+1}) \sum_{\pi_{i+2},\ldots,\pi_L \in X} p_{G_l\,\pi_{i+2}}$$

$$\times q_{\pi_{i+2}}(x_{i+2}) \ldots p_{\pi_{L-1}\,p_L} q_{\pi_L}(x_L) p_{\pi_L\,0} = \frac{f_k(i) p_{kl} q_l(x_{i+1}) b_l(i+1)}{P(x)},$$

where $f_k(i)$ and $b_l(i)$ are the quantities found from the forward and backward algorithms respectively. Analogous calculations for $i = 1$ and $i = L - 1$ show that the same result holds in these cases as well and thus we have

$$P(E_{i,(k,l)}|E(x)) = \frac{f_k(i) p_{kl} q_l(x_{i+1}) b_l(i+1)}{P(x)} \tag{3.19}$$

for $i = 1, \ldots, L - 1$ and $k, l = 1, \ldots, N$. The probability $P(E_{i,(k,l)}|E(x))$ is the probability with which x_i and x_{i+1} are emitted at states G_k and G_l

respectively and is also called the *posterior probability of states* G_k *and* G_l *at observations* i *and* $i+1$ *respectively given* x.

From formula (3.16) we also have

$$P(E_{1,l}|E(x)) = \frac{f_l(1)b_l(1)}{P(x)} = \frac{p_{0l}q_l(x_1)b_l(1)}{P(x)},$$

(3.20)

$$P(E_{L,k}|E(x)) = \frac{f_k(L)b_k(L)}{P(x)} = \frac{f_k(L)p_{k0}}{P(x)}.$$

If we assume for convenience that states $\mathcal{B}$ and $\mathcal{E}$ emit the symbols $\mathcal{B}$ and $\mathcal{E}$ respectively with probability 1 and define $E_{0,(\mathcal{B},l)} = E_{1,l}$, $E_{L,(k,\mathcal{E})} = E_{L,k}$, $f_{\mathcal{B}}(0) = 1$, $b_{\mathcal{E}}(L+1) = 1$, for $k,l = 1,\ldots,N$, it then follows from (3.20) that formulas (3.19) also hold for $E_{0,(\mathcal{B},l)}$ and $E_{L,(k,\mathcal{E})}$, $k,l = 1,\ldots,N$.

We are now ready to describe the Baum-Welch algorithm. The algorithm starts with some initial parameter values $p_{\alpha\beta}^{(0)}$, $q_k^{(0)}(a)$, $\alpha = \mathcal{B}, 1,\ldots,N$, $\beta = 1,\ldots,N,\mathcal{E}$, $a \in \mathcal{Q}$, and these values are chosen in such a way that the corresponding underlying Markov chain is non-trivially connected, its connectivity does not contradict the *a priori* connectivity, and the initial probabilities $P^{(0)}(x^r)$ of the training sequences are positive for all r. One can also incorporate prior knowledge about the training data into the initial values.

We will now describe the recursion step. Let $x^r = x_1^r \ldots x_{m_r}^r$, $r = 1,\ldots,n$, and let $p_{\alpha\beta}^{(s)}$, $q_k^{(s)}(a)$, $\alpha = \mathcal{B}, 1,\ldots,N$, $\beta = 1,\ldots,N,\mathcal{E}$, $a \in \mathcal{Q}$, be the parameter values after s steps of the algorithm. Denote by $P^{(s)}(x^r)$, $f_k^{r\,(s)}(i)$ and $b_l^{r\,(s)}(i)$ the probabilities and the quantities found from the forward and backward algorithm for the sequence x^r, $i = 1,\ldots,m_r$, $r = 1,\ldots,n$, $k,l = 1,\ldots,N$, with the above parameter values. The recursion step for the transition probabilities is based on formulas (3.19) and (3.20). For $k,l = 1,\ldots,N$ set

$$H_{kl}^{(s)} = \sum_{r=1}^{n} \frac{1}{P^{(s)}(x^r)} \sum_{i=1}^{m_r-1} f_k^{r\,(s)}(i)p_{kl}^{(s)}q_l^{(s)}(x_{i+1}^r)b_l^{r\,(s)}(i+1),$$

$$H_{\mathcal{B}l}^{(s)} = \sum_{r=1}^{n} \frac{p_{0l}^{(s)}q_l^{(s)}(x_1^r)b_l^{r\,(s)}(1)}{P^{(s)}(x^r)},$$

(3.21)

$$H_{k\mathcal{E}}^{(s)} = \sum_{r=1}^{n} \frac{f_k^{r\,(s)}(m_r)p_{k0}^{(s)}}{P^{(s)}(x^r)},$$

and the transition probabilities $p_{\alpha\beta}^{(s+1)}$ are produced from $H_{\alpha\beta}^{(s)}$ using the first formula in (3.17), $\alpha = \mathcal{B}, 1,\ldots,N$, $\beta = 1,\ldots,N,\mathcal{E}$. Note that if for some α,β we had $p_{\alpha\beta}^{(0)} = 0$, then the above estimation procedure ensures that $p_{\alpha\beta}^{(s)} = 0$ for all s. Therefore, the connectivity of the underlying Markov chain obtained on every iteration step does not contradict the *a priori* connectivity. It is also possible to show that the underlying Markov chain obtained on every step is non-trivially connected (see Exercise 3.11). Note that the first formula in (3.17) can only be used if the corresponding denominator is non-zero; otherwise the new values of the relevant transition probabilities are set arbitrarily.

Next, we will estimate the emission probabilities. For this purpose we will use the posterior probabilities $P(E_{i,k}|E(x))$ calculated for any sequence $x = x_1 \ldots x_L$ in the previous section, $i = 1, \ldots, L$, $k = 1, \ldots, N$. The recursion step for the emission probabilities is based on formula (3.16). For $l = 1, \ldots, N$ and $a \in \mathcal{Q}$ we set

$$J_l^{(s)}(a) = \sum_{r=1}^{n} \frac{1}{P^{(s)}(x^r)} \sum_{\{i=1,\ldots,m_r \,:\, x_i^r = a\}} f_l^{r\,(s)}(i) b_l^{r\,(s)}(i), \qquad (3.22)$$

and the emission probabilities $q_l^{(s+1)}(a)$ are produced from $J_l^{(s)}$ using the second formula in (3.17). Note that the second formula in (3.17) can only be used if the corresponding denominator is non-zero; otherwise the new values of the relevant emission probabilities are set arbitrarily.

Since the likelihood of the training data does not decrease with every step of the algorithm, $P^{(s)}(x^r) > 0$ for all r and s.

We will now give an example of applying the Baum-Welch algorithm.

Example 3.8. Consider the connectivity shown in Fig. 3.5. For the purposes of this example re-denote the state "g" by G_1 and the state "ig" by G_2. For simplicity assume that $\mathcal{Q}$ is the two-letter alphabet $\{A, B\}$ and suppose that we are given the following training data

$$x^1 : ABA$$
$$x^2 : ABB$$
$$x^3 : AB.$$

Set the initial parameter values as follows

$$p_{01}^{(0)} = 1, \qquad p_{02}^{(0)} = 0,$$

$$p_{11}^{(0)} = \frac{1}{2}, \qquad p_{12}^{(0)} = \frac{1}{2}, \qquad p_{10}^{(0)} = 0,$$

$$p_{21}^{(0)} = 0, \qquad p_{22}^{(0)} = 0, \qquad p_{20}^{(0)} = 1,$$

$$q_1^{(0)}(A) = \frac{1}{4}, q_1^{(0)}(B) = \frac{3}{4},$$

$$q_2^{(0)}(A) = \frac{1}{2}, q_2^{(0)}(B) = \frac{1}{2}.$$

Applying the forward and backward algorithms to each of the three training sequences we obtain

$$f_1^{1(0)}(1) = \frac{1}{4}, \quad f_2^{1(0)}(1) = 0, \quad f_1^{1(0)}(2) = \frac{3}{32}, \quad f_2^{1(0)}(2) = \frac{1}{16},$$

$$f_1^{1(0)}(3) = \frac{3}{256}, \quad f_2^{1(0)}(3) = \frac{3}{128}, \quad f_1^{2(0)}(1) = \frac{1}{4}, \quad f_2^{2(0)}(1) = 0,$$

$$f_1^{2(0)}(2) = \frac{3}{32}, \quad f_2^{2(0)}(2) = \frac{1}{16}, \quad f_1^{2(0)}(3) = \frac{9}{256}, \quad f_2^{2(0)}(3) = \frac{3}{128},$$

$$f_1^{3(0)}(1) = \frac{1}{4}, \quad f_2^{3(0)}(1) = 0, \quad f_1^{3(0)}(2) = \frac{3}{32}, \quad f_2^{3(0)}(2) = \frac{1}{16},$$

$$b_1^{1(0)}(1) = \frac{3}{32}, \quad b_2^{1(0)}(1) = 0, \quad b_1^{1(0)}(2) = \frac{1}{4}, \quad b_2^{1(0)}(2) = 0,$$

$$b_1^{1(0)}(3) = 0, \quad b_2^{1(0)}(3) = 1, \quad b_1^{2(0)}(1) = \frac{3}{32}, \quad b_2^{2(0)}(1) = 0,$$

$$b_1^{2(0)}(2) = \frac{1}{4}, \quad b_2^{2(0)}(2) = 0, \quad b_1^{2(0)}(3) = 0, \quad b_2^{2(0)}(3) = 1,$$

$$b_1^{3(0)}(1) = \frac{1}{4}, \quad b_2^{3(0)}(1) = 0, \quad b_1^{3(0)}(2) = 0, \quad b_2^{3(0)}(2) = 1,$$

$$P^{(0)}(x^1) = \frac{3}{128}, \quad P^{(0)}(x^2) = \frac{3}{128}, \quad P^{(0)}(x^3) = \frac{1}{16}.$$

The initial value of the likelihood therefore is

$$P^{(0)}(x^1)P^{(0)}(x^2)P^{(0)}(x^3) = \frac{9}{262144}.$$

Next, formulas (3.21) and (3.22) give

$$H_{11}^{(0)} = 2, \quad H_{12}^{(0)} = 3, \quad H_{21}^{(0)} = 0, \quad H_{22}^{(0)} = 0,$$

$$H_{\mathcal{B}1}^{(0)} = 3, \quad H_{\mathcal{B}2}^{(0)} = 0, \quad H_{1\mathcal{E}}^{(0)} = 0, \quad H_{2\mathcal{E}}^{(0)} = 3,$$

$$J_1^{(0)}(A) = 3, \ J_1^{(0)}(B) = 2, \ J_2^{(0)}(A) = 1, \ J_2^{(0)}(B) = 2,$$

and therefore from formulas (3.17) we obtain

$$p_{01}^{(1)} = 1, \qquad p_{02}^{(1)} = 0,$$

$$p_{11}^{(1)} = \frac{2}{5}, \qquad p_{12}^{(1)} = \frac{3}{5}, \qquad p_{10}^{(1)} = 0,$$

$$p_{21}^{(1)} = 0, \qquad p_{22}^{(1)} = 0, \qquad p_{20}^{(1)} = 1,$$

$$q_1^{(1)}(A) = \frac{3}{5}, \, q_1^{(1)}(B) = \frac{2}{5},$$

$$q_2^{(1)}(A) = \frac{1}{3}, \, q_2^{(1)}(B) = \frac{2}{3}.$$

Further, applying the forward algorithm to the sequences x^1, x^2, x^3 with the new parameter values we get

$$f_1^{1(1)}(1) = \frac{3}{5}, \qquad f_2^{1(1)}(1) = 0, \qquad f_1^{1(1)}(2) = \frac{12}{125}, \quad f_2^{1(1)}(2) = \frac{6}{25},$$

$$f_1^{1(1)}(3) = \frac{72}{3125}, \, f_2^{1(1)}(3) = \frac{12}{625}, \; f_1^{2(1)}(1) = \frac{3}{5}, \qquad f_2^{2(1)}(1) = 0,$$

$$f_1^{2(1)}(2) = \frac{12}{125}, \; f_2^{2(1)}(2) = \frac{6}{25}, \; f_1^{2(1)}(3) = \frac{48}{3125}, \, f_2^{2(1)}(3) = \frac{24}{625},$$

$$f_1^{3(1)}(1) = \frac{3}{5}, \qquad f_2^{3(1)}(1) = 0, \qquad f_1^{3(1)}(2) = \frac{12}{125}, \; f_2^{3(1)}(2) = \frac{6}{25},$$

$$P^{(1)}(x^1) = \frac{12}{625}, \; P^{(1)}(x^2) = \frac{24}{625}, \; P^{(1)}(x^3) = \frac{6}{25}.$$

Hence the value of the likelihood after one step of the Baum-Welch algorithm is

$$P^{(1)}(x^1)P^{(1)}(x^2)P^{(1)}(x^3) = \frac{1728}{9765625},$$

which is indeed larger than the previously determined initial value $P^{(0)}(x^1)P^{(0)}(x^2)P^{(0)}(x^3)$.

There is a frequently used alternative to the Baum-Welch training called the *Viterbi training algorithm*. It is also an iterative algorithm that starts with some initial parameter values satisfying the same conditions as in the case of the Baum-Welch training. On each step all Viterbi paths $\pi^*(x^1), \ldots \pi^*(x^n)$ for the training sequences are found and then the parameters are re-estimated as in the case when paths were known. We will show below that on each step of the Viterbi training $P(x^1, \pi^*(x^1)) \times \ldots \times P(x^n, \pi^*(x^n))$ does not decrease, that is, this algorithm is designed to attempt to maximize $P(x^1, \pi^*(x^1)) \times \ldots \times P(x^n, \pi^*(x^n))$, not the likelihood $P(x^1) \times \ldots \times P(x^n)$ which may in fact decrease. Perhaps for this reason, the Viterbi training does not generally perform as well as the Baum-Welch training.

We will now describe the recursion step in detail. As before, let $x^r = x_1^r \ldots x_{m_r}^r$, $r = 1, \ldots, n$, be the training sequences, and let $p_{\alpha\beta}^{(s)}$, $q_k^{(s)}(a)$, $\alpha = \mathcal{B}, 1, \ldots, N, \beta = 1, \ldots, N, \mathcal{E}, a \in \mathcal{Q}$, be the parameter values after s steps of the algorithm. All quantities calculated using these values will bear the superscript (s). With these parameter values we will now determine all the Viterbi paths for each training sequence. For simplicity we assume that there is in fact only one Viterbi path for each sequence on each step of the algorithm (the general case is completely analogous to this special one, but is harder notationally). Let $\pi^{*(s)}(x^r)$ be the Viterbi path for x^r, $r = 1, \ldots, n$. We further assume that every state of the underlying Markov chain is encountered in the paths $\pi^{*(s)}(x^1), \ldots, \pi^{*(s)}(x^n)$ for every s. We will now re-estimate the parameters from formulas (3.17) for the training dataset that consists of the pairs of sequences $(x^1, \pi^{*(s)}(x^1)), \ldots, (x^n, \pi^{*(s)}(x^n))$, as in the case when paths are known. This can be done since the Viterbi paths clearly agree with the *a priori* connectivity of the model. The HMM arising on each step has an underlying Markov chain that is non-trivially connected and whose connectivity does not contradict the *a priori* connectivity.

As we remarked above, the estimation procedure in the case when paths are known maximizes the corresponding likelihood of the training dataset, and therefore

$$\prod_{r=1}^{n} P^{(s)}(x^r, \pi^{*(s)}(x^r)) \leq \prod_{r=1}^{n} P^{(s+1)}(x^r, \pi^{*(s)}(x^r)). \tag{3.23}$$

Further, we clearly have

$$P^{(s+1)}(x^r, \pi^{*(s)}(x^r)) \leq P^{(s+1)}(x^r, \pi^{*(s+1)}(x^r)), \tag{3.24}$$

for all s and r. Together with (3.23) this gives

$$\prod_{r=1}^{n} P^{(s)}(x^r, \pi^{*(s)}(x^r)) \leq \prod_{r=1}^{n} P^{(s+1)}(x^r, \pi^{*(s+1)}(x^r)).$$

that is, $\prod_{r=1}^{n} P^{(s)}(x^r, \pi^{*(s)}(x^r))$ does not decrease from step to step. In particular, $P^{(s)}(x^r, \pi^{*(s)}(x^r)) > 0$ for all s and r.

Suppose that for some $s_0 \geq 0$ we have

$$\prod_{r=1}^{n} P^{(s_0+1)}(x^r, \pi^{*(s_0+1)}(x^r)) = \prod_{r=1}^{n} P^{(s_0)}(x^r, \pi^{*(s_0)}(x^r)). \tag{3.25}$$

Then formulas (3.23) and (3.24) imply that

$$P^{(s_0+1)}(x^r, \pi^{*(s_0+1)}(x^r)) = P^{(s_0+1)}(x^r, \pi^{*(s_0)}(x^r)),$$

for all r, and hence $\pi^{*(s_0+1)}(x^r) = \pi^{*(s_0)}(x^r)$ for all r. Therefore the parameter values after $s_0 + 2$ steps coincide with those after $s_0 + 1$ steps. This clearly implies that the parameter values after $s_0 + k$ steps for any $k \geq 2$ coincide with those after $s_0 + 1$ steps. Therefore the natural stopping criterion for the Viterbi training is based on condition (3.25) and regards step $s_0 + 1$ as the last one. Condition (3.25) is satisfied, for example, if $\pi^{*(s_0)}(x^r) = \pi^{*(s_0-1)}(x^r)$ for all r (here $s_0 \geq 1$), in which case one may even stop after s_0 steps, since the parameter values will not change afterwards.

The parameter values on each step are obtained from paths through the graph that represents the *a priori* connectivity, such that every state of the underlying Markov chain is encountered in at least one of the paths. Clearly, there are only finitely many sets of parameter values that can be obtained in this way. Therefore $\prod_{r=1}^{n} P^{(s)}(x^r, \pi^{*(s)}(x^r))$ cannot increase infinitely many times, and thus the Viterbi training always stops after finitely many steps. Note, however, that the algorithm may stop before reaching the maximal value of $P(x^1, \pi^*(x^1)) \times \ldots \times P(x^n, \pi^*(x^n))$.

We will now give an example of applying the Viterbi training. Below we denote by $v_k^{r(s)}(i)$, $V_k^{r(s)}(i)$, $\mathcal{V}^{r(s)}(m_r)$ the quantities and sets calculated from the Viterbi algorithm for the sequence x^r, $i = 1, \ldots, m_r$, $r = 1, \ldots, n$, $k = 1, \ldots, N$, using the parameter values after s steps of the training algorithm.

Example 3.9. Consider the situation of Example 3.8, but set the initial parameter values in a slightly different way

$$p_{01}^{(0)} = 1, \qquad p_{02}^{(0)} = 0,$$

$$p_{11}^{(0)} = \frac{1}{2}, \qquad p_{12}^{(0)} = \frac{1}{2}, \qquad p_{10}^{(0)} = 0,$$

$$p_{21}^{(0)} = 0, \qquad p_{22}^{(0)} = \frac{1}{2}, \qquad p_{20}^{(0)} = \frac{1}{2},$$

$$q_1^{(0)}(A) = \frac{1}{4}, \, q_1^{(0)}(B) = \frac{3}{4},$$

$$q_2^{(0)}(A) = \frac{1}{2}, \, q_2^{(0)}(B) = \frac{1}{2}.$$

We will now re-estimate the parameters using the Viterbi training. We have

$$v_1^{1(0)}(1) = \frac{1}{4}, \quad v_2^{1(0)}(1) = 0, \quad v_1^{1(0)}(2) = \frac{3}{32}, \quad \mathcal{V}_1^{1(0)}(1) = \{1\},$$

$$v_2^{1(0)}(2) = \frac{1}{16}, \quad \mathcal{V}_2^{1(0)}(1) = \{1\}, v_1^{1(0)}(3) = \frac{3}{256}, \quad \mathcal{V}_1^{1(0)}(2) = \{1\},$$

$$v_2^{1(0)}(3) = \frac{3}{128}, \quad \mathcal{V}_2^{1(0)}(2) = \{1\}, \mathcal{V}^{1(0)}(3) = \{2\},$$

$$v_1^{2(0)}(1) = \frac{1}{4}, \quad v_2^{2(0)}(1) = 0, \quad v_1^{2(0)}(2) = \frac{3}{32}, \quad \mathcal{V}_1^{2(0)}(1) = \{1\},$$

$$v_2^{2(0)}(2) = \frac{1}{16}, \quad \mathcal{V}_2^{2(0)}(1) = \{1\}, v_1^{2(0)}(3) = \frac{9}{256}, \quad \mathcal{V}_1^{2(0)}(2) = \{1\},$$

$$v_2^{2(0)}(3) = \frac{3}{128}, \quad \mathcal{V}_2^{2(0)}(2) = \{1\}, \mathcal{V}^{2(0)}(3) = \{2\},$$

$$v_1^{3(0)}(1) = \frac{1}{4}, \quad v_2^{3(0)}(1) = 0, \quad v_1^{3(0)}(2) = \frac{3}{32}, \quad \mathcal{V}_1^{3(0)}(1) = \{1\},$$

$$v_2^{3(0)}(2) = \frac{1}{16}, \quad \mathcal{V}_2^{3(0)}(1) = \{1\}, \mathcal{V}^{3(0)}(2) = \{2\}.$$

For each sequence there is a single Viterbi path. These paths $\pi^{*(0)}(x^1)$, $\pi^{*(0)}(x^2)$, $\pi^{*(0)}(x^3)$ and the corresponding probabilities are shown below

$$\pi^{*(0)}(x^1) : G_1G_1G_2, \ P^{(0)}(x^1, \pi^{*(0)}(x^1)) = \frac{3}{256},$$

$$\pi^{*(0)}(x^2) : G_1G_1G_2, \ P^{(0)}(x^2, \pi^{*(0)}(x^2)) = \frac{3}{256},$$

$$\pi^{*(0)}(x^3) : G_1G_2, \quad P^{(0)}(x^3, \pi^{*(0)}(x^3)) = \frac{1}{32}.$$

which, in particular, gives

$$P^{(0)}(x^1, \pi^{*(0)}(x^1))P^{(0)}(x^2, \pi^{*(0)}(x^2))P^{(0)}(x^3, \pi^{*(0)}(x^3)) = \frac{9}{2097152}.$$

We will now consider the dataset $(x^1, \pi^{*(0)}(x^1))$, $(x^2, \pi^{*(0)}(x^2))$, $(x^3, \pi^{*(0)}(x^3))$ and determine $H_{\alpha\beta}^{(0)}$ and $J_l^{(0)}(a)$ as in the case when paths are known. We obtain

$$H_{11}^{(0)} = 2, \quad H_{12}^{(0)} = 3, \quad H_{21}^{(0)} = 0, \quad H_{22}^{(0)} = 0,$$

$$H_{\mathcal{B}1}^{(0)} = 3, \quad H_{\mathcal{B}2}^{(0)} = 0, \quad H_{1\mathcal{E}}^{(0)} = 0, \quad H_{2\mathcal{E}}^{(0)} = 3,$$

$$J_1^{(0)}(A) = 3, \, J_1^{(0)}(B) = 2, \, J_2^{(0)}(A) = 1, \, J_2^{(0)}(B) = 2,$$

and from formulas (3.17) we get

$$p_{01}^{(1)} = 1, \quad p_{02}^{(1)} = 0,$$

$$p_{11}^{(1)} = \frac{2}{5}, \quad p_{12}^{(1)} = \frac{3}{5}, \quad p_{10}^{(1)} = 0,$$

$$p_{21}^{(1)} = 0, \quad p_{22}^{(1)} = 0, \quad p_{20}^{(1)} = 1,$$

$$q_1^{(1)}(A) = \frac{3}{5}, \, q_1^{(1)}(B) = \frac{2}{5},$$

$$q_2^{(1)}(A) = \frac{1}{3}, \, q_2^{(1)}(B) = \frac{2}{3}.$$

It is easy to see that $\pi^{*(1)}(x^r) = \pi^{*(0)}(x^r)$ for $r = 1, 2, 3$. Hence in this example the Viterbi training stops after one iteration. It is easy to calculate the probabilities $P^{(1)}(x^1, \pi^{*(1)}(x^1))$, $P^{(1)}(x^2, \pi^{*(1)}(x^2))$, $P^{(1)}(x^3, \pi^{*(1)}(x^3))$, and we obtain

$$P^{(1)}(x^1, \pi^{*(1)}(x^1)) \ P^{(1)}(x^2, \pi^{*(1)}(x^2)) P^{(1)}(x^3, \pi^{*(1)}(x^3)) =$$

$$12/625 \times 24/625 \times 6/25 = 1728/9765625,$$

which is indeed larger than the initial value $P^{(0)}(x^1, \pi^{*(0)}(x^1)) \times P^{(0)}(x^2, \pi^{*(0)}(x^2)) P^{(0)}(x^3, \pi^{*(0)}(x^3))$.

3.7 HMMs with Silent States

So far we have assumed that an *a priori* connectivity was fixed in advance. Specifying it, however, is an important part of modeling by means of HMMs. It may look reasonable at first sight to allow all possible transitions and hope that a parameter estimation process will eventually lead to a good model. Unfortunately, this approach almost never works in practice. It usually produces very bad models, even if training datasets are large. The main problem here is local maxima. The less constrained the model is, the more local maxima the corresponding likelihood function tends to have and hence the harder it becomes for an optimization algorithm to distinguish them from global ones. Well-performing HMMs are built by carefully deciding which transitions should be allowed in the model, and these decisions are based on prior knowledge about the problem of interest. Choosing a model structure is not an exact science, and it is difficult to give general recommendations on how to select the best possible design. In this section we will describe one trick that is often used in model building. The trick is to introduce states of a special kind, so-called *silent states*. These states do not emit any symbols, just like the begin and end states.

An *HMM with silent states* is a discrete-time finite Markov chain (that we always assume to be non-trivially connected) for which all the states (not including $\mathcal{B}$ and $\mathcal{E}$) are divided into two groups $X = \{G_1, \ldots, G_N\}$, $Y = \{D_1, \ldots, D_M\}$, $Y \neq \emptyset$, and for each state G_k and each symbol a in a finite alphabet $\mathcal{Q}$ an emission probability is specified. The corresponding transition and emission probabilities will be denoted as $p_{\mathcal{B} G_i} = p_{0 G_i}$, $p_{\mathcal{B} D_k} = p_{0 D_k}$, $p_{G_i G_j}$, $p_{G_i D_k}$, $p_{D_k G_i}$, $p_{D_k D_l}$, $p_{G_i \mathcal{E}} = p_{G_i 0}$, $p_{D_k \mathcal{E}} = p_{D_k 0}$, $q_{G_i}(a)$, for $i, j = 1, \ldots, N$, $k, l = 1, \ldots, M$, $a \in \mathcal{Q}$. As before, the Markov chain will be called the *underlying Markov chain of the HMM*. The states in Y do not emit any symbols and are called *silent*. The states in X are called *non-silent*.

HMMs with silent states are useful, for example, in the following situation. Suppose we have a very long chain of states in which each state is required to be connected to any state further in the chain, as in the connectivity shown in Fig. 3.7. The number of parameters required to be estimated quickly be-

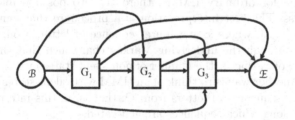

Fig. 3.7.

comes very large as the length of the chain increases, and all the transition probabilities cannot be well estimated from realistic datasets (note that for the connectivity in Fig. 3.7 the number of free parameters corresponding to the transition probabilities is 5). In this situation one can use silent states to reduce the number of parameters. Consider, for example, the connectivity shown in Fig. 3.8, where D_1, D_2, D_3 are silent states.

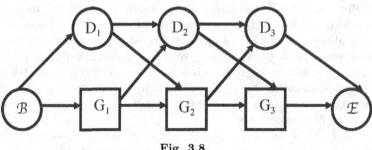

Fig. 3.8.

Since the silent states do not emit any letters, it is possible to go from any G_i to any G_j with $j > i$ without emitting any letters. Hence the collections of sequences that are generated with non-zero probabilities by an HMM with connectivity in Fig. 3.7 and an HMM with connectivity in Fig. 3.8 coincide. Although for the connectivity shown in Fig. 3.8 the number of free parameters corresponding to the transition probabilities is also 5, it is easy to see that connectivities analogous to that in Fig. 3.8 will require a much smaller number of parameters than those analogous to the connectivity in Fig. 3.7, as the length of the chains grows.

One can think of an HMM with silent states as a process of sequence generation. As for ordinary HMMs, there are two possible interpretations of this process. The first interpretation is a procedure that generates pairs of sequences (x, π), where x is a finite sequence of letters from $\mathcal{Q}$ and π is a finite path through the underlying Markov chain such that the *non-silent length* $L_{n/s}(\pi)$ of π, that is, the number of non-silent states in π, is equal to the length of x. Another way is to think of an HMM with silent states as a process that generates sequences of letters from $\mathcal{Q}$, that is, in this interpretation we ignore paths along which sequences are generated.

All concepts introduced in the preceding sections for ordinary HMMs have analogues for HMMs with silent states. Let $x = x_1 \ldots x_L$ be a sequence of letters from $\mathcal{Q}$ and $\pi = \pi_1 \ldots \pi_T$, for $T \geq L$, be a path with $L_{n/s}(\pi) = L$. Let $\pi_{i_1}, \ldots, \pi_{i_L} \in X$ for some $1 \leq i_1 < \ldots < i_L \leq T$. We will then define the probability $P(x, \pi)$ of the pair (x, π) as follows

$$P(x, \pi) = p_{0\,\pi_1} p_{\pi_1 \pi_2} \times \ldots \times p_{\pi_{T-1} \pi_T} p_{\pi_T 0} \times q_{\pi_{i_1}}(x_1) q_{\pi_{i_2}}(x_2) \times \ldots \times q_{\pi_{i_L}}(x_L).$$

It is also natural to define "the total probability" $P(x)$ of x as

$$P(x) = \sum_{\text{all } \pi:\, L_{n/s}(\pi) = L} P(x, \pi). \qquad (3.26)$$

It follows from Exercise 3.14 that $\sum_x P(x) = 1$, where the summation is taken over all sequences of all finite lengths, including length 0 (sequences of length 0 correspond to paths that consist only of silent states).

We will be interested in generalizing the algorithms discussed in the preceding sections for HMMs to the case of HMMs with silent states. It can be done under the following additional condition: we say that an HMM *does not have silent loops* if the silent states can be enumerated in such a way that transitions with non-zero probabilities between any two silent states are only possible from the lower numbered state to the higher numbered one, and the probability of any transition from a silent state into itself is equal to 0. From now on we will assume that all HMMs do not have silent loops and that the enumeration of silent states $D_1, \ldots, D_M$ agrees with this condition. We will also apply this terminology to connectivities and speak about *a priori* connectivities that do not have silent loops.

We will first generalize the Viterbi algorithm. Fix a sequence $x = x_1 \ldots x_L$ of letters from the alphabet Q and define for $k = 1, \ldots, N$

$$v_{G_k}(1) = q_{G_k}(x_1) \max \Big\{ p_{0\,G_k},$$
$$\max_{\substack{1 \le s \le M, \\ 1 \le i_1 < \ldots < i_s \le M}} p_{0\,D_{i_1}} p_{D_{i_1}\,D_{i_2}} \cdots p_{D_{i_{s-1}}\,D_{i_s}} p_{D_{i_s}\,G_k} \Big\}.$$

For every k let $\mathcal{V}_{G_k}(0)$ contain all paths among $0\,G_k$ and $0\,D_{i_1} D_{i_2} \ldots D_{i_s} G_k$ that realize the value $v_{G_k}(1)$. Next, for $i = 1, \ldots, L$ and $m = 1, \ldots, M$ set

$$v_{D_m}(i) = \max_{k=1,\ldots,N} \Big\{ v_{G_k}(i) \max \Big\{ p_{G_k\,D_m},$$
$$\max_{\substack{1 \le s \le m-1, \\ 1 \le i_1 < \ldots < i_s \le m-1}} p_{G_k\,D_{i_1}} p_{D_{i_1}\,D_{i_2}} \cdots p_{D_{i_{s-1}}\,D_{i_s}} p_{D_{i_s}\,D_m} \Big\} \Big\},$$

and let $\mathcal{V}_{D_m}(i-1)$ be the collection of all paths among $G_k D_m$ and $G_k D_{i_1} D_{i_2} \ldots D_{i_s} D_m$ that realize the value $v_{D_m}(i)$.

In order to perform a recursion step, for $i = 1, \ldots, L-1$ and $k = 1, \ldots, N$ set

$$v_{G_k}(i+1) = q_{G_k}(x_{i+1}) \max \Big\{ \max_{l=1,\ldots,N} v_{G_l}(i) p_{G_l\,G_k}, \ \max_{m=1,\ldots,M} v_{D_m}(i) p_{D_m\,G_k} \Big\},$$

and define $\mathcal{V}_{G_k}(i)$ to be the collection of all paths among $G_l G_k$ and $D_m G_k$ that realize the value $v_{G_k}(i+1)$. It is not difficult to show that

$$\max_{\text{all } \pi: \ L_{n/s}(\pi) = L} P(x, \pi) = \max \Big\{ \max_{k=1,\ldots,N} v_{G_k}(L) p_{G_k\,0}, \\ \max_{m=1,\ldots,M} v_{D_m}(L) p_{D_m\,0} \Big\} \tag{3.27}$$

(see Exercise 3.15).

We also define $\mathcal{V}(L)$ to be the collection of all states that realize the value $\max_{\text{all } \pi: \, L_{n/s}(\pi) = L} P(x, \pi)$. Traceback starts at any element of $\mathcal{V}(L)$, goes through the relevant paths in some $\mathcal{V}_{G_k}(i)$ and $\mathcal{V}_{D_m}(i)$ with $0 \leq i \leq L - 1$, ending with a path in a certain $\mathcal{V}_{G_k}(0)$.

We will now give an example of applying the Viterbi algorithm in the case of HMMs with silent states.

Example 3.10. Consider the HMM for which $\mathcal{Q}$ is the two-letter alphabet $\{A, B\}$, whose underlying Markov chain is shown in Fig. 3.9 and whose emis-

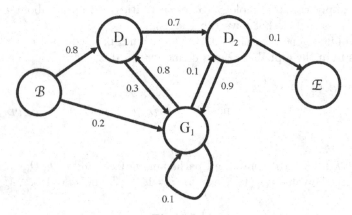

Fig. 3.9.

sion probabilities are as follows

$$q_{G_1}(A) = 0.2, \qquad q_{G_1}(B) = 0.8.$$

We let $x = ABB$ and find the (unique) Viterbi path π^* for it. We have

$$
\begin{aligned}
v_{G_1}(1) &= 0.2 \times 0.8 \times 0.7 \times 0.9 = 0.1008, & \mathcal{V}_{G_1}(0) &= \{0 \, D_1 D_2 G_1\}, \\
v_{D_1}(1) &= 0.1008 \times 0.8 = 0.08064, & \mathcal{V}_{D_1}(0) &= \{G_1 D_1\}, \\
v_{D_2}(1) &= 0.1008 \times 0.8 \times 0.7 = 0.056448, & \mathcal{V}_{D_2}(0) &= \{G_1 D_1 D_2\}, \\
v_{G_1}(2) &= 0.8 \times 0.056448 \times 0.9 = 0.04064256, & \mathcal{V}_{G_1}(1) &= \{D_2 G_1\}, \\
v_{D_1}(2) &= 0.04064256 \times 0.8 = 0.03251405, & \mathcal{V}_{D_1}(1) &= \{G_1 D_1\}, \\
v_{D_2}(2) &= 0.04064256 \times 0.8 \times 0.7 = 0.02275983, & \mathcal{V}_{D_2}(1) &= \{G_1 D_1 D_2\}, \\
v_{G_1}(3) &= 0.8 \times 0.02275983 \times 0.9 = 0.01638708, & \mathcal{V}_{G_1}(2) &= \{D_2 G_1\}, \\
v_{D_1}(3) &= 0.01638708 \times 0.8 = 0.01310966, & \mathcal{V}_{D_1}(2) &= \{G_1 D_1\}, \\
v_{D_2}(3) &= 0.01638708 \times 0.8 \times 0.7 = 0.009176765, & \mathcal{V}_{D_2}(2) &= \{G_1 D_1 D_2\}.
\end{aligned}
$$

This gives

$$\max_{\text{all } \pi:\, L_{n/s}(\pi)\, =\, 3} P(x,\pi) = 0.009176765 \times 0.1 = 0.0009176765,$$

and

$$\mathcal{V}(3) = \{D_2\}.$$

Tracing back we obtain the Viterbi path

$$\pi^* = 0\, D_1 D_2 G_1 D_1 D_2 G_1 D_1 D_2 G_1 D_1 D_2\, 0.$$

We will now generalize the forward algorithm. Fix a sequence $x = x_1 \ldots x_L$ and define for $k = 1, \ldots, N$

$$f_{G_k}(1) = q_{G_k}(x_1)\left(p_{0\,G_k} + \sum_{\substack{1 \le s \le M, \\ 1 \le i_1 < \ldots < i_s \le M}} p_{0\,D_{i_1}} p_{D_{i_1}\,D_{i_2}} \cdots p_{D_{i_{s-1}}\,D_{i_s}} p_{D_{i_s}\,G_k} \right).$$

Next, for $i = 1, \ldots, L$ and $m = 1, \ldots, M$ set

$$f_{D_m}(i) = \sum_{k=1,\ldots,N} \left(f_{G_k}(i)\left(p_{G_k\,D_m} \right. \right.$$

$$\left. \left. + \sum_{\substack{1 \le s \le m-1, \\ 1 \le i_1 < \ldots < i_s \le m-1}} p_{G_k\,D_{i_1}} p_{D_{i_1}\,D_{i_2}} \cdots p_{D_{i_{s-1}}\,D_{i_s}} p_{D_{i_s}\,D_m} \right) \right).$$

In order to perform a recursion step, for $i = 1, \ldots, L-1$ and $k = 1, \ldots, N$ set

$$f_{G_k}(i+1) = q_{G_k}(x_{i+1})\left(\sum_{l=1,\ldots,N} f_{G_l}(i) p_{G_l\,G_k} + \sum_{m=1,\ldots,M} f_{D_m}(i) p_{D_m\,G_k} \right).$$

It is not difficult to show that

$$P(x) = \left(\sum_{k=1,\ldots,N} f_{G_k}(L) p_{G_k\,0} + \sum_{m=1,\ldots,M} f_{D_m}(L) p_{D_m\,0} \right) \tag{3.28}$$

(see Exercise 3.16).

We will now give an example of applying the forward algorithm in the case of HMMs with silent states.

Example 3.11. We will find $P(x)$ for the sequence $x = ABB$ in Example 3.10. We have

$f_{G_1}(1) = 0.2 \times (0.2 + 0.8 \times 0.3 + 0.8 \times 0.7 \times 0.9) = 0.1888,$
$f_{D_1}(1) = 0.1888 \times 0.8 = 0.15104,$
$f_{D_2}(1) = 0.1888 \times (0.1 + 0.8 \times 0.7) = 0.124608,$
$f_{G_1}(2) = 0.8 \times (0.1888 \times 0.1 + 0.15104 \times 0.3 + 0.124608 \times 0.9) = 0.1410714,$
$f_{D_1}(2) = 0.1410714 \times 0.8 = 0.1128571,$
$f_{D_2}(2) = 0.1410714 \times (0.1 + 0.8 \times 0.7) = 0.09310712,$
$f_{G_1}(3) = 0.8 \times (0.1410714 \times 0.1 + 0.1128571 \times 0.3 + 0.09310712 \times 0.9)$
$\qquad = 0.1054085,$
$f_{D_1}(3) = 0.1054085 \times 0.8 = 0.0843268,$
$f_{D_2}(3) = 0.1054085 \times (0.1 + 0.8 \times 0.7) = 0.06956961,$

which gives

$$P(x) = 0.06956961 \times 0.1 = 0.006956961.$$

It is also possible to generalize the backward algorithm as well as the Baum-Welch and Viterbi training algorithms to the case of HMMs with silent states (see Exercises 3.17–3.19). These generalizations follow the principles that we have utilized generalizing the Viterbi and forward algorithms: one must include in consideration chains of silent states in an appropriate way. To illustrate this point once again, below we give an example of parameter estimation for HMMs with silent states in the case when paths are known.

Example 3.12. Let the *a priori* connectivity be as shown in Fig. 3.10, and the

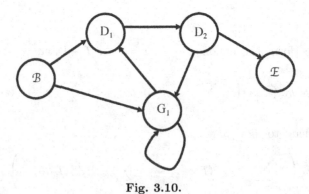

Fig. 3.10.

training dataset be (x^1, π^1), (x^2, π^2), (x^3, π^3) with

$$x^1 : ABBA, \ \pi^1 : 0\,G_1G_1G_1D_1D_2G_1D_1D_2\,0$$
$$x^2 : AB, \qquad \pi^2 : 0\,D_1D_2G_1G_1D_1D_2\,0$$
$$x^3 : BAB, \qquad \pi^3 : 0\,D_1D_2G_1D_1D_2G_1D_1D_2G_1D_1D_2\,0.$$

Clearly, the path component of the dataset agrees with the *a priori* connectivity.

From formulas (3.17) we obtain

$$p_{0\,G_1} = \frac{1}{3}, \quad p_{0\,D_1} = \frac{2}{3},$$

$$p_{G_1\,G_1} = \frac{1}{3}, \quad p_{G_1\,D_1} = \frac{2}{3},$$

$$p_{D_2\,G_1} = \frac{5}{8}, \quad p_{D_2\,0} = \frac{3}{8},$$

$$q_{G_1}(A) = \frac{4}{9}, \quad q_{G_1}(B) = \frac{5}{9}.$$

It is also clear that $p_{D_1\,D_2} = 1$.

3.8 Profile HMMs

Recall that Markov chains and HMMs can be used for searching with the purpose of detecting potential membership of a query sequence in a sequence family. Suppose, for example, that we are given a family $\mathcal{F}$ of related sequences, and we wish to know whether or not a query sequence x shares some of the common features of the sequences in $\mathcal{F}$. A straightforward approach is to first choose a connectivity that looks suitable (for example, the simple one shown in Fig. 3.1, if we are dealing with DNA sequences and Markov models), to use the sequences in $\mathcal{F}$ as the training data to estimate the parameters and, finally, assess a probabilistic quantity arising from the resulting model (for instance, $P(x)$) to decide whether or not x should be accepted as a new member of the family $\mathcal{F}$. The problem with this approach, however, is the choice of connectivity. An arbitrarily picked connectivity will almost certainly lead to a very poorly performing model, and there is no comprehensive theory that could help to choose a "good" connectivity.

In this section we will present a particular connectivity for the underlying Markov models of HMMs that performs very well in various searches. To use it for searching, however, a multiple alignment of sequences in $\mathcal{F}$ is required. As we explained in Chap. 2, producing a "good" multiple alignment is a difficult problem in itself, but for the purposes of this section we will assume that a multiple alignment is given. Certainly, in this situation one may try to avoid constructing a probabilistic model altogether and to attempt to detect directly the presence of common features identifiable from the alignment in the sequence x. However, having a probabilistic way of doing so has numerous advantages; in particular, it allows to compute, at least in some sense, the probability of x belonging to $\mathcal{F}$. There are special HMMs called *profile HMMs* that are commonly used for modeling multiple alignments. They are one of the most popular applications of HMMs in molecular biology.

We will now explain how a profile HMM is built from a multiple alignment. Consider the *a priori* connectivity of the type shown in Fig. 3.11.

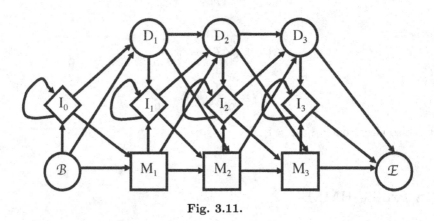

Fig. 3.11.

Here M_j and I_j are non-silent, and D_j are silent states; M_j are called *match states*, I_j *insert states*, and D_j *delete states*. The connectivity shown in Fig. 3.11 has length (the number of match states) equal to 3. In general, the length is derived from the given multiple alignment. One simple rule is to set the length to be equal to the number of columns in the alignment where gaps occupy no more than half the positions. Such columns will be called *match columns*, and all the other ones will be called *insert columns*. There are also other methods for selecting the length, but everywhere below we will assume that this rule is applied.

We wish to use the family $\mathcal{F}$ as a training dataset for this *a priori* connectivity. Training will be done as in the case when paths are known, so for each sequence in $\mathcal{F}$ we will first produce a path through the connectivity graph. Paths come from the multiple alignment. For each sequence in $\mathcal{F}$ consider the corresponding row in the alignment. The row contains letters from $\mathcal{Q}$ and may contain gaps. If a letter is found in a match column, we assign a match state to its position; if a letter is found in an insert column, we assign an insert state to its position; if a gap is found in a match column, we assign a delete state to its position; if a gap is found in an insert column, we skip this position and do not assign any state to it. Reading the row from left to right then produces a path through the connectivity graph for the sequence.

We will illustrate the above procedure by the following example.

Example 3.13. Let Q be the two-letter alphabet $\{A, B\}$, $\mathcal{F} = \{BAAB, AABA, BBB, AA\}$, and suppose we are given the multiple alignment

$$
\begin{array}{l}
\quad\ 1\ 2\ 3\ 4\ 5\ 6\ 7 \\
x^1 : -\ B\ A\ -\ A\ B\ - \\
x^2 : A\ -\ A\ B\ -\ A\ - \\
x^3 : -\ -\ B\ -\ B\ -\ B \\
x^4 : -\ -\ A\ -\ -\ A\ -.
\end{array}
$$

Here columns 3,5,6 are match columns, and the other ones are insert columns. Hence, the length of the connectivity is chosen to be equal to 3, as in the example in Fig. 3.11. The corresponding paths for the sequences x^1, x^2, x^3, x^4 are as follows

$$
\begin{array}{l}
\pi^1 : 0\ I_0 M_1 M_2 M_3\ 0 \\
\pi^2 : 0\ I_0 M_1 I_1 D_2 M_3\ 0 \\
\pi^3 : 0\ M_1 M_2 D_3 I_3\ 0 \\
\pi^4 : 0\ M_1 D_2 M_3\ 0.
\end{array}
$$

We will now consider the dataset (x^1, π^1), (x^2, π^2), (x^3, π^3), (x^4, π^4) and estimate the parameters using formulas (3.17), as in Example 3.12. We obtain

$$
p_{0\,I_0} = \frac{1}{2}, \quad p_{0\,D_1} = 0, \quad p_{0\,M_1} = \frac{1}{2},
$$

$$
p_{M_1\,I_1} = \frac{1}{4},\ p_{M_1\,D_2} = \frac{1}{4},\ p_{M_1\,M_2} = \frac{1}{2},
$$

$$
p_{M_2\,I_2} = 0,\ p_{M_2\,D_3} = \frac{1}{2},\ p_{M_2\,M_3} = \frac{1}{2},
$$

$$
p_{M_3\,I_3} = 0,\ p_{M_3\,0} = 1,
$$

$$
p_{I_0\,I_0} = 0,\ p_{I_0\,D_1} = 0,\ p_{I_0\,M_1} = 1,
$$

$$
p_{I_1\,I_1} = 0,\ p_{I_1\,D_2} = 1,\ p_{I_1\,M_2} = 0,
$$

$$
p_{I_3\,I_3} = 0,\ p_{I_3\,0} = 1,
$$

$$p_{D_2\,I_2} = 0, \quad p_{D_2\,M_3} = 1, \quad p_{D_2\,D_3} = 0,$$

$$p_{D_3\,I_3} = 1, \quad p_{D_3\,0} = 0,$$

$$q_{M_1}(A) = \frac{3}{4}, \, q_{M_1}(B) = \frac{1}{4},$$

$$q_{M_2}(A) = \frac{1}{2}, \, q_{M_2}(B) = \frac{1}{2},$$

$$q_{M_3}(A) = \frac{2}{3}, \, q_{M_3}(B) = \frac{1}{3},$$

$$q_{I_0}(A) = \frac{1}{2}, \quad q_{I_0}(B) = \frac{1}{2},$$

$$q_{I_1}(A) = 0, \quad q_{I_1}(B) = 1,$$

$$q_{I_3}(A) = 0, \quad q_{I_3}(B) = 1.$$

Note that the given data does not allow us to estimate the transition and emission probabilities $p_{I_2\,I_2}$, $p_{I_2\,D_3}$, $p_{I_2\,M_3}$, $q_{I_2}(A)$, $q_{I_2}(B)$, since I_2 does not occur anywhere in the sequences π^1, π^2, π^3, π^4. This is a common problem with parameter estimation for profile HMMs. To set the remaining transition and emission probabilities, heuristic procedures (such as adding *pseudocounts*) are frequently used. These procedures may incorporate prior knowledge about the dataset in question.

For an example of parameter estimation from a family of globin sequences and subsequent search of a protein sequence database by means of the resulting profile HMM see [DEKM].

3.9 Multiple Sequence Alignment by Profile HMMs

In the previous section we showed how one could construct a profile HMM from a given multiple alignment. In this section we will assume that the sequences in a family $\mathcal{F}$ are unaligned and explain how a profile HMM can be used to produce a multiple alignment for these sequences. This procedure is heuristic, but has shown good results.

Consider the *a priori* connectivity of a profile HMM as introduced in the previous section. First of all, we must choose its length. The commonly used rule is to set it to the average length of the sequences in $\mathcal{F}$. Next, we use the Baum-Welch training or the Viterbi training to estimate the transition and emission probabilities from the training sequences. Note that since the sequences are unaligned, we do not have paths for the training data and hence cannot estimate the parameters directly as it was done in the preceding section. Next, by the Viterbi algorithm, for each training sequence we find the most probable paths through the resulting model. Finally, we build a multiple alignment from these paths. Letters are aligned if they are emitted by the same match states in their paths. Gaps are then inserted appropriately. Since there may be several ways of positioning the letters emitted at the insert states with respect to the letters emitted at the match states, we may obtain more than one multiple alignment, as in Example 3.14 below. We may obtain more than one alignment also because some sequences can have more than one Viterbi path. The HMM does not attempt to align subsequences that came from the insert states in the paths, hence we may not obtain a complete alignment.

The following example illustrates the last step of the multiple alignment process.

Example 3.14. Let $\mathcal{Q}$ be the DNA alphabet $\{A, C, G, T\}$ and $\mathcal{F} = \{x^1 = ACCG, x^2 = TGCG, x^3 = AT, x^4 = CG, x^5 = ATC, x^6 = TTG\}$. The average length of the sequences is 3, so we will consider the profile HMM connectivity of length 3, as shown in Fig. 3.11. Suppose we have estimated the parameters by either the Baum-Welch or Viterbi training procedure, and suppose that we have found the Viterbi paths for each sequence from the estimated parameter values. Assume further that there is only one Viterbi path $\pi^*(x^j)$ for each x^j, and that the paths are

$$\pi^*(x^1) : 0 \, M_1 M_2 M_3 I_3 \, 0$$
$$\pi^*(x^2) : 0 \, I_0 D_1 I_1 M_2 D_3 I_3 \, 0$$
$$\pi^*(x^3) : 0 \, D_1 M_2 M_3 \, 0$$
$$\pi^*(x^4) : 0 \, M_1 M_2 D_3 \, 0$$
$$\pi^*(x^5) : 0 \, D_1 M_2 D_3 I_3 I_3 \, 0$$
$$\pi^*(x^6) : 0 \, D_1 D_2 M_3 I_3 I_3 \, 0.$$

These paths lead to the following three multiple alignments

$$\begin{aligned}
x^1 &: - - - A \, C \, C \, g \\
x^2 &: T \, G - C - g \\
x^3 &: - - - - A \, T - - \\
x^4 &: - - - C \, G - - - \\
x^5 &: - - - - A - t \ c \\
x^6 &: - - - - - T \, t \ g,
\end{aligned}$$

$$\begin{aligned}
x^1 &: - A - C \, C \, g \\
x^2 &: T - G \, C - g \\
x^3 &: - - - - A \, T - - \\
x^4 &: - C - G - - - \\
x^5 &: - - - - A - t \ c \\
x^6 &: - - - - - T \, t \ g,
\end{aligned}$$

$$x^1 : A - - C\,C\,g$$
$$x^2 : - T\,G\,C - g$$
$$x^3 : - - - A\,T - -$$
$$x^4 : C - - G - - -$$
$$x^5 : - - - A - t\;\;c$$
$$x^6 : - - - - T\,t\;\;g,$$

where we do not attempt to align the letters emitted at I_3 (shown in lower case).

The multiple alignment method described in this section was successfully tested in [KH] on a family of globin sequences. It is implemented in the software package HMMER [E] (see http://hmmer.wustl.edu). More details on using profile HMMs for sequence alignment can be found in [DEKM].

HMMs in general and profile HMMs in particular can be applied to modeling datasets of many types, both in biology and other areas. In the next chapter we will mention an application of profile HMMs to the problem of protein folding.

Exercises

3.1. Consider a Markov chain with state set $X = \{G_1, \ldots, G_N\}$, transition probabilities $p_{ij} = p_{G_i G_j}$ and initialization probabilities $P(G_j)$, $i, j = 1, \ldots, N$. Fix $L \in \mathbb{N}$ and for a sequence $x = x_1 \ldots x_L$ with $x_j \in X$ for all j, define $P(x)$ by formula (3.1). Show that $\sum_x P(x) = 1$, where the summation is taken over all sequences of length L.

3.2. Consider a Markov chain with state set $X = \{G_1, \ldots, G_N\}$, end state and transition probabilities $p_{0 G_j}$, $p_{G_i G_j}$, $p_{G_j 0}$, $i, j = 1, \ldots, N$. Suppose that the chain is non-trivially connected. For a sequence $x = x_1 \ldots x_L$ with $x_j \in X$ for all j, define $P(x)$ by formula (3.4). Show that $\sum_x P(x) = 1$, where the summation is taken over all sequences of finite positive length (the length is not fixed). [If you find it hard to prove this statement in general, do it in the case when X contains only two elements.]

3.3. Estimate the transition probabilities for the *a priori* connectivity shown in Fig. 3.12 from the following training sequences

$$x^1 : 0\,G_1 G_2 G_3 G_3 G_2\,0$$
$$x^2 : 0\,G_2 G_3 G_3 G_3 G_2 G_3\,0$$
$$x^3 : 0\,G_1 G_2 G_3\,0$$
$$x^4 : 0\,G_2 G_3 G_2 G_3 G_2\,0$$

(note that the training data agrees with the above connectivity). What is the connectivity of the resulting Markov model?

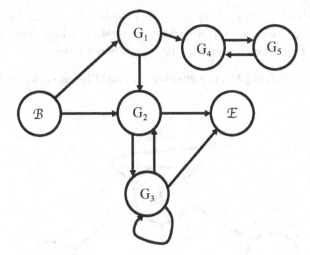

Fig. 3.12.

3.4. Consider the DNA model obtained in Example 3.1. Let $y = AAA\textbf{ATG}CTATTGATGGATGAAATGAAAATT\textbf{TAA}AAG$ be a stretch of prokaryotic DNA, where the start and stop codons present in y are shown in bold. This sequence contains three candidate gene sequences

$$y^1 : \textbf{ATG}CTATTGATGGATGAAATGAAAATT\textbf{TAA}$$
$$y^2 : \textbf{ATG}GATGAAATGAAAATT\textbf{TAA}$$
$$y^3 : \textbf{ATG}AAAATT\textbf{TAA}.$$

The ORFs corresponding to these sequences are

$$z^1 : CTATTGATGGATGAAATGAAAATT$$
$$z^2 : GATGAAATGAAAATT$$
$$z^3 : AAAATT.$$

respectively. Suppose that an ORF x is accepted as the ORF of a prokaryotic gene, if $P(x) \geq 0.0003$, with $P(x)$ calculated from formula (3.4). Assess on this basis z^1, z^2 and z^3. Will any of y^1, y^2, y^3 be accepted as prokaryotic genes?

3.5. Consider an HMM whose underlying Markov chain has a state set $X = \{G_1, \ldots, G_N\}$, end state and transition probabilities $p_{0\,G_j}$, $p_{G_i\,G_j}$, $p_{G_j\,0}$, $i, j = 1, \ldots, N$. Suppose that the chain is non-trivially connected, and let $q_{G_j}(a)$ be the emission probabilities of the HMM, $j = 1, \ldots, N$, $a \in \mathcal{Q}$. For a pair of sequences $(x, \pi) = (x_1 \ldots x_L, \pi_1 \ldots \pi_L)$ define $P(x, \pi)$ by formula (3.8). Show that $\sum_{(x,\pi)} P(x, \pi) = 1$, where the summation is taken over all pairs of sequences of finite positive length (the length is not fixed).

3.6. Let the underlying Markov chain of an HMM be as shown in Fig. 3.13.

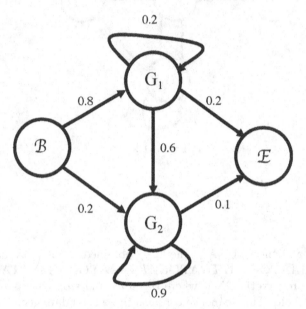

Fig. 3.13.

Assume that Q is the two-letter alphabet $\{A, B\}$ and let the emission probabilities be as follows

$$q_1(A) = 0.5, \ q_1(B) = 0.5,$$
$$q_2(A) = 0.9, \ q_2(B) = 0.1.$$

Suppose that a sequence x is accepted as a member of the family $\mathcal{F}$ of training sequences, if $P(x) \geq 0.02$, with $P(x)$ calculated from formula (3.9). Assess on this basis the sequences $y^1 = ABB$ and $y^2 = BAA$. Will either of y^1, y^2 be accepted as a member of $\mathcal{F}$? What are the most probable paths $\pi^*(y^1)$ and $\pi^*(y^2)$ for these sequences? [In this case there is only one most probable path for each sequence.]

3.7. Consider the HMM for which Q is the two-letter alphabet $\{A, B\}$, whose underlying Markov chain is shown in Fig. 3.14 and whose emission probabilities are as follows

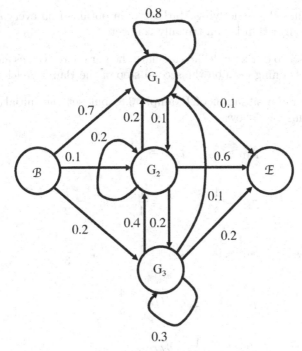

Fig. 3.14.

$$q_1(A) = 0.7, \; q_1(B) = 0.3,$$
$$q_2(A) = 0.2, \; q_2(B) = 0.8,$$
$$q_3(A) = 0.4, \; q_3(B) = 0.6.$$

For the sequence $x = ABA$ find all the Viterbi paths and $\max_{\text{all } \pi \text{ of length 3}}$ $P(x, \pi)$ by applying the Viterbi algorithm.

3.8. Consider the HMM from Exercise 3.7 and find $P(x)$, where $x = BAB$, by applying the forward algorithm.

3.9. In Exercise 3.8 find $P(x)$ by the backward algorithm and perform posterior decoding for the sequence x. Also find the (unique) Viterbi path for x and compare it with the one provided by the posterior decoding.

3.10. For the *a priori* connectivity shown in Fig. 3.4 in Example 3.2 and the two-letter alphabet $\{A, B\}$ estimate the transition and emission probabilities from the following training data

$$x^1 : AAAB, \quad \pi^1 : G_1G_1G_1G_1,$$
$$x^2 : BBA, \quad \pi^2 : G_1G_1G_2,$$
$$x^3 : BB, \quad \pi^3 : G_1G_2.$$

3.11. Show that the underlying Markov chain obtained on every step of the Baum-Welch algorithm is non-trivially connected.

3.12. In the setup of Example 3.8 calculate the parameter values and the likelihood of the training data on the second step of the Baum-Welch algorithm.

3.13. Consider the situation of Example 3.8, but set the initial parameter values differently as follows

$$p_{01}^{(0)} = \frac{1}{4}, \quad p_{02}^{(0)} = \frac{3}{4},$$

$$p_{11}^{(0)} = \frac{1}{4}, \quad p_{12}^{(0)} = \frac{1}{4}, \quad p_{10}^{(0)} = \frac{1}{2},$$

$$p_{21}^{(0)} = 0, \quad p_{22}^{(0)} = \frac{1}{3}, \quad p_{20}^{(0)} = \frac{2}{3},$$

$$q_1^{(0)}(A) = \frac{1}{5}, q_1^{(0)}(B) = \frac{4}{5},$$

$$q_2^{(0)}(A) = \frac{1}{3}, q_2^{(0)}(B) = \frac{2}{3}.$$

Re-estimate the parameters once using the Viterbi training algorithm.

3.14. Consider an HMM with silent states, and for every finite sequence x of letters from the corresponding alphabet Q define $P(x)$ by formula (3.26). Show that $\sum_x P(x) = 1$, where the summation is taken over all sequences of all finite lengths, including length 0 (the length is not fixed).

3.15. Prove identity (3.27).

3.16. Prove identity (3.28).

3.17. Generalize the backward algorithm to the case of HMMs with silent states which do not have silent loops.

3.18. Generalize the Baum-Welch training algorithm to the case of HMMs with silent states, where the *a priori* connectivities do not have silent loops. For the *a priori* connectivity from Example 3.12 shown in Fig. 3.10 make one step of the generalized Baum-Welch algorithm for the training data

$$x^1 : AB$$
$$x^2 : BB$$
$$x^3 : BA$$

starting with the initial values

$$p^{(0)}_{0\,G_1} = \frac{1}{4}, \quad p^{(0)}_{0\,D_1} = \frac{3}{4},$$

$$p^{(0)}_{G_1\,G_1} = \frac{3}{5}, \quad p^{(0)}_{G_1\,D_1} = \frac{2}{5},$$

$$p^{(0)}_{D_2\,G_1} = \frac{5}{7}, \quad p^{(0)}_{D_2\,0} = \frac{2}{7},$$

$$q^{(0)}_{G_1}(A) = \frac{1}{8}, \quad q^{(0)}_{G_1}(B) = \frac{7}{8}.$$

3.19. Generalize the Viterbi training algorithm to the case of HMMs with silent states, where the *a priori* connectivities do not have silent loops. In the situation of Exercise 3.18 make one step of the generalized Viterbi training algorithm.

3.20. As in Example 3.13, estimate the transition and emission probabilities for the corresponding connectivity from the family $\mathcal{F} = \{ABAAA, BAAABB, BABA, AAB, ABABBA\}$ of sequences of letters from the two letter alphabet $\{A, B\}$, given the following multiple alignment

$$
\begin{aligned}
&x^1 : A\ B\ A\ -\ -\ A\ A\ - \\
&x^2 : B\ A\ A\ A\ -\ B\ -\ B \\
&x^3 : -\ B\ -\ A\ B\ A\ -\ - \\
&x^4 : -\ -\ -\ -\ A\ A\ B\ -\ - \\
&x^5 : -\ A\ -\ B\ A\ B\ B\ A.
\end{aligned}
$$

Have you been able to estimate all the parameters?

3.21. In the situation of Example 3.14 let the only Viterbi paths for x^j, $j = 1, 2, 3, 4, 5, 6$ be

$$
\begin{aligned}
&\pi^*(x^1) : 0\,D_1 I_1 I_1 M_2 D_3 I_3\,0 \\
&\pi^*(x^2) : 0\,M_1 D_2 I_2 I_2 M_3\,0 \\
&\pi^*(x^3) : 0\,M_1 D_2 M_3\,0 \\
&\pi^*(x^4) : 0\,I_0 D_1 D_2 D_3 I_3\,0 \\
&\pi^*(x^5) : 0\,M_1 M_2 M_3\,0 \\
&\pi^*(x^6) : 0\,M_1 M_2 D_3 I_3\,0.
\end{aligned}
$$

Find all multiple alignments of the sequences arising from these paths.

4

Protein Folding

One of the most important problems in molecular biology is the protein folding problem: given the amino acid sequence of a protein, what is the protein's structure in three dimensions? This problem is important since the structure of a protein provides a key to understanding its biological function. In this chapter we will only slightly touch on the problem. For a more in-depth discussion see, e.g., the surveys [N], [DC].

4.1 Levels of Protein Structure

It is widely assumed that the amino acid sequence contains all information about the native three-dimensional structure of a protein molecule under given physiological conditions. One reason to believe this is the following thermodynamic principle established in the 1950's by Christian Anfinsen's denaturation-renaturation experiments on ribonuclease. If one changes the values of environmental parameters such as temperature or solvent conditions, the protein will undergo a transition from the native state to an unfolded state and become inactive. When the parameter values are reset to the values corresponding to the physiological conditions, the protein refolds and becomes active again. This thermodynamic principle was later confirmed by similar experiments on other small globular proteins.

Determining the tree-dimensional structure of a protein is a very hard task. The most reliable results are produced by experimental approaches such as *nuclear magnetic resonance (NMR)* and *X-ray crystallography*. However, such approaches are expensive and can require years to produce the structure of a single protein. Therefore the number of known protein sequences is much larger than the number of known three-dimensional protein structures, and this gap constantly grows as a result of various sequencing projects. Thus, mathematical, statistical and computational methods that may give some indication of structure (and as a result, function) are becoming increasingly important.

There are different levels of protein structure.

(i) *Primary Structure:* the amino acid sequence of a protein.

(ii) *Secondary Structure:* local regular structures commonly found within proteins. There are two main types of such structures: *α-helices* and *β-sheets*. In an α-helix amino acids are arranged into a right-handed spiral with 3.6 residues per turn. An α-helix is traditionally pictured either as a ribbon (see Fig. 4.1) or as a cylinder.

Fig. 4.1.

A β-sheet consists of two or more *β-strands* connected by hydrogen bonds. Atoms in a β-strand are arranged in a particular way as shown in Fig. 4.2.

Fig. 4.2.

If the orientations of all β-strands in a β-sheet coincide, it is called *parallel*, if the orientations alternate, it is called *antiparallel*, and otherwise it is called of *mixed type*. A β-sheet is pictured as a collection of oriented strips, where each of the strips represents a single β-strand. Figure 4.3 shows a simple two-strand antiparallel β-sheet.

Fig. 4.3.

Other secondary structure elements are less pronounced. There are, for example, *turns* and *loops*.

(iii) *Super-Secondary Structure or Motif:* local folding patterns built up from particular secondary structures. Examples are the *EF-hand motif* that consists of an α-helix, followed by a turn, followed by another α-helix, the *β-helix motif* that consists of three β-sheets, *coiled coil motifs* that consist of two or more α-helices wrapped around each other. Figure 4.4 shows a protein containing two *β-barrels*, which are "closed" β-sheets (PDB ID: 1CBI, [TBB]).

Fig. 4.4.

(iv) *Fold:* the three-dimensional structure of an independently folding fragment of the polypeptide chain that often has an associated function; such fragments are called *domains*. A protein can have several folds connected by loops. Figure 4.5 shows a two-domain protein (PDB ID: 1QRJ, [KS]).

(v) *Tertiary Structure:* the full three-dimensional structure of a protein. Figure 4.6 shows the tertiary structure of sperm whale myoglobin (PDB ID: 108M, [S]).

Fig. 4.5.

Fig. 4.6.

(vi) *Quaternary Structure:* an arrangement of several protein chains in space. Figure 4.7 shows the bovine mitochondrial Cytochrome Bc1 complex that consists of 11 chains (PDB ID: 1BE3, [IJ]).

Fig. 4.7.

In these terms the protein folding problem can be stated as follows: predict the secondary, super-secondary, tertiary, quaternary structures and folds from the primary structure. In practice, one can only hope to predict secondary structure, motifs and in some cases folds and tertiary structures. In the following sections we will discuss several methods used for prediction.

4.2 Prediction by Profile HMMs

To predict secondary structure elements, motifs or folds from the primary structure one can directly use profile HMMs as defined in Sect. 3.8. To construct a profile HMM, one has to start with a family of aligned sequences. If one aims at predicting a particular structural feature, then as the sequence family a collection of protein sequences that contain the feature must be taken. The sequences are aligned and a profile HMM is derived from the alignments as in Sect. 3.8. For a query protein sequence x search is performed with the resulting HMM as was described in Sect. 3.2. If $P(x, \pi^*)$ or $P(x)$ is greater than a certain cutoff, the feature is reported in the sequence x. We also mention that other learning machines such as neural networks can be used in a similar way instead of HMMs.

This principle is implemented in the web-based resource Pfam (see http://www.sanger.ac.uk/Pfam/) that concentrates on fold prediction. Pfam contains over 2000 folds and for each of them a profile HMM has been estimated from a family of proteins containing the fold. A query sequence is run past each of these HMMs. If a part of the sequence has probability greater than a certain cutoff for a particular HMM, the corresponding domain is reported in the sequence. For more information the reader is referred to the Pfam web site.

4.3 Threading

Threading is an approach to the protein folding problem based on the observation that many protein folds in existing databases are very similar. As a result of this, it has been conjectured that there are only a limited number of folds in nature. Some predict that there are fewer than 1000 different folds. Hence it is natural to attempt to determine the structure of a query sequence by finding the "optimal fit" of fragments of the sequence to some folds in a library of folds, that is, to "thread" it through known structures.

There are many ways to do threading, and we will only describe one of them. Threading can be done by aligning sequences to structures. Suppose we have an amino acid sequence x and a structure S. We have to define what aligning x to S means and how one can score such an alignment. A straightforward approach would be to choose an amino acid sequence y that is known to give rise to the structure S and align x to y. Such an alignment can be scored by using standard PAM or BLOSUM matrices described in Chap. 9. The major problem with this method is that similar structures have significant sequence variability, so it is not clear which sequence y corresponding to the structure S one should choose. Selecting different sequences y may produce significantly different alignments of x to S.

We will now describe a more realistic approach to threading that is sometimes called the *environmental template method*. The idea behind it is that instead of aligning a sequence to a sequence, one aligns a sequence to a string of descriptors that describe the environment of each residue in the structure S. Specifically, for each residue in S one determines how buried it is, how polar/hydrophobic its environment is, and what secondary structure it is in. Figure 4.8 shows six *burial/polarity environments* introduced in [BLE]: B_1 (buried, hydrophobic), B_2 (buried, moderately polar), B_3 (buried, polar), P_1 (partially buried, moderately polar), P_2 (partially buried, polar), E (exposed).

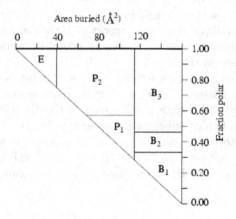

Fig. 4.8.

In addition to the six environments each residue in S is categorized into three classes corresponding to the secondary structure it is in: α (for α-helices), β (for β-sheets) and γ (for all other secondary structures). In total there are 18 *environmental descriptors*. We denote the collection of these descriptors by $\mathcal{C}$.

One can now convert the known structure S into a string: $S = e_1, \ldots, e_n$, where e_j are elements of $\mathcal{C}$. To align x to this string a scoring scheme is required. A scoring matrix is obtained in the following way. For a from the amino acid alphabet and $e \in \mathcal{C}$ set

$$s(a, e) = \log \left(\frac{p_{a\,e}}{p_a} \right),$$

where p_{ae} and p_a are derived from a database of known protein structures: p_{ae} is the frequency of the occurrence of the amino acid a in the environment e in the database and p_a is the frequency of the occurrence of a anywhere in the database. We remark that this scoring matrix is produced in the spirit of the general approach to constructing scoring matrices discussed in Sect. 9.1 (note that p_{ae} can be treated as the conditional probability of a given e). Figure 4.9 shows a scoring matrix which is a slight improvement of that derived in [BLE].

	W	F	Y	L	I	V	M	A	G	P	C	T	S	Q	N	E	D	H	K	R
B,α	1.11	1.28	0.27	1.3	1.11	0.74	1.26	-0.77	-2.22	-1.56	-0.43	-1.72	-2.43	-1.38	-1.76	-2.15	-2.48	-0.34	-1.37	-1.8
B,β	0.92	0.96	0.17	1.07	1.5	1.18	0.51	-1.05	-2.35	-0.77	-0.45	-1.27	-2.56	-2.03	-2.18	-1.59	-1.8	-2.26	-3.04	-1.52
B,γ	0.96	1.40	0.52	1.06	0.93	1	0.91	-0.54	-2.78	0.59	-0.59	-1.41	-2.99	-0.84	-2.61	-2.01	-2.63	-0.61	-2.78	-2.35
B,α	1.01	0.87	0.86	0.71	0.55	0.41	1.02	-0.65	-2.04	-0.97	0.15	-0.67	-1.33	0.16	-0.48	-0.58	-0.8	0.82	-0.94	-0.11
B,β	0.83	1.32	1.30	0.36	1.07	0.71	0.49	-1.52	-2.22	-0.86	-0.72	-1.14	-0.82	-0.79	-0.26	-0.2	-2.08	-0.05	-0.83	-0.41
B,γ	1.62	1.04	1.14	0.77	0.81	0.66	1	-0.81	-1.71	-0.07	-0.62	-1.03	-1.23	-0.87	-0.56	-1.13	-1.7	0.54	-2.12	-0.44
B,α	0.86	-0.22	0.5	0.16	-0.02	-0.29	0.87	-0.44	-1.09	-1.11	-1.38	-0.69	-1.01	0.16	-0.07	0.09	-0.43	0.61	0.56	1.1
B,β	0.07	0.37	1.09	0.14	0.26	0.16	-0.68	-1.08	-2.29	-0.01	-0.79	-0.1	-0.71	0.52	-0.33	-0.42	-0.76	0.8	0.35	0.84
B,γ	1.12	0.71	1.25	0.29	-0.54	-0.4	0.23	-0.87	-0.61	-0.11	-0.98	-0.48	-0.61	0.1	0.09	-0.46	-0.83	1.04	0.08	0.71
P,α	-1.29	-0.85	-0.88	-0.3	-0.06	0.3	-0.42	0.76	-0.46	-0.41	0.95	0.39	0.47	-0.32	-0.58	-0.43	-0.28	-0.91	-0.5	-0.51
P,β	0.34	-0.61	-0.09	-0.81	0.09	0.44	-0.4	0.59	-0.22	-0.65	1.28	0.95	0.49	-2.38	-0.92	-0.68	-0.61	-0.53	-2.01	-0.89
P,γ	-1.25	-1.29	-1.4	-0.33	-0.28	-0.09	-0.9	0.49	-0.39	0.64	1.29	0.55	0.59	-0.57	-0.26	-0.59	0.34	-1.21	-0.72	-0.88
P,α	-1.09	-1.35	-0.55	-0.46	-0.59	-0.62	-0.27	-0.02	-0.58	-0.25	-0.7	-0.13	-0.38	0.62	-0.02	0.2	0.29	0.17	0.660	0.56
P,β	-0.71	-0.56	-0.3	-1.33	-0.35	0.08	-0.76	-0.52	-0.87	-1.01	-0.87	0.79	0.49	0.1	0	0.41	-0.03	-0.49	0.55	0.19
P,γ	-0.42	-0.84	-0.43	-0.68	-0.94	-0.74	-0.83	-0.25	-0.42	0.44	-0.81	0.08	0.17	0.25	0.51	0.28	0.51	0.2	0.47	0.24
$E\alpha$	-1.26	-1.81	-1.7	-1.37	-2.36	-1.25	-0.9	0.44	0.63	0.05	-0.17	-0.2	0.16	0.29	0.32	0.6	0.44	-0.06	0.07	-0.2
$E\beta$	0.81	-0.83	-0.03	-1.6	-1.39	-1.66	-0.62	0.14	1.75	-0.88	-0.04	-0.17	0.65	-0.12	0.01	-0.37	-0.3	-0.76	-1.54	-1.12
$E\gamma$	-2.06	-1.63	-1.04	-1.14	-1.63	-0.8	-1.3	0.14	1.1	0.25	-0.35	0.08	0.34	-0.03	0.41	0.04	0.23	-0.41	-0.1	-0.41

Fig. 4.9.

We note that assigning a residue to a particular burial/polarity class is a non-trivial task and further improvements in the scoring matrix can be made by obtaining more precise estimates on the buried area and polarity of residues in known structures.

When x is aligned to S, gaps in x are allowed. Therefore one also has to select a model for aligning gap regions to sequences of the elements of C. Usually smaller gap penalties are selected for the α- and β-environments. The rationale for this strategy is that insertions and deletions are less likely to occur in regions of regular secondary structure. The gap model is usually either linear or affine. Similarly, gaps in S may be allowed, in which case one of the standard gap models is selected to score gap regions aligned to a sequence of amino acid.

Once a scoring system has been selected, one can find all optimal local alignments between x and S. This can be done by applying either the Smith-Waterman algorithm (see Sect. 2.3) or a faster heuristic algorithm. Aligning x to all structures in a database, one obtains overall optimal local alignments that hopefully characterize the three-dimensional structures of the corresponding subsequences in x. Of course, as with any sequence alignment procedure, one also must assess the statistical significance of the score.

4.4 Molecular Modeling

We will now turn to a biophysical method that attempts to predict the structure of a protein *ab initio*, that is, by trying to apply laws of physics to describe the protein molecule, rather than by using databases of known structures. Due to a large number of atoms involved, it is very hard to give a complete mathematical description of a protein molecule, including both quantum mechanical and relativistic effects. Therefore, in applications the classical approach is utilized. Specifically, we will attempt to model the *potential energy* of a protein molecule under normal physiological conditions, as well as the energy of the surrounding media. As we will see, the energy function depends on a number of parameters whose values determine the three-dimensional structure of the molecule. Under this approach, it is assumed that the native state of the molecule is given by those parameter values that minimize the energy function (assuming that a point of minimum exists). Note that the energy function may have multiple minima; in this case the molecule is assumed to have multiple native states, which indeed occurs in reality.

To convey the general ideas of the approach, we will only discuss one rather simplistic way to model the potential energy function of a protein molecule. For more sophisticated energy functions and more detailed discussion of molecular modeling of proteins see the survey paper [N]. Energy functions take into account *all* atoms in a protein molecule. Since the number of amino acids in proteins ranges from 25 to 3000 and the number of atoms ranges from around 500 to more than 10000, dealing even with the simplest energy functions may be a difficult computational task.

The potential energy function that we present here is broken up into two parts: the bonded interaction component and non-bonded interaction component. The bonded interaction component itself consists of three parts corresponding to bonds, bond angles and dihedral angles as shown in Fig. 4.10, where circles represent atoms and segments represent bonds.

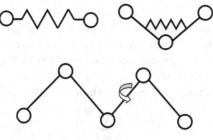

Fig. 4.10.

All bonds are modeled as harmonic oscillators (springs), and therefore the *bond energy* is given by

$$E_B = \sum_i \frac{1}{2} K_{B_i} (r_i - r_i^0)^2,$$

where for the ith bond, K_{B_i} is the *spring constant*, r_i is the length of the bond, r_i^0 is its *equilibrium length*, and the summation is taken over all bonds.

Bond angles are also modeled as harmonic oscillators, and therefore the *bond angle energy* is given by

$$E_{BA} = \sum_j \frac{1}{2} K_{A_j} (\theta_j - \theta_j^0)^2,$$

where for the jth angle, K_{A_j} is the spring constant, θ_j is the size of the angle, θ_j^0 is its *equilibrium size*, and the summation is taken over all bond angles.

It is natural to model the *dihedral angle energy* by using a periodic function. We allow this energy for each dihedral angle to have multiple minima which corresponds to potential multiple barriers encountered when rotating the molecule by 2π around the relevant bond. Hence one way to model the dihedral angle energy is as follows

$$E_{DA} = \sum_k \frac{1}{2} K_{D_k} (1 + \cos(n_k \chi_k - \chi_k^0)),$$

where for the kth dihedral angle, K_{D_k} is the *dihedral energy barrier*, χ_k is the size of the angle, n_k is introduced to allow for multiple minima, χ_k^0 is the *phase angle*, and the summation is taken over all dihedral angles.

The non-bonded interaction component consists of two parts: *Van der Waals energy* and *electrostatic energy*.

The Van der Waals energy between two atoms is given in terms of an attractive interaction and repulsive interaction as follows

$$E_{\mathrm{VDW}} = \sum_{i<j} \left(\frac{A_{ij}}{r_{ij}^{12}} - \frac{B_{ij}}{r_{ij}^{6}} \right),$$

where the summation is taken over all pairs of atoms, and for the (i,j)th pair r_{ij} is the distance between the atoms, A_{ij} is a constant associated with the repulsion between the atoms, and B_{ij} is a constant associated with the attraction between the atoms.

The electrostatic energy arises from the Coulomb law

$$E_{\mathrm{elec}} = \sum_{i<j} \frac{q_i q_j}{\varepsilon r_{ij}^{2}},$$

where, as before, the summation is taken over all pairs of atoms i, j, r_{ij} is the distance between the atoms, q_i, q_j are the electrostatic charges on the atoms and ε is the dielectric constant of the surrounding medium in which the protein is being modeled.

Combining all the above terms we can calculate the total potential energy of the molecule as

$$E_{\mathrm{total}} = E_B + E_{BA} + E_{DA} + E_{\mathrm{VDW}} + E_{\mathrm{elec}}.$$

The constants appearing in the terms of E_{total} are determined based on experiments with smaller molecules from which the molecules under study are built. However, there is usually not enough experimental information to completely determine the constants, and therefore certain *ab initio* quantum mechanical calculations are done as well.

So far we have modeled the energy function in the absence of any external factors like solvents. In practice, however, it is important to take into account the medium surrounding the protein molecule, usually water (in the case of globular proteins). One approach is to randomly distribute thousands of water molecules around the protein and to explicitly take into account the contributions made by the atoms that make up these molecules to the potential energy function. This way we obtain the final formula for the energy. This approach, however, is computationally expensive because of a large number of additional atoms. Therefore, simplified ways to model the surrounding medium are usually utilized.

4.5 Lattice HP-Model

The complexity of the protein folding problem has led to various simplified approaches that concentrate only on some known major factors that affect protein structure. One such major factor (in the case of globular proteins) is the presence of *hydrophobic* and *hydrophilic* (also called *polar*) residues at particular sites in the molecule. Hydrophobic amino acids tend to stay away from water and therefore are buried in the core of a protein molecule, whereas polar ones tend to be exposed to water and therefore stay close to the surface of the protein. The structure of a protein molecule is very strongly affected by the proportions of hydrophobic and polar residues and their positions in the sequence. The *lattice HP-model* (see, e.g., [Di], [DC], [LD], [CD], [LTW]) singles out this particular factor and attempts to deduce the structure of a protein molecule solely from its hydrophobic/polar composition.

For the purposes of this model the amino acid sequence of a protein is first converted into a binary string, called the *HP-sequence* of the amino acid sequence. Namely, each amino acid is classified as either hydrophobic and replaced with H, or polar and replaced with P. Table 4.1 shows the *HP-classification* of amino acids.

Table 4.1.

Single letter code	HP-code
A	H
R	P
N	P
D	P
C	H
Q	P
E	P
G	H
H	P
I	H
L	H
K	P
M	H
F	H
P	H
S	P
T	P
W	H
Y	H
V	H

Further, space is discretized into a three-dimensional unit lattice, and the residues of a molecule are only allowed to occupy positions at the vertices of the lattice in such a way that there is no more than one residue per vertex and adjacent residues in the sequence occupy adjacent lattice points. Every configuration obtained in this way corresponds to a possible tertiary structure of the protein. There are, of course, many structures that can be obtained from "threading" the amino acid sequence through the lattice. In order to choose the ones that we believe may have something in common with the native structure of the protein, an "energy function" E is introduced. Then, as in the preceding section, optimal structures, that is, structures for which E takes its minimal value, are taken as the native ones.

Many energy functions are considered. One popular choice is as follows: E is equal to minus the number of adjacent HH-pairs in the lattice. Hence, for this choice of E optimal structures are those for which the H residues are packed most densely.

Example 4.1. Consider the following sequence of amino acids: $IPTGEC$. Its HP-sequence is $HHPHPH$, and one possible configuration for this sequence in the lattice is shown in Fig. 4.11.

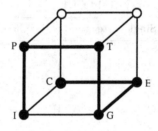

Fig. 4.11.

The value of E for this configuration is -3. Figure 4.12 shows another possible configuration. The value of E for it is equal to -1. Therefore, from the point of view of the lattice HP-model with the energy function defined above, the second configuration is less optimal than the first one.

Certainly, optimal structures found by the model depend on the choice of energy function. Another frequently used function is as follows: E is equal to minus the number of adjacent HH-pairs in the lattice that are not adjacent in the sequence. Using this definition for the configurations shown in Figs. 4.11 and 4.12 we obtain $E = -2$ and $E = 0$ respectively; thus the configuration in Fig. 4.11 is more optimal for this choice of energy function as well.

Although many biophysicists work with the lattice HP-model, it is in fact NP-*complete* which indicates that an algorithm for finding optimal configurations in polynomial time may not exist [BL]. However, there are fast algorithms that obtain structures with values of the energy function not exceeding particular fractions of the optimal value, for example, not exceeding

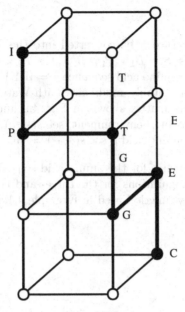

Fig. 4.12.

3/8 of the optimal energy value [HI]. There have also been many computer simulations with this model.

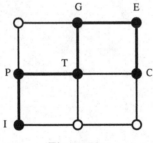

Fig. 4.13.

Other lattice models are used as well; for instance, one popular simplification of the model considered above is the *two-dimensional lattice HP-model*, where residues are allowed to occupy the vertices of a planar lattice. The two-dimensional model is simpler, but, unfortunately, is also *NP*-complete [CY]. One possible two-dimensional configuration for the amino acid sequence from Example 4.1 is shown in Fig. 4.13. The value of the energy function used in Example 4.1 for this configuration is -1.

Exercises

4.1. Let a protein structure S be converted into the following string of environmental descriptors: $S = B_1\alpha \; P_1\gamma \; B_2\beta \; E\alpha \; P_2\beta$. Find all optimal local alignments between the amino acid sequence $x = WYVARK$ and this string. For this purpose use a suitably modified Smith-Waterman algorithm from Section 2.3 with scoring matrix shown in Fig. 4.9 and the following linear gap model: if e is an α- or β-environment, set $s(-, e) = -50$, otherwise set $s(-, e) = -10$; for any amino acid a set $s(a, -) = -5$.

4.2. Write the HP-sequence for the amino acid sequence $LASVEGAS$ and find all its optimal configurations for the three- and two-dimensional lattice HP-models with energy function used in Example 4.1.

5

Phylogenetic Reconstruction

5.1 Phylogenetic Trees

Trees have been used in biology for a long time to graphically represent evolutionary relationships among species and genes. A *rooted tree* by definition descends in two directions from a single node called the *root*, bifurcates at lower nodes and ends at terminal nodes called *tips* or *leaves*. The tips are labeled by the names of the species or sequences being considered; the latter are called *operational taxonomic units (OTUs)* or simply *taxa*.

A rooted tree represents evolution directed from the common ancestor of all the OTUs (the root) towards the OTUs. The other internal nodes of the tree represent the ancestors of particular groups of the OTUs. By removing the root from a rooted tree and joining the two branches descending from the root into a single branch, one obtains an *unrooted tree*. Such trees do not contain information about the direction of evolution and specify only evolutionary relationships among the OTUs. Figure 5.1 shows a rooted tree for four mammalian species on the left and the corresponding unrooted tree on the right.

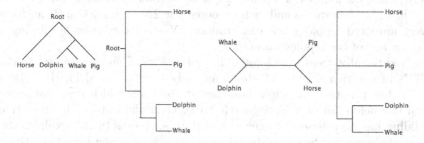

Fig. 5.1.

Two representations are given for each tree. Note that the two internal nodes of the rooted tree represent respectively the common ancestor of the group {dolphin, whale, pig} and the common ancestor of the smaller group {dolphin, whale}.

The length of each branch of a tree is (ideally) a positive number that represents the degree of relatedness between the species or sequences corresponding to the nodes at the endpoints of the branch and is often computed as the product of the length of the time interval that historically separates the species or sequences and a particular value of the *evolutionary rate*, which attempts to take into account the fact that some species or genes evolve faster than others (evolutionary rates will be formally introduced in Sect. 5.4). Note that Fig. 5.1 gives only the correct branching pattern, not branch lengths. Branch lengths are often shown as labels next to the corresponding branches (see, e.g., Fig. 5.10). The branching pattern of a tree (without any reference to the branch lengths, but with a fixed assignment of the OTUs to the leaves) is called the *(labeled) tree topology*. Trees as described above are called *phylogenetic trees*. Any phylogenetic tree whose leaves are labeled by particular OTUs is said to *relate* the OTUs. In this chapter we will consider methods for identifying trees that relate species or sequences in an "optimal" or "likely" way. These methods produce trees that we hope reflect, at least to some extent, the real evolutionary relationships among the OTUs.

One should bear in mind that representing species or gene divergence by means of a rooted phylogenetic tree contains the hidden assumption that the divergence occurred "instantaneously" or, rather, that the product of the period of time that divergence had taken and a "divergence rate" is small compared to the branch lengths. In reality it may take a substantial amount of time for divergence to manifest itself on the population level.

For N OTUs the number of topologies of all possible rooted trees that relate the OTUs is $(2N - 3)!/(2^{N-2}(N - 2)!)$ and the number of topologies of all possible unrooted ones is $(2N - 5)!/(2^{N-3}(N - 3)!)$ (see Exercise 5.1). For example, there is only one unrooted topology for $N = 3$ (see Fig. 5.5) and only three unrooted tree topologies for $N = 4$ (see Example 5.10). However, as N grows, the number of topologies quickly becomes very large. Each rooted tree has $2N - 2$ branches and each unrooted one $2N - 3$ branches. Hence from every unrooted topology one can produce $2N - 3$ rooted ones by placing a root on any of the topology branches.

Traditionally, trees were built from morphological similarities among the OTUs in question. Figure 5.2 shows an early tree published in [H] in 1866 (note that this tree has three branches at the root, which does not agree with our definition of a phylogenetic tree). Over the last few decades tree building has been based on gene and protein sequences (which explains the term *phylogenetic*). Sequence based methods are more sensitive than those based on morphological similarities, since sequence divergence precedes species divergence. One should also bear in mind that gene divergence can occur not just because of speciation (in which case the diverged genes are called

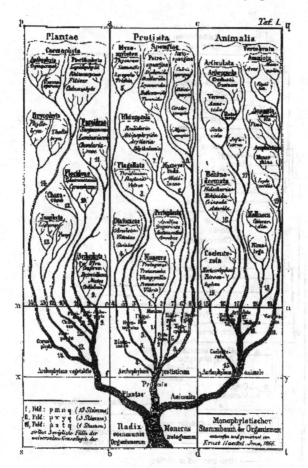

Fig. 5.2.

orthologues), but also because of gene duplication (in which case the diverged genes are called *paralogues*). Therefore, in principle phylogenetic trees can be built for families of gene sequences some of which come from the same organism. Hence it is natural to think of phylogenetic trees as *gene trees*, not species trees. Figure 5.3 shows a rooted phylogenetic tree for 52 mammalian species determined by considering the DNA sequences of the breast and ovarian cancer susceptibility gene 1 (*BRCA1*). The figure is reprinted with permission from Nature (see [MS]). Copyright 2001 Macmillan Magazines Limited.

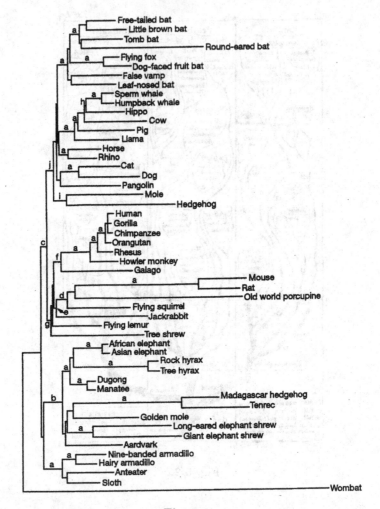

Fig. 5.3.

It frequently happens that choosing different sets of sequences for the same family of species results in different gene trees, and even from a single dataset multiple trees can be produced. However, historically there is only one sequence of events that had led to the formation of the species being studied, and hence only one tree represents the true evolutionary relationships. The problem of combining different gene trees into a single "consensus tree" relating the actual species is still largely open (see, e.g., [BHV]).

The process of building phylogenetic trees that are in some sense optimal for a given set of OTUs is called *phylogenetic reconstruction* and involves the following basic steps.

(i) Choosing a family of homologous sequences as OTUs (see Sect. 2.1 for a brief discussion of homology). One must attempt to select the sequences in such a way that there is a sufficiently strong "phylogenetic signal" in the family. If the signal is weak (that is, if the sequences are extremely diverged), the existing phylogenetic methods will still produce some tree, but it probably will not be very informative (note that assessing the reliability of phylogenetic trees is an important problem in its own right, but we do not discuss it in this book – see, e.g., [F2] and a brief survey in [EG]). Choosing suitable OTUs is a kind of art and is often constrained by the availability of sequence data for particular organisms.

(ii) Putting the sequences into a multiple alignment and obtaining a *reduced multiple alignment* by discarding the columns that contain gaps (discarding the gaps, in fact, is not necessary and there are ways to incorporate them in phylogenetic reconstruction, but here we will only discuss simplified methods that produce trees from reduced multiple alignments). For example, suppose that for the four species from Fig. 5.1 we are given the DNA sequences (of course, in reality sequences are much longer)

$$\begin{array}{ll} \text{pig:} & GCTGCA \\ \text{horse:} & GCTGA \\ \text{whale:} & GTCC \\ \text{dolphin:} & GCTCCC, \end{array}$$

and suppose that by some method we have produced the following multiple alignment for these sequences

$$\begin{array}{ll} \text{pig:} & G\ C\ T\ G\ C\ A \\ \text{horse:} & G\ C\ T\ G\ -\ A \\ \text{whale:} & G\ -\ T\ C\ C\ - \\ \text{dolphin:} & G\ C\ T\ C\ C\ C. \end{array}$$

Then, after discarding the columns containing gaps, we obtain the reduced multiple alignment

$$\begin{array}{ll} \text{pig:} & G\ T\ G \\ \text{horse:} & G\ T\ G \\ \text{whale:} & G\ T\ C \\ \text{dolphin:} & G\ T\ C. \end{array}$$

(iii) Inferring a phylogenetic tree from the reduced multiple alignment. Usually, the hardest part of any such inference is the determination of tree topology. In the example considered above it is clear that pig should be clustered with horse and whale with dolphin, as in the unrooted topology shown in Fig. 5.1. The determination of branch lengths and a root position requires further analysis (and in some cases additional information).

In reality, when the sequences involved are much longer and the number of OTUs is much larger, the resulting reduced multiple alignment is analyzed by a particular method of phylogenetic reconstruction. Such methods can be classified into three groups:

(a) Parsimony methods,
(b) Distance methods,
(c) Probabilistic methods arising from the maximum likelihood approach.

Methods of each type make specific assumptions about evolution, and we discuss each type separately in the forthcoming sections. We will see that some of the methods concentrate on determining only tree topologies, whereas others also produce branch lengths. Further, some of the methods give only unrooted trees, whereas others also indicate where the roots could be. We will also see methods that attempt to reconstruct ancestral sequences at the root and internal nodes. We remark that methods of each type may produce several trees from a single reduced multiple alignment. If there are more than one optimal multiple alignments for the sequences of interest (and hence more than one reduced multiple alignments), one should deal with each alignment separately, and thus the number of phylogenetic trees for a single sequence family can be quite large.

In reality methods of phylogenetic reconstruction are applied to either DNA or protein sequences. However, all the procedures make sense for sequences of letters from any finite alphabet Q, and, as before, we will sometimes use the artificial alphabets for illustration purposes.

5.2 Parsimony Methods

Parsimony methods find rooted tree topologies, not branch lengths. They also attempt to reconstruct ancestral sequences at the root and internal nodes. Under this approach a total *cost* is calculated for each rooted tree topology, and optimal topologies are defined as the ones that have the smallest total cost. They are called the *most parsimonious topologies*. Thus, the parsimony approach is based on the assumption that sequences always evolve in the "most economic way" in the sense that the total cost is minimized.

There are many specific *cost functions*. As an example, we will consider the simplest one where unit cost is made for each substitution. For a fixed rooted topology, we assign sequences to the root and all internal nodes; the length of these sequences is equal to that of the reduced multiple alignment. For any such sequence assignment the cost is the sum of costs of all branches, where the cost of the branch joining two nodes is the minimal number of substitutions required to move from the sequence at one node to the sequence at the other one. The cost of the topology then is the minimal cost over all such sequence assignments.

We will illustrate this method by the following example.

Example 5.1. Let Q be the three-letter alphabet $\{A, B, C\}$, and suppose that we are given the following reduced multiple alignment of three sequences

$$x^1 : A \ A \ B$$
$$x^2 : A \ C \ B$$
$$x^3 : C \ C \ B.$$

There are only three possible rooted topologies for three OTUs, and it is easy to observe that each topology is most parsimonious for x^1, x^2, x^3. Figure 5.4 shows the topologies together with all optimal sequence assignments. The cost of each topology is equal to 2.

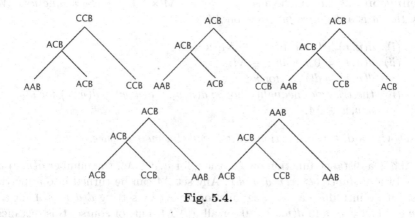

Fig. 5.4.

The cost calculation for a fixed topology can be done by simple *Fitch's algorithm*. This algorithm is quick, and if the number N of OTUs is moderate, it is realistic to compute the cost of each topology and then select the most parsimonious ones. If, however, N is very large, doing so is a huge computational task. In such cases a non-comprehensive search of the topology space may be conducted, but with this approach some or all of the most parsimonious topologies may be missed. As we will see later, the problem of searching the topology space for large values of N is common for many phylogenetic methods. However, there is a specific algorithm for enumerating tree topologies called the *branch and bound algorithm* that, when coupled with Fitch's algorithm, is guaranteed to find all the most parsimonious trees in an acceptable amount of time, even if N is quite large.

The branch and bound algorithm is only useful for the parsimony methods, which makes them the fastest available methods of phylogenetic reconstruction. Despite the fact that they are perhaps the least accurate ones, they are still sometimes used because of their computational speed. We do not concentrate on them here. The interested reader can find the descriptions of Fitch's algorithm and the branch and bound algorithm in [DEKM].

5.3 Distance Methods

Distance methods reconstruct trees (rooted or unrooted, depending on the method) from a set of pairwise distances between the sequences in a fixed reduced multiple alignment. In this section we do not explain how such distances are obtained and assume that they are given. One popular procedure for producing distances, or, rather, certain substitutes for distances that we call "pseudodistances", will be discussed in Sect. 5.5.

We start with a formal definition.

Definition 5.2. *Let $\mathcal{M}$ be a set and let $d : \mathcal{M} \times \mathcal{M} \to \mathbb{R}$ be a function. We say that d is a distance function on $\mathcal{M}$ if*

    (i) $d(u, v) > 0$ for all $u, v \in \mathcal{M}$, $u \neq v$,
    (ii) $d(u, u) = 0$ for all $u \in \mathcal{M}$,
    (iii) $d(u, v) = d(v, u)$ for all $u, v \in \mathcal{M}$,
    (iv) the triangle inequality holds: $d(u, v) \leq d(u, w) + d(w, v)$ for all $u, v, w \in \mathcal{M}$.

A set with a distance function on it is called a metric space.

If d is a distance function on $\mathcal{M}$, then, for $u, v \in \mathcal{M}$, the number $d(u, v)$ is called the *distance between u and v*. Any set $\mathcal{M}$ can be turned into a metric space if we introduce a distance function on $\mathcal{M}$ by setting $d(u, v) = 1$ for all $u, v \in \mathcal{M}$, $u \neq v$, and $d(u, u) = 0$ for all $u \in \mathcal{M}$, but, of course, this distance function is not very informative.

We will be interested in the special case of distance functions on a finite set $\mathcal{M} = \{x^1, \ldots, x^N\}$ of sequences (OTUs) for which we would like to build a phylogenetic tree. Assume that a distance function d is defined on $\mathcal{M}$ and that d is *biologically relevant*, that is, it incorporates some information about the degree of divergence among the sequences in $\mathcal{M}$: for example, $d(x^i, x^j) > d(x^k, x^l)$, if x^i and x^j have diverged from their common ancestor more than x^k and x^l have diverged from theirs. For simplicity we will write d_{ij} instead of $d(x^i, x^j)$. It will be convenient for us to represent d by the symmetric *distance matrix* $M_d = (d_{ij})$.

If we fix an unrooted tree $\mathcal{T}$ relating the OTUs, we obtain a *tree-generated distance function* $d^{\mathcal{T}}$ on $\mathcal{M}$ by declaring $d^{\mathcal{T}}(x^i, x^j) = d_{ij}^{\mathcal{T}}$ to be the length of the shortest path from x^i to x^j in $\mathcal{T}$. It is not hard to check that under very general assumptions $d^{\mathcal{T}}$ is indeed a distance function on $\mathcal{M}$ (see Exercise 5.3). The broad aim of distance methods for phylogenetic reconstruction is to determine all trees $\mathcal{T}$ for which $d^{\mathcal{T}}$ is in some sense as close as possible to d. Any such tree is considered to be optimal from the point of view of distance methods. Thus, the emphasis of distance methods is on determining unrooted trees together with branch lengths (although we will also describe one distance method that produces rooted trees).

The following natural question arises: is any distance function d on $\mathcal{M}$ *additive*, that is, does there exist a tree $\mathcal{T}$ that *generates* d, which means that $d^{\mathcal{T}} = d$ ($d_{ij}^{\mathcal{T}} = d_{ij}$ for all i, j)? The answer to this question is obviously positive for $N = 2$. Suppose now that $N = 3$. In this case we are looking for three positive numbers x, y, z such that

$$
\begin{aligned}
x + y &= d_{12}, \\
x + z &= d_{13}, \\
y + z &= d_{23}
\end{aligned}
\tag{5.1}
$$

(see Fig. 5.5).

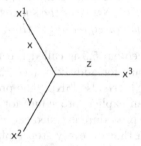

Fig. 5.5.

The solution to equations (5.1) is

$$x = \frac{1}{2}(d_{12} + d_{13} - d_{23}),$$

$$y = \frac{1}{2}(d_{12} + d_{23} - d_{13}), \tag{5.2}$$

$$z = \frac{1}{2}(d_{13} + d_{23} - d_{12}).$$

We note that the numbers in the right-hand side of identities (5.2) are *non-negative* due to the triangle inequality. However, since this inequality is not strict, they do not have to be *positive*, some of them may be equal to 0. Hence, it will be convenient for us to allow zero branch lengths, and we will assume that all branch lengths are non-negative, rather than positive numbers from now on. In biological applications branches of zero length are interpreted as "very short" branches. The definition of additivity in this case is identical to the previously given one, and thus (5.2) shows that, if $N = 3$, any distance function is additive on $\mathcal{M}$ in this generalized sense. If some of the branch lengths of a tree $\mathcal{T}$ are zero, $d^{\mathcal{T}}$ may not satisfy condition (i) of Definition 5.2, and we will sometimes impose this condition separately (see, e.g., Exercise 5.3). Note also that if zero branch lengths are allowed, phylogenetic trees are no longer required to have the bifurcating pattern discussed in the preceding section, but may have any branching pattern at any of its internal nodes.

As we have seen, for $N = 2, 3$ there is a *unique* tree that generates a given distance function. The uniqueness of such a tree is a general fact for additive distance functions (see Exercise 5.4).

Later we will observe (see, e.g., Example 5.6), that for $N \geq 4$ not every distance function on $\mathcal{M}$ is additive. In fact, additive distance functions can be characterized in the following way.

Theorem 5.3. *Let d be a distance function on a $\mathcal{M}$ and $N \geq 4$. Then d is additive if and only if the following condition holds: for every set of four distinct numbers $1 \leq i, j, k, l \leq N$ two of the sums $d_{ij} + d_{kl}$, $d_{ik} + d_{jl}$, $d_{il} + d_{jk}$ coincide and are greater than or equal to the third one.*

The condition from Theorem 5.3 is called the *four-point condition*. This condition is clearly necessary for additivity (see Exercise 5.5). The proof of sufficiency was given in [SN] (see also [SK]). The proof in [SN] is constructive in the sense that it gives an explicit procedure for determining the (unique) tree that generates d. This procedure is called the *neighbor-joining algorithm*. It is an iterative algorithm that on every step replaces a pair of OTUs with a single new OTU, and iterates until there are only three OTUs left. For $N = 3$ there is just one unrooted tree topology, and the corresponding branch lengths are found from formulas (5.2). The tree is then built by a traceback procedure that works by recalling which single OTU on a particular step arose from which pair of OTUs available on the preceding step.

We will now describe the algorithm in detail. For every $i = 1, \ldots, N$ define

$$r_i = \frac{1}{N-2} \sum_{k=1}^{N} d_{ik}.$$

Further, for all $i, j = 1, \ldots, N$, $i < j$, set

$$D_{ij} = d_{ij} - (r_i + r_j).$$

It will be convenient for us to write D_{ij} into an upper-triangular matrix $D = (D_{ij})$. Now, pick a pair $1 \leq i, j \leq N$ for which D_{ij} is minimal (such a pair may not be unique). We will now group together the OTUs x^i, x^j, that is, replace them with a single element that we call the OTU x^{N+1}. The new OTU x^{N+1} represents an internal node of the future tree connected to x^i and x^j, and is placed at the following distances from them respectively

$$d_{N+1\,i} = \frac{1}{2}(d_{ij} + r_i - r_j),$$

$$d_{N+1\,j} = \frac{1}{2}(d_{ij} + r_j - r_i).$$

We will now define the distances between x^{N+1} and any x^m with $m \neq i, j$ as follows

$$d_{N+1\,m} = \frac{1}{2}(d_{im} + d_{jm} - d_{ij}).$$

Now we have the new collection of $N-1$ OTUs $\mathcal{M}' = \{x^m, x^{N+1}, m \neq i, j\}$ and can repeat the above procedure once again. The algorithm is iterated until only three OTUs are left, in which case there is just one unrooted tree topology, and the corresponding branch lengths are found from formulas (5.2). The tree is then built by a traceback procedure.

We will now give an example of applying the neighbor-joining algorithm.

Example 5.4. Let $N = 6$ and suppose that we are given the following distance matrix

$$
\begin{array}{c|cccccc}
M_d & x^1 & x^2 & x^3 & x^4 & x^5 & x^6 \\
\hline
x^1 & 0 & 8 & 3 & 14 & 10 & 12 \\
x^2 & 8 & 0 & 9 & 10 & 6 & 8 \\
x^3 & 3 & 9 & 0 & 15 & 11 & 13 \\
x^4 & 14 & 10 & 15 & 0 & 10 & 8 \\
x^5 & 10 & 6 & 11 & 10 & 0 & 8 \\
x^6 & 12 & 8 & 13 & 8 & 8 & 0
\end{array}
$$

It is not hard to verify that d is indeed a distance function. Formally, before applying the neighbor-joining algorithm, we need to check that d satisfies the four-point condition. However, we will apply the neighbor-joining algorithm to d without doing such verification. We will obtain a tree $\mathcal{T}$ and compare the corresponding function $d^{\mathcal{T}}$ with d. We will see that $d^{\mathcal{T}} = d$, and hence, indeed, d satisfies the four-point condition.

We have

$$r_1 = \frac{47}{4}, r_2 = \frac{41}{4}, r_3 = \frac{51}{4},$$

$$r_4 = \frac{57}{4}, r_5 = \frac{45}{4}, r_6 = \frac{49}{4}.$$

This gives the following matrix

$$
\begin{array}{c c c c c c c}
D & x^1 & x^2 & x^3 & x^4 & x^5 & x^6 \\
\end{array}
$$

$$
x^1 \quad -14 \quad -\dfrac{43}{2} \quad -12 \quad -13 \quad -12
$$

$$
x^2 \qquad\qquad -14 \quad -\dfrac{29}{2} \quad -\dfrac{31}{2} \quad -\dfrac{29}{2}
$$

$$
x^3 \qquad\qquad\qquad\qquad -12 \quad -13 \quad -12
$$

$$
x^4 \qquad\qquad\qquad\qquad\qquad -\dfrac{31}{2} \quad -\dfrac{37}{2}
$$

$$
x^5 \qquad\qquad\qquad\qquad\qquad\qquad -\dfrac{31}{2}
$$

In the above matrix the minimal value is $D_{13} = -43/2$. We now introduce a new OTU x^7 that will replace the pair x^1, x^3. We place x^7 at the distance

$$
d_{71} = \frac{1}{2}(d_{31} + r_1 - r_3) = 1
$$

from x^1 and at the distance

$$
d_{73} = \frac{1}{2}(d_{31} + r_3 - r_1) = 2
$$

from x^3, as shown in Fig. 5.6.

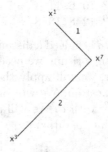

Fig. 5.6.

We will now compute distances between x^7 and each of x^2, x^4, x^5, x^6. We have

$$d_{72} = \frac{1}{2}(d_{12} + d_{32} - d_{13}) = 7,$$

$$d_{74} = \frac{1}{2}(d_{14} + d_{34} - d_{13}) = 13,$$

$$d_{75} = \frac{1}{2}(d_{15} + d_{35} - d_{13}) = 9,$$

$$d_{76} = \frac{1}{2}(d_{16} + d_{36} - d_{13}) = 11,$$

which gives the following distance matrix for the OTUs x^2, x^4, x^5, x^6, x^7

M_d	x^2	x^4	x^5	x^6	x^7
x^2	0	10	6	8	7
x^4	10	0	10	8	13
x^5	6	10	0	8	9
x^6	8	8	8	0	11
x^7	7	13	9	11	0

For this new distance matrix we will repeat the process again and obtain

$$r_2 = \frac{31}{3}, \, r_4 = \frac{41}{3}, \, r_5 = 11,$$

$$r_6 = \frac{35}{3}, \, r_7 = \frac{40}{3},$$

which gives

D	x^2	x^4	x^5	x^6	x^7
x^2		-14	$-\frac{46}{3}$	-14	$-\frac{50}{3}$
x^4			$-\frac{44}{3}$	$-\frac{52}{3}$	-14
x^5				$-\frac{44}{3}$	$-\frac{46}{3}$
x^6					-14

We now introduce a new OTU x^8 that will replace the pair x^4, x^6 (note that D_{46} is minimal in the above matrix). We place x^8 at the distance 5 from x^4 and at the distance 3 from x^6 as shown in Fig. 5.7.

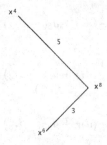

Fig. 5.7.

The distance matrix for the OTUs x^2, x^5, x^7, x^8 is

$$
\begin{array}{c|cccc}
M_d & x^2 & x^5 & x^7 & x^8 \\
x^2 & 0 & 6 & 7 & 5 \\
x^5 & 6 & 0 & 9 & 5 \\
x^7 & 7 & 9 & 0 & 8 \\
x^8 & 5 & 5 & 8 & 0
\end{array}
$$

On the next step of the algorithm we obtain

$$r_2 = 9,\ r_5 = 10,\ r_7 = 12,\ r_8 = 9,$$

and

$$
\begin{array}{c|cccc}
D & x^2 & x^5 & x^7 & x^8 \\
x^2 & & -13 & -14 & -13 \\
x^5 & & & -13 & -14 \\
x^7 & & & & -13
\end{array}
$$

At this point we can group together either x^2 and x^7, or x^5 and x^8, since both D_{27} and D_{58} are minimal in the above matrix (the resulting tree will not depend on our choice). We group together x^5 and x^8, that is, we introduce a new OTU x^9, place it at the distance 3 from x^5 and at the distance 2 from x^8 as shown in Fig. 5.8, and calculate distances from x^9 to x^2 and x^7 which gives the following distance matrix for the three OTUs

$$
\begin{array}{c|ccc}
M_d & x^2 & x^7 & x^9 \\
x^2 & 0 & 7 & 3 \\
x^7 & 7 & 0 & 6 \\
x^9 & 3 & 6 & 0
\end{array}
$$

It follows from formulas (5.2) that the above distance function is generated by the tree shown in Fig. 5.9.

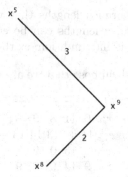

Fig. 5.8.

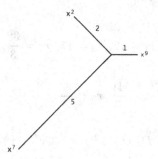

Fig. 5.9.

Next, merging the trees from Figs. 5.6–5.9, we obtain the tree T shown in Fig. 5.10.

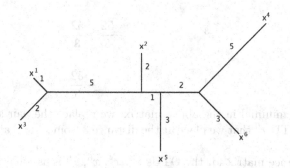

Fig. 5.10.

It is easy to verify that T generates d and therefore the distance function d indeed satisfies the four-point condition.

In Example 5.4 all the branch lengths that we computed were positive numbers. In fact, some branch lengths can be equal to zero for a distance function satisfying the four-point condition, as the following example shows.

Example 5.5. Let $N = 5$ and suppose that we are given the following distance matrix

$$
\begin{array}{c|ccccc}
M_d & x^1 & x^2 & x^3 & x^4 & x^5 \\
\hline
x^1 & 0 & 11 & 8 & 9 & 8 \\
x^2 & 11 & 0 & 13 & 14 & 13 \\
x^3 & 8 & 13 & 0 & 9 & 8 \\
x^4 & 9 & 14 & 9 & 0 & 9 \\
x^5 & 8 & 13 & 8 & 9 & 0
\end{array}
$$

It is not hard to verify that d is indeed a distance function. From the matrix M_d we obtain

$$ r_1 = 12, \; r_2 = 17, \; r_3 = \frac{38}{3}, $$

$$ r_4 = \frac{41}{3}, \; r_5 = \frac{38}{3}, $$

and

$$
\begin{array}{c|ccccc}
D & x^1 & x^2 & x^3 & x^4 & x^5 \\
\hline
x^1 & & -18 & -\dfrac{50}{3} & -\dfrac{50}{3} & -\dfrac{50}{3} \\
x^2 & & & -\dfrac{50}{3} & -\dfrac{50}{3} & -\dfrac{50}{3} \\
x^3 & & & & -\dfrac{52}{3} & -\dfrac{52}{3} \\
x^4 & & & & & -\dfrac{52}{3}
\end{array}
$$

Since D_{12} is minimal in the above matrix, we replace the pair x^1, x^2 with a single new OTU x^6 that we place at the distance 3 from x^1 and at the distance 8 from x^2.

The distance matrix for the OTUs x^3, x^4, x^5, x^6 is as follows

$$
\begin{array}{c|cccc}
M_d & x^3 & x^4 & x^5 & x^6 \\
\hline
x^3 & 0 & 9 & 8 & 5 \\
x^4 & 9 & 0 & 9 & 6 \\
x^5 & 8 & 9 & 0 & 5 \\
x^6 & 5 & 6 & 5 & 0
\end{array}
$$

and therefore we obtain

$$r_3 = 11,\ r_4 = 12,\ r_5 = 11,\ r_6 = 8,$$

and

$$
\begin{array}{c|cccc}
D & x^3 & x^4 & x^5 & x^6 \\
\hline
x^3 & & -14 & -14 & -14 \\
x^4 & & & -14 & -14 \\
x^5 & & & & -14 \\
\end{array}
$$

At this point we can group together any pair of OTUs, say, x^4, x^5. A new OTU x^7 is placed at the distance 5 from x^4 and at the distance 4 from x^5.

The distance matrix for the OTUs x^3, x^6, x^7 is

$$
\begin{array}{c|ccc}
M_d & x^3 & x^6 & x^7 \\
\hline
x^3 & 0 & 5 & 4 \\
x^6 & 5 & 0 & 1 \\
x^7 & 4 & 1 & 0 \\
\end{array}
$$

and formulas (5.2) give for these OTUs the tree shown in Fig. 5.11 (note that

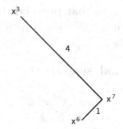

Fig. 5.11.

the length of the branch leading to x^7 is equal to 0).

Then the tree T produced by the neighbor-joining algorithm is as shown in Fig. 5.12.

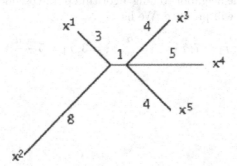

Fig. 5.12.

It is easy to verify that $\mathcal{T}$ generates d and therefore the distance function d indeed satisfies the four-point condition.

Often the neighbor-joining algorithm is applied in the situation when the distance function in question does not satisfy the four-point condition. This is done because distance functions obtained in reality almost never satisfy it. Moreover, in practice the triangle inequality (see (iv) in Definition 5.2) is hard to satisfy either, and what one usually deals with in reality is a *pseudodistance function*, rather than a distance function. The definition of pseudodistance function is obtained from Definition 5.2 by removing requirement (iv), and, as before, for a set of OTUs we will usually write a pseudodistance function d as the corresponding *pseudodistance matrix* $M_d = (d_{ij})$.

If the neighbor-joining algorithm is formally applied to a pseudodistance function that does not satisfy the four-point condition, various anomalies may occur: the algorithm may produce more than one tree, these trees may have branches of negative lengths (note that biologists are sometimes able to interpret negative branch lengths), the functions $d^{\mathcal{T}}$ obtained from these trees may not coincide with the original pseudodistance function d.

We illustrate these anomalies by the following example (see also Exercise 5.7). Before proceeding, we note that pseudodistance functions may lead to negative branch lengths even for $N = 3$, which can be seen from formulas (5.2).

Example 5.6. Let $N = 4$ and suppose that we are given the following pseudodistance matrix

$$
\begin{array}{c|cccc}
M_d & x^1 & x^2 & x^3 & x^4 \\
\hline
x^1 & 0 & 5 & 2 & 7 \\
x^2 & 5 & 0 & 1 & a \\
x^3 & 2 & 1 & 0 & 3 \\
x^4 & 7 & a & 3 & 0 \\
\end{array}
$$

where $a > 6$. The function d does not satisfy the triangle inequality since $d_{14} > d_{13} + d_{34}$. It does not satisfy the four-point condition either because $d_{13} + d_{24} = 2 + a > 8$, $d_{12} + d_{34} = 8$, $d_{14} + d_{23} = 8$.

Let us apply the neighbor-joining algorithm to the pseudodistance function d and see what it will produce. We have

$$
r_1 = 7, \ r_2 = 3 + \frac{a}{2}, \ r_3 = 3, \ r_4 = 5 + \frac{a}{2},
$$

and

$$D \quad x^1 \qquad x^2 \qquad x^3 \qquad x^4$$

$$x^1 \qquad -5-\frac{a}{2} \qquad -8 \ -5-\frac{a}{2}$$

$$x^2 \qquad\qquad\qquad -5-\frac{a}{2} \qquad -8$$

$$x^3 \qquad\qquad\qquad\qquad\qquad -5-\frac{a}{2}$$

Since $a > 6$, each of D_{12}, D_{14}, D_{23}, D_{34} is minimal. As we will see, grouping together different pairs of OTUs will lead in some cases to different trees.

First, let us group together x^1, x^2, and introduce a new OTU x^5 that we place at the distance $9/2 - a/4$ from x^1 and at the distance $1/2 + a/4$ from x^2 (note that the first of these branch lengths is negative if $a > 18$). The new "pseudodistance matrix" for x^3, x^4, x^5 is as follows

$$M_d \quad x^3 \quad x^4 \qquad x^5$$

$$x^3 \quad 0 \quad\ 3 \qquad -1$$

$$x^4 \quad 3 \quad\ 0 \qquad 1+\frac{a}{2}$$

$$x^5 \quad -1 \ 1+\frac{a}{2} \quad 0$$

The above matrix is not in fact that of a pseudodistance function since some of its entries are negative. However, we can formally apply formulas (5.2) to it to obtain the tree shown in Fig. 5.13 (note that one of the branch lengths is negative).

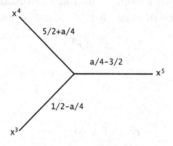

Fig. 5.13.

This tree gives rise to the tree $\mathcal{T}_1$ shown in Fig. 5.14.

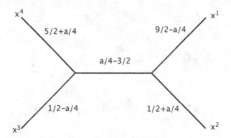

Fig. 5.14.

Although some of the branches of $\mathcal{T}_1$ have negative lengths, we can still attempt to sum up the relevant branch lengths and calculate $d^{\mathcal{T}_1}$. The resulting matrix is

$$
\begin{array}{c|cccc}
M_{d^{\mathcal{T}_1}} & x^1 & x^2 & x^3 & x^4 \\[1em]
x^1 & 0 & 5 & \dfrac{7}{2} - \dfrac{a}{4} & \dfrac{11}{2} + \dfrac{a}{4} \\[1.5em]
x^2 & 5 & 0 & \dfrac{a}{4} - \dfrac{1}{2} & \dfrac{3}{2} + \dfrac{3a}{4} \\[1.5em]
x^3 & \dfrac{7}{2} - \dfrac{a}{4} & \dfrac{a}{4} - \dfrac{1}{2} & 0 & 3 \\[1.5em]
x^4 & \dfrac{11}{2} + \dfrac{a}{4} & \dfrac{3}{2} + \dfrac{3a}{4} & 3 & 0
\end{array}
$$

which is clearly different from M_d.

If we group together x^3, x^4, it is easy to see that we will again obtain the tree $\mathcal{T}_1$ in Fig. 5.14.

Let us now group together x^1, x^4. We introduce a new OTU x^5 and place it at the distance $9/2 - a/4$ from x^1 and at the distance $5/2 + a/4$ from x^4. The new "pseudodistance matrix" for x^2, x^3, x^5 is as follows

$$
\begin{array}{c|ccc}
M_d & x^2 & x^3 & x^5 \\[1em]
x^2 & 0 & 1 & \dfrac{a}{2} - 1 \\[1.5em]
x^3 & 1 & 0 & -1 \\[1.5em]
x^5 & \dfrac{a}{2} - 1 & -1 & 0
\end{array}
$$

Again, the above matrix is not in fact that of a pseudodistance function since some of its entries are negative. However, as we have already done earlier, we can formally apply formulas (5.2) to it to obtain the tree shown in Fig. 5.15.

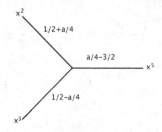

Fig. 5.15.

The above tree gives rise to the tree T_2 shown in Fig. 5.16.

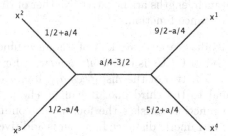

Fig. 5.16.

Although some of the branches of T_2 have negative lengths, we, as before, can still attempt to sum up the relevant branch lengths and calculate d^{T_2}. The resulting matrix is

$$
\begin{array}{c|cccc}
M_{d^{T_2}} & x^1 & x^2 & x^3 & x^4 \\[2mm]
x^1 & 0 & \dfrac{7}{2}+\dfrac{a}{4} & \dfrac{7}{2}-\dfrac{a}{4} & 7 \\[4mm]
x^2 & \dfrac{7}{2}+\dfrac{a}{4} & 0 & 1 & \dfrac{3}{2}+\dfrac{3a}{4} \\[4mm]
x^3 & \dfrac{7}{2}-\dfrac{a}{4} & 1 & 0 & \dfrac{3}{2}+\dfrac{a}{4} \\[4mm]
x^4 & 7 & \dfrac{3}{2}+\dfrac{3a}{4} & \dfrac{3}{2}+\dfrac{a}{4} & 0
\end{array}
$$

which is different from M_d as well.

Finally, if we group together x_2, x_3, we again obtain the tree T_2 in Fig. 5.16. Thus, the neighbor-joining algorithm in this example produces two trees, and some of their branch lengths are negative. Neither of these trees generates the original pseudodistance function.

We will now introduce a special class of distance functions. A distance function d on a set $\mathcal{M}$ of OTUs is called *ultrameric*, if for any three distinct elements $x^i, x^j, x^k \in \mathcal{M}$, two of the distances d_{ij}, d_{ik}, d_{jk} coincide and are greater than or equal to the third one. It can be checked directly that an ultrameric distance function satisfies the four point condition (see Exercise 5.8). Therefore, any ultrameric distance function is additive and the tree that generates it can be recovered by the neighbor-joining algorithm.

We will now give an example of applying the neighbor-joining algorithm to an ultrameric distance function.

Example 5.7. Let $N = 5$ and suppose that we are given the following distance matrix

$$
\begin{array}{c|ccccc}
M_d & x^1 & x^2 & x^3 & x^4 & x^5 \\
x^1 & 0 & 16 & 6 & 16 & 6 \\
x^2 & 16 & 0 & 16 & 8 & 16 \\
x^3 & 6 & 16 & 0 & 16 & 2 \\
x^4 & 16 & 8 & 16 & 0 & 16 \\
x^5 & 6 & 16 & 2 & 16 & 0
\end{array}
$$

It is easy to verify directly that d is indeed a distance function and is ultrameric (see Exercise 5.9). From the matrix M_d we obtain

$$r_1 = \frac{44}{3}, r_2 = \frac{56}{3}, r_3 = \frac{40}{3},$$

$$r_4 = \frac{56}{3}, r_5 = \frac{40}{3},$$

and

$$D \quad x^1 \quad x^2 \quad x^3 \quad x^4 \quad x^5$$

$$x^1 \qquad -\frac{52}{3} \quad -22 \quad -\frac{52}{3} \quad -22$$

$$x^2 \qquad\qquad -16 \quad -\frac{88}{3} \quad -16$$

$$x^3 \qquad\qquad\qquad -16 \quad -\frac{74}{3}$$

$$x^4 \qquad\qquad\qquad\qquad -16$$

Since D_{24} is minimal in the above matrix, we replace the pair x^2, x^4 with a single new OTU x^6 that we place at the distance 4 from each of x^2 and x^4.

The distance matrix for the OTUs x^1, x^3, x^5, x^6 is as follows

$$
\begin{array}{c|cccc}
M_d & x^1 & x^3 & x^5 & x^6 \\
x^1 & 0 & 6 & 6 & 12 \\
x^3 & 6 & 0 & 2 & 12 \\
x^5 & 6 & 2 & 0 & 12 \\
x^6 & 12 & 12 & 12 & 0
\end{array}
$$

and therefore we obtain

$$r_1 = 12, \ r_3 = 10, \ r_5 = 10, \ r_6 = 18,$$

and

$$
\begin{array}{c|cccc}
D & x^1 & x^3 & x^5 & x^6 \\
x^1 & & -16 & -16 & -18 \\
x^3 & & & -18 & -16 \\
x^5 & & & & -16
\end{array}
$$

We now group together x^1, x^6. A new OTU x^7 is placed at the distance 3 from x^1 and at the distance 9 from x^6.

The distance matrix for the OTUs x^3, x^5, x^7 is

$$
\begin{array}{c|ccc}
M_d & x^3 & x^5 & x^7 \\
x^3 & 0 & 2 & 3 \\
x^5 & 2 & 0 & 3 \\
x^7 & 3 & 3 & 0
\end{array}
$$

and formulas (5.2) give for these OTUs the tree shown in Fig. 5.17.

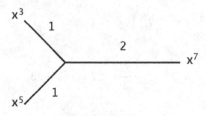

Fig. 5.17.

Then the tree T produced by the neighbor-joining algorithm is as shown in Fig. 5.18.

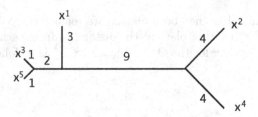

Fig. 5.18.

The tree T found in Example 5.7 possesses an interesting property: there is a way to place a root on this tree so that for any node the lengths of all paths descending from this node to the tips lying below the node are equal. This root position in shown in Fig. 5.19.

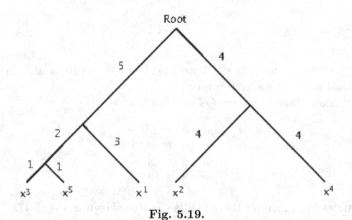

Fig. 5.19.

Rooted trees satisfying the above property describe the evolution of species or sequences whose evolutionary rates do not change through time, and hence the branch lengths are proportional to the real periods of time with the proportionality coefficient being independent of the branch. This condition is called the *molecular clock condition* and trees satisfying this condition are called *molecular clock trees*. It should be noted that the evolution of any family of species or sequences can be represented by a molecular clock tree, if one sets the branch lengths to be equal to the corresponding time intervals elapsed between various separation events. In practice, however, branch lengths are always computed by scaling the time intervals, which reflects the observation that evolution along different branches often occurs with different rates. Therefore, in biology molecular clock trees are usually reserved to represent the evolution of species or sequences with constant evolutionary rates.

If $\mathcal{T}$ is a molecular clock tree, then $d^{\mathcal{T}}$ is an ultrameric distance function under very general assumptions (see Exercise 5.10). Conversely, any ultrameric distance function is generated by a molecular clock tree. The proof of this statement can be derived from the arguments in [SN], [SK] that show that if one applies the neighbor-joining algorithm to an ultrameric distance function, then the resulting tree can be turned into a molecular clock tree by appropriately placing a root, just as we did in Example 5.7. However, for an ultrameric distance function there is a simpler algorithm that recovers the corresponding molecular clock tree. It is called the *Unweighted Pair Group Method Using Arithmetic Averages* or *UPGMA* [SM].

We will now describe the UPGMA algorithm. Suppose we are given an ultrameric distance function d on a set $\mathcal{M} = \{x^1, \ldots, x^N\}$ of OTUs. We place the OTUs at height 0 and will build the tree from bottom to top by introducing new OTUs representing the interior nodes and the root of the future tree, and placing them at particular heights. UPGMA combines the OTUs in clusters. If C^i and C^j are two clusters of OTUs from $\mathcal{M}$, define the distance between them as

$$d(C^i, C^j) = \frac{1}{N(C^i)N(C^j)} \sum_{a \in C^i, b \in C^j} d_{ab}, \qquad (5.3)$$

where $N(C^i)$ and $N(C^j)$ denote the numbers of OTUs in the clusters C^i and C^j respectively.

Assign initially each OTU x^i to its own single-element cluster C^i; we say that x^i is *associated with* C^i. Next, choose two clusters C^i and C^j for which $d(C^i, C^j)$ is minimal. Define a new cluster $C^{N+1} = C^i \cup C^j$ and set the distances from C^{N+1} to the remaining clusters by formula (5.3). Introduce now a new OTU x^{N+1} and place it at the *total* height $d(C^i, C^j)/2$ above x^i and x^j. The OTU x^{N+1} is associated with the cluster C^{N+1} and represents an interior node of the future tree connected to x^i and x^j. We replace the pair x^i, x^j with x^{N+1} and set the distance between x^{N+1} and any other OTU as the distance between the associated clusters. Thus we now have $N - 1$ clusters, for which we repeat the above procedure. We iterate the algorithm until only two clusters remain, say, C^m associated with an OTU x^m and C^l associated with an OTU x^l. We then place the root of the tree above x^m and x^l at the total height $d(C^m, C^l)/2$.

We will now apply the UPGMA algorithm to the distance function from Example 5.7.

Example 5.8. Since d_{35} is minimal, we form a new cluster C^6 as $C^6 = \{x^3, x^5\}$. From formula (5.3) we obtain

$$d(C^1, C^6) = \frac{1}{2}(d_{13} + d_{15}) = 6,$$

$$d(C^2, C^6) = \frac{1}{2}(d_{23} + d_{25}) = 16,$$

$$d(C^4, C^6) = \frac{1}{2}(d_{43} + d_{45}) = 16.$$

We now introduce a new OTU x^6 and place it at the height $d_{35}/2 = 1$ above x^3 and x^5, as shown in Fig. 5.20.

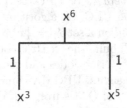

Fig. 5.20.

For the OTUs x^1, x^2, x^4 and x^6 we have the following distance matrix

$$
\begin{array}{c|cccc}
M_d & x^1 & x^2 & x^4 & x^6 \\
x^1 & 0 & 16 & 16 & 6 \\
x^2 & 16 & 0 & 8 & 16 \\
x^4 & 16 & 8 & 0 & 16 \\
x^6 & 6 & 16 & 16 & 0 \\
\end{array}
$$

Here d_{16} is minimal, and we form a new cluster C^7 as $C^7 = \{x^1, x^3, x^5\}$ (recall that x^6 is associated with the cluster $C^6 = \{x^3, x^5\}$). From formula (5.3) we obtain

$$d(C^2, C^7) = \frac{1}{3}(d_{21} + d_{23} + d_{25}) = 16,$$

$$d(C^4, C^7) = \frac{1}{3}(d_{41} + d_{43} + d_{45}) = 16.$$

We now introduce a new OTU x^7 and place it at the total height $d_{16}/2 = 3$ above x^1 and x^6, as shown in Fig. 5.21.

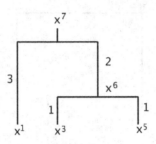

Fig. 5.21.

For the OTUs x^2, x^4 and x^7 we have the following distance matrix

$$\begin{array}{c|ccc} M_d & x^2 & x^4 & x^7 \\ x^2 & 0 & 8 & 16 \\ x^4 & 8 & 0 & 16 \\ x^7 & 16 & 16 & 0 \end{array}$$

Here d_{24} is minimal, and we form a new cluster C^8 as $C^8 = \{x^2, x^4\}$. From formula (5.3) we obtain

$$d(C^7, C^8) = \frac{1}{3 \times 2}(d_{12} + d_{32} + d_{52} + d_{14} + d_{34} + d_{54}) = 16.$$

We now introduce a new OTU x^8 and place it at the height $d_{24}/2 = 4$ above x^2 and x^4, as shown in Fig. 5.22.

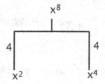

Fig. 5.22.

Finally, we place the root at the total height $d(C^7, C^8)/2 = 8$ above x^7 and x^8 which gives the tree shown Fig. 5.23.

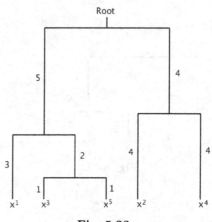

Fig. 5.23.

This tree is identical to the tree in Fig. 5.19, except that it has a "squared" shape which is the preferred shape for UPGMA.

We also remark that we did not need to check that d is ultrameric at the beginning. We could simply apply the UPGMA algorithm to it and notice that the resulting tree generates d.

If a distance function satisfies the four-point condition, but is not ultrameric, the UPGMA algorithm will produce a wrong tree (that is, a tree different from the one produced by the neighbor-joining algorithm), since it always finds a molecular clock tree. Even the topology of the tree derived by UPGMA may be wrong. We will illustrate this effect by the following example (see also Exercise 5.13).

Example 5.9. Let $N = 4$ and suppose that we are given the following distance matrix

$$M_d\ x^1\ x^2\ x^3\ x^4$$

$$
\begin{array}{c c c c c}
x^1 & 0 & 3 & 9 & 9 \\
x^2 & 3 & 0 & 10 & 8 \\
x^3 & 9 & 10 & 0 & 16 \\
x^4 & 9 & 8 & 16 & 0
\end{array}
$$

It is easy to check that d is indeed a distance function, that it satisfies the four-point condition, and that it is not ultrameric (see Exercise 5.12).

We will first apply the UPGMA algorithm to d. Since d_{12} is minimal, we form a new cluster C^5 as $C^5 = \{x^1, x^2\}$. The new OTU x^5 associated with C^5 is placed at the height $3/2$ above x^1, x^2, and the distance matrix for the OTUs x^3, x^4, x^5 is as follows

$$M_d\ x^3\ x^4\ x^5$$

$$
x^3 \quad 0 \quad 16 \quad \frac{19}{2}
$$

$$
x^4 \quad 16 \quad 0 \quad \frac{17}{2}
$$

$$
x^5 \quad \frac{19}{2} \quad \frac{17}{2} \quad 0
$$

Here d_{45} is minimal, and we form a new cluster $C^6 = \{x^1, x^2, x^4\}$ and associate with it a new OTU x^6 that we place at the total height $17/4$ above x^4 and x^6. The distance from C^3 to C^7 is $35/3$, and we therefore place the root above x^3 and x^7 at the total height $35/6$. The resulting tree T_1 is shown in Fig. 5.24. It is easy to observe that $d^{T_1} \neq d$.

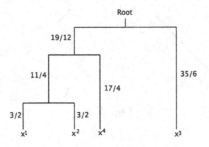

Fig. 5.24.

Let us now apply the neighbor-joining algorithm to d. We obtain

$$
r_1 = r_2 \frac{21}{2},\ r_3 = \frac{35}{2},\ r_4 = \frac{33}{2},
$$

and

$$
\begin{array}{c|cccc}
D & x^1 & x^2 & x^3 & x^4 \\
\hline
x^1 & & -18 & -19 & -18 \\
x^2 & & & -18 & -19 \\
x^3 & & & & -18
\end{array}
$$

Since D_{13} is minimal, we group together the OTUs x^1, x^3, and introduce a new OTU x^5 that replaces them and that is placed at the distance 1 from x^1 and at the distance 8 from x^3. The distance matrix for the OTUs x^2, x^4, x^5 is as follows

$$
\begin{array}{c|ccc}
M_d & x^2 & x^4 & x^5 \\
\hline
x^2 & 0 & 8 & 2 \\
x^4 & 8 & 0 & 8 \\
x^5 & 2 & 8 & 0
\end{array}
$$

and from formulas (5.2) we obtain for them the tree shown in Fig. 5.25.

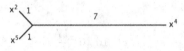

Fig. 5.25.

Hence, the tree $\mathcal{T}_2$ produced by the neighbor-joining algorithm is as shown Fig. 5.26.

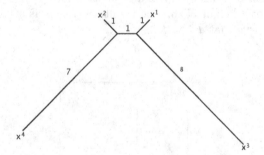

Fig. 5.26.

Observe that the tree $\mathcal{T}_2$ and the unrooted variant of the tree $\mathcal{T}_1$ are very different. Not only the branch lengths do not match, but the topologies do not coincide either.

UPGMA, as the neighbor-joining algorithm, is often applied in the situation when d does not satisfy either the triangle inequality or the four point condition, that is, when d is just a pseudodistance function. We have seen above that using the neighbor-joining method with a pseudodistance function may lead to various anomalies (see Example 5.6). In contrast, if we use UP-GMA with a pseudodistance function, we will always obtain a (not necessarily unique) molecular clock tree with positive branch lengths, but the reliability of such a tree may be equally low.

Certainly, if one applies either the neighbor-joining algorithm or UPGMA to a pseudodistance function d and obtains a tree T_0, one cannot in general hope for the identity $d^{T_0} = d$, but one may ask the natural questions: how close is d^{T_0} to d and is d^{T_0} the closest function to d among all functions d^T calculated from all possible trees T relating the OTUs from $\mathcal{M}$ (possibly, even allowing negative branch lengths)? To attempt to answer these questions one first has to define what "close" means, that is, to introduce a way of comparing two pseudodistance functions. We will compare two pseudodistance functions in the spirit of the *sum of squares* [C-SE] and its variants (see, e.g., [FM]).

For two pseudodistance functions d and d' on the same set of OTUs $\mathcal{M} = \{x^1, \ldots, x^N\}$ define the sum of squares as

$$\varrho(d, d') = \sum_{1 \le i < j \le N} (d_{ij} - d'_{ij})^2.$$

We will be interested in the special case when $d' = d^T$, where an unrooted tree T relates the OTUs from $\mathcal{M}$. We set

$$ss_d(T) = \varrho(d, d^T).$$

There is a method of phylogenetic reconstruction associated with ss_d called the *least squares method*. With this method, for a given pseudodistance function d one looks for all unrooted trees T with the property that $ss_d(T)$ is minimal (assuming that a point of minimum of ss_d exists in the space of all trees). Every such a tree is optimal from the point of view of the least squares method. Hence, the second question from the previous paragraph can be reformulated as follows: if T is a tree produced from a pseudodistance function d by either the neighbor-joining or UPGMA algorithm, is it optimal from the point of view of the least squares method? Certainly, if d is a distance function that respectively either satisfies the four-point condition or is ultrameric, then the answer is positive. For general pseudodistance functions, however, it is not always positive, as we will see in Example 5.10 below for the case of the neighbor-joining algorithm (one can give an analogous example for the case of the UPGMA algorithm as well – see Exercise 5.14).

The least squares method ideally requires minimizing $ss_d(T)$ over *all* possible unrooted trees, but, of course, in real software packages only *some* unrooted topologies with *some* branch length assignments can be examined, which reduces the sensitivity of the method. Branch lengths can be either constrained to be non-negative numbers, or be allowed to be any real numbers. In Example 5.10 below we determine the (unique) optimal tree for the pseudodistance function from Example 5.6. Since for $N = 4$ there are only three unrooted topologies, we will be able to give an *analytic solution* to the least square method in this case. Recall that the neighbor-joining method in Example 5.6 produced trees where some of the branch lengths were negative; therefore in order to compare the two methods, we do not constrain branch lengths for the least square method in Example 5.10.

Example 5.10. We will consider three types of trees corresponding to the three possible unrooted topologies.

Type I. A general tree $T_I(\alpha, \beta, \gamma, \delta, \varepsilon)$ of this type is shown in Fig. 5.27.

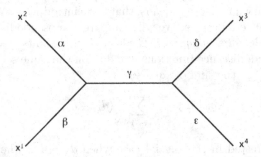

Fig. 5.27.

We have

$$\varphi(\alpha, \beta, \gamma, \delta, \varepsilon) = ss_d\Big((T_I(\alpha, \beta, \gamma, \delta, \varepsilon))\Big) = (\alpha + \beta - 5)^2 + (\beta + \gamma + \delta - 2)^2$$
$$+ (\beta + \gamma + \varepsilon - 7)^2 + (\alpha + \gamma + \delta - 1)^2 + (\alpha + \gamma + \varepsilon - a)^2 + (\delta + \varepsilon - 3)^2.$$

We will attempt to find the points of minimum of φ by determining its critical points. Differentiating φ with respect to each of the five variables and setting all partial derivatives to zero gives the following linear system of equations

$$3\alpha + \beta + 2\gamma + \delta + \varepsilon = 6 + a,$$
$$\alpha + 3\beta + 2\gamma + \delta + \varepsilon = 14,$$
$$2\alpha + 2\beta + 4\gamma + 2\delta + 2\varepsilon = 10 + a,$$
$$\alpha + \beta + 2\gamma + 3\delta + \varepsilon = 6,$$
$$\alpha + \beta + 2\gamma + \delta + 3\varepsilon = 10 + a.$$

The solution to this system is not hard to find and we obtain

$$\alpha = \frac{1}{2} + \frac{a}{4}, \ \beta = \frac{9}{2} - \frac{a}{4}, \ \gamma = \frac{a}{4} - \frac{3}{2},$$

$$\delta = \frac{1}{2} - \frac{a}{4}, \ \varepsilon = \frac{5}{2} + \frac{a}{4}.$$

It is easy to see that these values of branch lengths give the unique point of minimum for φ, and the value of φ at this point is $4(3/2 - a/4)^2$. Observe now that the resulting tree $T_I(\alpha, \beta, \gamma, \delta, \varepsilon)$ is precisely the tree T_1 from Fig. 5.14 found in Example 5.6.

Type II. A general tree $T_{II}(\alpha, \beta, \gamma, \delta, \varepsilon)$ of this type is shown in Fig. 5.28.

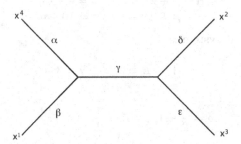

Fig. 5.28.

It can be proved (see Exercise 5.15) that $\psi(\alpha, \beta, \gamma, \delta, \varepsilon) = ss_d\Big((T_{II}(\alpha, \beta, \gamma, \delta, \varepsilon)\Big)$ is minimized by the tree T_2 from Fig. 5.16 found in Example 5.6, that the point of minimum is unique and that the minimal value of ψ is also $4(3/2 - a/4)^2$.

Type III. A general tree $T_{III}(\alpha, \beta, \gamma, \delta, \varepsilon)$ of this type is shown in Fig. 5.29.

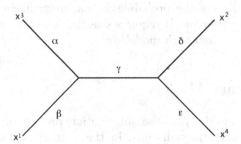

Fig. 5.29.

We have

$$\eta(\alpha, \beta, \gamma, \delta, \varepsilon) = ss_d\Big((T_{III}(\alpha, \beta, \gamma, \delta, \varepsilon)\Big) = (\beta + \gamma + \delta - 5)^2 + (\alpha + \beta - 2)^2$$
$$+ (\beta + \gamma + \varepsilon - 7)^2 + (\alpha + \gamma + \delta - 1)^2 + (\delta + \varepsilon - a)^2 + (\alpha + \gamma + \varepsilon - 3)^2.$$

As above, we will attempt to find the points of minimum of η by determining its critical points. Differentiating η with respect to each of the five variables and setting all partial derivatives to zero gives the following linear system of equations

$$3\alpha + \beta + 2\gamma + \delta + \varepsilon = 6,$$
$$\alpha + 3\beta + 2\gamma + \delta + \varepsilon = 14,$$
$$2\alpha + 2\beta + 4\gamma + 2\delta + 2\varepsilon = 16,$$
$$\alpha + \beta + 2\gamma + 3\delta + \varepsilon = 6 + a,$$
$$\alpha + \beta + 2\gamma + \delta + 3\varepsilon = 10 + a.$$

The solution to this system is

$$\alpha = -1, \quad \beta = 3, \quad \gamma = 3 - \frac{a}{2},$$

$$\delta = \frac{a}{2} - 1, \, \varepsilon = 1 + \frac{a}{2}.$$

It is easy to see that these values of branch lengths give the unique point of minimum for η, and that the value of η at this point is equal to 0.

Hence, if negative branch lengths are allowed, the least squares method in this example performs better than the neighbor-joining algorithm. The least squares method finds a tree $\mathcal{T}$ (where some branch lengths are negative) for which $d^{\mathcal{T}} = d$. In general, however, even if negative branch lengths are allowed, one cannot hope to obtain a tree $\mathcal{T}$ such that $ss_d(\mathcal{T}) = 0$, as in this example.

Often variants of the sum of squares are used. For example, the *weighted sum of squares* introduced in [FM] is quite popular, and the corresponding method of phylogenetic reconstruction is implemented in a program called Fitch which is part of the software package PHYLIP [F3].

We will now discuss the probabilistic maximum likelihood approach to phylogenetic reconstruction. It requires selecting an evolutionary model, and therefore we will consider such models first.

5.4 Evolutionary Models

Evolutionary models describe the substitution process in DNA, RNA and amino acid sequences through time. In the next section they will be used to model this process along the branches of a phylogenetic tree, and model parameters will be allowed to change from branch to branch. We will concentrate on DNA sequences, since RNA and amino acid sequences are treated similarly. Most currently available evolutionary models assume independence among nucleotide sites, and it is therefore sufficient to describe evolution at a single site. This is done by specifying a *continuous-time finite Markov chain* or *continuous-time finite Markov model* depending on the time parameter $t \geq 0$. For the purposes of this exposition a continuous-time Markov model will be understood as a family of ordinary discrete-time Markov models, one for each value of $t \geq 0$, where the states are labeled by the letters of the DNA alphabet $\mathcal{Q} = \{A, C, G, T\}$ and no constraints on the connectivity are imposed in advance. The corresponding matrices of transition probabilities will be written as

$$P(t) = \begin{pmatrix} p_{AA}(t) & p_{AC}(t) & p_{AG}(t) & p_{AT}(t) \\ p_{CA}(t) & p_{CC}(t) & p_{CG}(t) & p_{CT}(t) \\ p_{GA}(t) & p_{GC}(t) & p_{GG}(t) & p_{GT}(t) \\ p_{TA}(t) & p_{TC}(t) & p_{TG}(t) & p_{TT}(t) \end{pmatrix}.$$

In particular, for every value of $t \geq 0$ each element in $P(t)$ is non-negative and the entries in each row sum up to 1. As we will explain below, there is also a relation among the matrices $P(t)$ for different values of t. We further assume that each of the discrete-time Markov models has the same vector of initialization probabilities.

For comparison, we note that a discrete-time Markov chain with matrix of transition probabilities P can also be regarded as a family of Markov chains given by P^n, $n \in \mathbb{N}$. For every n the matrix P^n can be thought of as the matrix of transition probabilities after $n + 1$ steps of the Markov chain (see Sect. 6.11).

Modeling the substitution process at a nucleotide site is done by assuming that $P(t)$ gives the probabilities of all possible state changes in time t. Namely, at any moment the site can be in one of the four possible states: A, C, G and T, and the assumption is that the above matrix gives the probabilities of state change in time t. For example, $p_{AC}(t)$ is the probability of the site changing its state from A to C in time t.

If we suppose that the substitution process at the nucleotide site goes in accordance with such a Markov chain, we in fact make the following

Assumption. *If at some time t_0 the site was in state $i \in \{A, C, G, T\}$, then the probability of the event that at time $t_0 + t$ the site will be in state $j \in \{A, C, G, T\}$ depends only on i, j and t (and is exactly the element $p_{ij}(t)$ of the matrix $P(t)$).*

This assumption leads to the following important observation. Consider the probability $p_{ij}(t + \tau)$ for some $t, \tau \geq 0$. This is the probability of the site going from state i to state j in time $t + \tau$. Any such transition can be realized by first going from state i to any state k in time t and then going from state k to state j in time τ. In accordance with the above assumption it is natural to require that

$$p_{ij}(t + \tau) = \sum_{k \in \mathcal{Q}} p_{ik}(t) p_{kj}(\tau)$$

for all i, j. In the matrix form this identity can be written as follows

$$P(t + \tau) = P(t)P(\tau). \tag{5.4}$$

This identity is part of the definition of continuous-time Markov chain, and we will always assume that it is satisfied.

We remark that there is an analogue of identity (5.4) for ordinary discrete-time Markov chains as well. In this case (5.4) becomes the tautological identity

$$P^{n+m} = P^n P^m, \quad n, m = 0, 1, 2 \dots.$$

We will only consider *regular* continuous-time Markov chains, which means that $P(0)$ is the identity matrix E and that $P(t)$ is differentiable at every $t \geq 0$, that is, each element in the matrix $P(t)$ is differentiable at every $t \geq 0$ as a function of t (for $t = 0$ differentiability is understood as the existence of one-sided derivatives).

We will need the following theorem.

Theorem 5.11. *Under the above assumptions $P(t)$ has the form*

$$P(t) = \exp(tQ), \tag{5.5}$$

where Q is some 4×4-matrix.

Of course, we have to define what the *exponential of a matrix* is.

Definition 5.12. *Let A be a square $m \times m$-matrix. Then $\exp(A)$ is defined to be the $m \times m$-matrix given by the sum of the following series*

$$\exp(A) = E + A + \frac{A^2}{2!} + \frac{A^3}{3!} + \dots = \sum_{n=0}^{\infty} \frac{A^n}{n!}. \tag{5.6}$$

Identity (5.6) must be understood as a collection of m^2 scalar identities, one for each matrix element. Thus, the right-hand side of (5.6) consists of m^2 series. It is possible to prove (see Exercise 5.16) that each of these series is convergent for any matrix A.

Sketch of Proof of Theorem 5.11:
It follows from (5.4) that for $t \geq 0, h > 0$ the following holds

$$\frac{P(t + h) - P(t)}{h} = \frac{P(t)(P(h) - E)}{h} = \frac{P(t)(P(h) - P(0))}{h}.$$

When $h \to 0$ the above identity implies

$$P'(t) = P(t)P'(0),$$

which gives, as in the case of scalar functions (see Exercise 5.17), that

$$P(t) = \exp(tQ),$$

where $Q = P'(0)$.   ∎

Since in identity (5.5) we have $Q = P'(0)$, the matrix Q is sometimes called the *matrix of instantaneous change*. One important property of Q is that the elements in each row of Q sum up to 0 (see Exercise 5.18). By varying Q one obtains all popular models. Four such models will be described below.

Before considering the specific models we will briefly discuss two conditions that continuous-time Markov chains describing the substitution process in DNA are often required to satisfy.

(1) The uniqueness of a stationary probability distribution

Definition 5.13. *A vector $\varphi = (\varphi_A, \varphi_C, \varphi_G, \varphi_T)$ with $\varphi_i \geq 0$ and $\sum_{i \in \mathcal{Q}} \varphi_i = 1$ is called a stationary probability distribution of a Markov chain if $\varphi Q = 0$. Due to (5.5) this is equivalent to requiring that $\varphi P(t) \equiv \varphi$.*

The concept of stationary probability distribution also makes sense for ordinary discrete-time Markov chains. In this case a vector φ with $\varphi_i \geq 0$ and $\sum_{i=1}^{N} \varphi_i = 1$ is called a stationary probability distribution of a Markov chain whose matrix of transition probabilities is P, if $\varphi P = \varphi$ (see Exercise 5.19).

Any Markov chain (either continuous-time or discrete-time) possesses a stationary probability distribution (see [Do]), but it may not be unique. For example, for the trivial continuous-time Markov chain with $Q = 0$, any vector φ with $\varphi_i \geq 0$ and $\sum_{i=1}^{N} \varphi_i = 1$ is a stationary probability distribution. We will only consider continuous-time Markov chains for which a stationary probability distribution is unique, which is the case under certain assumptions. This holds, for example, if for some $t_0 > 0$ we have $p_{ij}(t_0) > 0$ for all $i, j \in \mathcal{Q}$ (see [Do]).

We will now discuss two important properties of stationary probability distributions.

(a) Define the probability $p_i(t)$ of the event that the nucleotide site is in state i at time t as follows

$$p_i(t) = \sum_{k \in \mathcal{Q}} \varrho_k p_{ki}(t),$$

where $\varrho = (\varrho_A, \varrho_C, \varrho_G, \varrho_T)$ is the vector of initialization probabilities for the Markov chain. Setting $\varrho = \varphi$ we obtain

$$p_i(t) = \sum_{k \in \mathcal{Q}} \varphi_k p_{ki}(t) = \varphi_i.$$

Hence, if the nucleotide site is assumed to evolve in accordance with a Markov chain with stationary probability distribution φ, and if φ is taken as the vector of initialization probabilities (which is commonly done), then $p_i(t)$

does not depend on t and is equal to φ_i for all i. This is of course a very strong (and hardly realistic) assumption on the way the nucleotide site evolves. It is used mainly because of its computational convenience.

(b) We have

$$P(t) \to \begin{pmatrix} \varphi_A & \varphi_C & \varphi_G & \varphi_T \\ \varphi_A & \varphi_C & \varphi_G & \varphi_T \\ \varphi_A & \varphi_C & \varphi_G & \varphi_T \\ \varphi_A & \varphi_C & \varphi_G & \varphi_T \end{pmatrix},$$

as $t \to \infty$, that is, for large t, $P(t)$ "stabilizes" in accordance with the stationary probability distribution. A similar property holds for ordinary discrete-time Markov chains as well. Namely, if a discrete-time Markov chain with matrix of transition probabilities P possesses a unique stationary probability distribution φ, then under certain additional assumptions we have

$$P^n \to \begin{pmatrix} \varphi_1 & \cdots & \varphi_N \\ \vdots & \vdots & \vdots \\ \varphi_1 & \cdots & \varphi_N \end{pmatrix}, \tag{5.7}$$

as $n \to \infty$ (see Exercise 5.20). For example, if for some $n_0 \in \mathbb{N}$ all elements of P^{n_0} are positive (and hence all elements of P^n for $n \geq n_0$ are positive as well), then a stationary probability distribution is unique and (5.7) holds (see [Do]).

For the general theory of stationary probability distributions the interested reader is referred to [Kar], [Do] (see also [EG]).

(2) Time reversibility

Definition 5.14. *Let ϱ be the vector of initialization probabilities of a continuous-time Markov chain, and suppose that $p_i(t) \neq 0$ for all $i \in \mathcal{Q}$ and $t \geq 0$. We define the reversed Markov chain as the continuous-time Markov chain given by the transition probability matrices $P^*(t)$ with*

$$p_{ij}^*(t) = \frac{\varrho_j p_{ji}(t)}{p_i(t)},$$

for all i, j.

Loosely speaking, replacing $P(t)$ with $P^*(t)$ corresponds to reversing the time parameter t. Assuming that each component of the (unique) stationary probability distribution φ is non-zero and setting $\varrho = \varphi$, we obtain

$$p_{ij}^*(t) = \frac{\varphi_j p_{ji}(t)}{\varphi_i},$$

for all i, j.

Definition 5.15. *A Markov chain satisfying the assumptions of Definition 5.14 is called time-reversible or simply reversible if $P^*(t) = P(t)$ for all $t \geq 0$.*

If each component of φ is non-zero, then for $\varrho = \varphi$ reversibility is equivalent to

$$\varphi_i p_{ij}(t) = \varphi_j p_{ji}(t), \tag{5.8}$$

for all i, j. Note that identity (5.8) makes sense even if some of the components of φ are equal to zero, and in what follows we will always understand reversibility in the sense of this identity. Note that due to (5.5), identity (5.8) is equivalent to the condition that the matrix

$$\begin{pmatrix} \varphi_A & 0 & 0 & 0 \\ 0 & \varphi_C & 0 & 0 \\ 0 & 0 & \varphi_G & 0 \\ 0 & 0 & 0 & \varphi_T \end{pmatrix} Q$$

is symmetric.

We will now describe four evolutionary models commonly used for phylogenetic reconstruction.

5.4.1 The Jukes-Cantor Model

The *Jukes-Cantor model* was introduced in [JC] and is given by setting

$$Q = \begin{pmatrix} -3\alpha/4 & \alpha/4 & \alpha/4 & \alpha/4 \\ \alpha/4 & -3\alpha/4 & \alpha/4 & \alpha/4 \\ \alpha/4 & \alpha/4 & -3\alpha/4 & \alpha/4 \\ \alpha/4 & \alpha/4 & \alpha/4 & -3\alpha/4 \end{pmatrix},$$

where α is a positive constant called the *evolutionary rate* (the role of evolutionary rates in phylogenetic reconstruction will be discussed in the next section).

We will now calculate the corresponding matrix $P(t) = \exp(tQ)$. Since we need to find the sum of the series

$$\sum_{n=0}^{\infty} \frac{t^n Q^n}{n!},$$

we will first determine the powers of Q. We will show by induction that

$$Q^n = (-\alpha)^{n-1} Q, \tag{5.9}$$

for all $n \in \mathbb{N}$.

Clearly, (5.9) holds for $n = 1$. Assume that $n_0 \geq 2$ and that (5.9) holds for all $n < n_0$. Then

$$Q^{n_0} = Q^{n_0-1}Q = (-\alpha)^{n_0-2}Q^2.$$

Calculating Q^2 we obtain

$$Q^2 = \begin{pmatrix} 3\alpha^2/4 & -\alpha^2/4 & -\alpha^2/4 & -\alpha^2/4 \\ -\alpha^2/4 & 3\alpha^2/4 & -\alpha^2/4 & -\alpha^2/4 \\ -\alpha^2/4 & -\alpha^2/4 & 3\alpha^2/4 & -\alpha^2/4 \\ -\alpha^2/4 & -\alpha^2/4 & -\alpha^2/4 & 3\alpha^2/4 \end{pmatrix} = -\alpha Q.$$

Hence

$$Q^{n_0} = (-\alpha)^{n_0-2}(-\alpha)Q = (-\alpha)^{n_0-1}Q,$$

and (5.9) is proved.

We can now find $P(t) = \exp(tQ)$. Indeed,

$$\sum_{n=0}^{\infty} \frac{t^n Q^n}{n!} = E + \sum_{n=1}^{\infty} \frac{t^n Q^n}{n!}$$

$$= E + \left(\sum_{n=1}^{\infty} \frac{t^n(-\alpha)^{n-1}}{n!} \right) Q = E - \frac{1}{\alpha} \left(\sum_{n=1}^{\infty} \frac{(-t\alpha)^n}{n!} \right) Q$$

$$= E - \frac{1}{\alpha}(\exp(-t\alpha) - 1)Q.$$

This implies

$$p_{ii}(t) = \frac{1}{4} + \frac{3}{4}\exp(-t\alpha) \quad \text{for all } i,$$

(5.10)

$$p_{ij}(t) = \frac{1}{4} - \frac{1}{4}\exp(-t\alpha) \quad \text{for all } i \neq j.$$

Let $\varphi = (1/4, 1/4, 1/4, 1/4)$. A simple calculation shows that $\varphi Q = 0$, and hence φ is a stationary probability distribution of the Jukes-Cantor model. It is not hard to show that φ is the only stationary probability distribution of the model and that the model is time-reversible (see Exercise 5.21).

The Jukes-Cantor model is one of the earliest models and is not very realistic. In particular, it assumes that the probabilities to find a nucleotide site in any of the four possible states are all equal to $1/4$ for all t.

5.4.2 The Kimura Model

The *Kimura model* [Ki] is a generalization of the Jukes-Cantor model and is given by setting

$$Q = \begin{pmatrix} -(2\beta+1)\alpha/4 & \beta\alpha/4 & \alpha/4 & \beta\alpha/4 \\ \beta\alpha/4 & -(2\beta+1)\alpha/4 & \beta\alpha/4 & \alpha/4 \\ \alpha/4 & \beta\alpha/4 & -(2\beta+1)\alpha/4 & \beta\alpha/4 \\ \beta\alpha/4 & \alpha/4 & \beta\alpha/4 & -(2\beta+1)\alpha/4 \end{pmatrix},$$

where, as before, $\alpha > 0$ is referred to as the evolutionary rate, and $\beta > 0$ is an additional parameter. The Kimura model turns into the Jukes-Cantor model for $\beta = 1$.

As before, we can find the corresponding matrix $P(t) = \exp(tQ)$ for the Kimura model (see Exercise 5.22). We have

$$p_{ii}(t) = \frac{1}{4} + \frac{1}{4}\exp(-t\beta\alpha) + \frac{1}{2}\exp\left(-t\frac{(\beta+1)\alpha}{2}\right),$$

$$p_{AC}(t) = p_{CA}(t) = p_{AT}(t) = p_{TA}(t) = p_{CG}(t) = p_{GC}(t)$$

$$= p_{GT}(t) = p_{TG}(t) = \frac{1}{4} - \frac{1}{4}\exp(-t\beta\alpha), \tag{5.11}$$

$$p_{AG}(t) = p_{GA}(t) = p_{CT}(t) = p_{TC}(t) = \frac{1}{4} + \frac{1}{4}\exp(-t\beta\alpha)$$

$$-\frac{1}{2}\exp\left(-t\frac{(\beta+1)\alpha}{2}\right).$$

The Kimura model thus incorporates a certain difference between two types of nucleotide substitutions: *transversions* ($A \to C$, $C \to A$, $A \to T$, $T \to A$, $C \to G$, $G \to C$, $G \to T$, $T \to G$) and *transitions* ($A \to G$, $G \to A$, $C \to T$, $T \to C$). Its stationary probability distribution is unique (and is identical to that of the Jukes-Cantor model), and the model is reversible (see Exercise 5.23).

5.4.3 The Felsenstein Model

The *Felsenstein model* [F1] is also a generalization of the Jukes-Cantor model and is given by the matrix

$$Q = \begin{pmatrix} -\alpha(\pi_C + \pi_G \\ +\pi_T) & \alpha\pi_C & \alpha\pi_G & \alpha\pi_T \\ \alpha\pi_A & -\alpha(\pi_A + \pi_G \\ +\pi_T) & \alpha\pi_G & \alpha\pi_T \\ \alpha\pi_A & \alpha\pi_C & -\alpha(\pi_A + \pi_C \\ +\pi_T) & \alpha\pi_T \\ \alpha\pi_A & \alpha\pi_C & \alpha\pi_G & -\alpha(\pi_A + \pi_C \\ +\pi_G) \end{pmatrix},$$

where, as before, $\alpha > 0$ is the evolutionary rate, and π_i for $i \in Q$ are non-negative parameters satisfying $\pi_A + \pi_C + \pi_G + \pi_T = 1$. The Jukes-Cantor model is a special case of the Felsenstein model for $\pi_A = \pi_C = \pi_G = \pi_T = 1/4$. The corresponding matrix $P(t) = \exp(tQ)$ is given by

$$p_{ii}(t) = \pi_i + (1 - \pi_i)\exp(-t\alpha), \text{ for all } i,$$

$$p_{ij}(t) = \pi_j - \exp(-t\alpha)\pi_j, \qquad \text{for all } i \neq j.$$

The stationary probability distribution of the Felsenstein model is unique and coincides with the vector $(\pi_A, \pi_C, \pi_G, \pi_T)$; this model is also reversible (see Exercise 5.24). The Felsenstein model is much more flexible than the Jukes-Cantor model since it allows to construct a continuous-time Markov model with a given stationary probability distribution.

5.4.4 The Hasegawa-Kishino-Yano (HKY) Model

The *HKY model* [HKY] generalizes the Felsenstein model in the same way as the Kimura model generalizes the Jukes-Cantor model. The corresponding matrix Q is as follows

$$Q = \begin{pmatrix} -\alpha(\beta\pi_C + \pi_G \\ +\beta\pi_T) & \beta\alpha\pi_C & \alpha\pi_G & \beta\alpha\pi_T \\ \beta\alpha\pi_A & -\alpha(\beta\pi_A + \beta\pi_G \\ +\pi_T) & \beta\alpha\pi_G & \alpha\pi_T \\ \alpha\pi_A & \beta\alpha\pi_C & -\alpha(\pi_A + \beta\pi_C \\ +\beta\pi_T) & \beta\alpha\pi_T \\ \beta\alpha\pi_A & \alpha\pi_C & \beta\alpha\pi_G & -\alpha(\beta\pi_A + \pi_C \\ +\beta\pi_G) \end{pmatrix},$$

where $\alpha > 0$ is the evolutionary rate, $\beta > 0$ is a parameter responsible for distinguishing between transitions and transversions, π_i for $i \in Q$ are non-negative parameters satisfying $\pi_A + \pi_C + \pi_G + \pi_T = 1$. The only stationary probability distribution of the HKY model is the vector $(\pi_A, \pi_C, \pi_G, \pi_T)$, and the model is reversible (see Exercise 5.25). This model generalizes the three models described above. It differentiates between transitions and transversions, and allows to construct a continuous-time Markov model with a given stationary probability distribution. The matrix $P(t) = \exp(tQ)$ has a rather complicated form and we omit it.

5.5 Maximum Likelihood Method

In this section we will show how evolutionary models can be used to infer a phylogenetic tree (or trees) from a reduced multiple alignment of DNA sequences. Let $\mathcal{M} = \{x^1, \ldots, x^N\}$ be the original set of OTUs and let $D = \{\hat{x}^1, \ldots, \hat{x}^N\}$ be the collection of the shorter sequences that form the reduced multiple alignment. First of all, we select an evolutionary model that we assume to be one of the four models described in the previous section (although everything that follows can be also applied to an arbitrary regular reversible evolutionary model). Here we do not explain how one can choose the "best" model for a particular dataset; this question will be addressed to some extent in the next section.

Suppose we have selected a model. We will now make the following

Evolutionary Assumption.

(i) The sequences in the dataset D evolved from their common ancestor along a molecular clock tree whose branches are measured as time intervals (we will refer to this tree as the true tree),

(ii) the evolution along the true tree involved only substitutions, not deletions or insertions,

(iii) each site evolved along the true tree independently of the others and identically to the others,

(iv) the evolution of each site along each branch of the true tree did not depend on its evolution along any other branch,

(v) the substitution process for each site along each branch of the true tree went in accordance with the selected Markov model,

(vi) the evolutionary rate parameter α is branch-specific, that is, allowed to change from branch to branch,

(vii) the parameter β (in the case of the Kimura and HKY models) is also branch-specific subject to the constraint that it has the same value on the two branches descending from the root.

Our ultimate goal is to determine the true tree. We will attempt to find it by calculating the *likelihood $L(D|T)$ of the dataset D given T* (a formula for $L(D|T)$ will be given later) for every molecular clock tree T relating the sequences in D, and by selecting the tree (or trees) for which $L(D|T)$ is maximal (assuming that a point of maximum exists). This procedure is the core of the maximum likelihood approach. The likelihood $L(D|T)$ is in fact the probability (calculated in accordance with the evolutionary assumption) of the event that the sequences in the dataset D are related by the tree T, and the maximal likelihood method selects the trees for which this probability has the largest value as optimal trees.

The maximum likelihood approach is probabilistic rather than deterministic which has many advantages. For instance, unlike most approaches to phylogenetic reconstruction discussed in the preceding sections, it comes with its own measure of error. If, for example, T is optimal for the maximal likelihood method, but $L(D|T)$ is small, one probably should not accept T as a reliable tree.

Observe that it is unreasonable to expect that the branch lengths of the true tree can be determined by any method whatsoever. Indeed, the true tree was assumed to be a molecular clock tree whose branch lengths are measured as time intervals. However, evolution might have been faster along some branches and slower along others. Since no information on the speed of evolution is usually available for the dataset D, no analysis can extract the actual time intervals from it. Instead, one can hope to determine time intervals scaled by particular values of the *evolutionary rate*. Biologically, the evolutionary rate is thought of as the speed of evolution along a particular branch of a tree. Mathematically, it is a model parameter (that we always denote by α) which is simply a scaling factor for the matrix $Q = P'(0)$. In the previous section we have seen evolutionary rates incorporated into the matrices of instantaneous change for the four specific models. In accordance with (vi) of the evolutionary assumption, the evolutionary rate is branch-specific. Later we will see from the explicit formula for the likelihood that, for a fixed tree topology, if t_j is the parameter for the length of the jth branch and α_j is value of the evolutionary rate along this branch, then $L(D|\mathcal{T})$ depends on α_j and t_j by way of the product $\alpha_j \times t_j$. Hence, when $L(D|\mathcal{T})$ is maximized for every topology and subsequently over all topologies, one only obtains optimal values of the product of time and the evolutionary rate, and these values are taken as the branch lengths of the corresponding optimal trees. Therefore, from now on we will assume that the tree space we are working on consists of all rooted (not necessarily molecular clock) trees relating the sequences from D, and that the evolutionary rate is set to 1 on each branch of every tree.

The likelihood $L(D|\mathcal{T})$ is a function of $\mathcal{T}$ (which includes the topology and branch lengths – constrained to be non-negative) as well as the branch-specific parameter β (in the case of the Kimura and HKY models), and therefore even for a moderate number N of OTUs $L(D|\mathcal{T})$ typically depends on a large number of variables. Ideally, one would like to maximize $L(D|\mathcal{T})$ over all these variables, but in practice doing so is a slow process even for a fixed topology. Therefore, in reality a very limited number of topologies can be examined, which decreases the sensitivity of the method. Luckily, as will be explained below, the reversibility of the evolutionary model helps to slightly reduce the topology space that needs to be searched, but one should keep in mind that, firstly, reversibility imposes biologically unjustifiable constraints on the evolution of each nucleotide site, and, secondly, the reduced topology space is still very large in cases of interest.

The maximum likelihood method is implemented, for example, as part of the PHYLIP software package [F3], and the interested reader can learn about various topology space search algorithms utilized there as well as other features of this package from the PHYLIP website http://evolution.genetics.washington.edu/phylip.html.

We will now explain how the likelihood of a dataset given a tree is calculated. Rather than writing a general formula, we will consider a specific example that will convey the idea. Suppose that the reduced multiple alignment is as follows

$$\hat{x}^1 : A\,A\,G$$
$$\hat{x}^2 : C\,T\,C$$
$$\hat{x}^3 : C\,G\,G$$
$$\hat{x}^4 : G\,G\,A,$$

and let T be the rooted tree shown in Fig. 5.30.

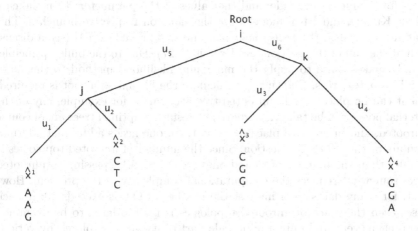

Fig. 5.30.

First of all, we define site-specific likelihoods as

$$L_1(D|T) = \sum_{ijk \in Q} \varphi_i p_{ij}(u_5) p_{ik}(u_6) p_{jA}(u_1) p_{jC}(u_2) p_{kC}(u_3) p_{kG}(u_4),$$

$$L_2(D|T) = \sum_{ijk \in Q} \varphi_i p_{ij}(u_5) p_{ik}(u_6) p_{jA}(u_1) p_{jT}(u_2) p_{kG}(u_3) p_{kG}(u_4),$$

$$L_3(D|T) = \sum_{ijk \in Q} \varphi_i p_{ij}(u_5) p_{ik}(u_6) p_{jG}(u_1) p_{jC}(u_2) p_{kG}(u_3) p_{kA}(u_4),$$

where φ is the stationary probability distribution of the model. In the above formulas the parameter β (in the case of the Kimura and HKY models) is branch-specific. More precisely, the lth branch carries its own parameter β_l for $l = 1, \ldots, 6$, with the constraint $\beta_5 = \beta_6$, which agrees with (vii) of the evolutionary assumption. For example, $p_{jA}(u_1)$ is calculated using β_1 and $p_{ij}(u_5)$, $p_{ik}(u_6)$ using $\beta_5 = \beta_6$. The *likelihood of the dataset* is then defined as the product of the site-specific likelihoods

$$L(D|T) = L_1(D|T)L_2(D|T)L_3(D|T).$$

An analogous definition can be given for the case of molecular clock trees and arbitrary values of the evolutionary rate, which was our original setup. It is clear that in this case, for a fixed tree topology, $L(D|T)$ indeed depends on α_j and t_j by way of the product $\alpha_j \times t_j$, as we stated above.

Further, one can show that the reversibility of the model implies that $L(D|T)$ does not in fact depend on the root position, that is, $L(D|T_1) = L(D|T_2)$, if the topologies and branch lengths of the corresponding unrooted trees for T_1 and T_2 coincide, and the values of the parameter β (in the case of the Kimura and HKY models) are the same on respective branches. This statement is called the *pulley principle* and was proved in [F1] (see a discussion at the end of this section and Exercise 5.27). Due to the pulley principle, it only makes sense to apply the maximum likelihood method to the space of all *unrooted* trees. Certainly, to calculate the likelihood a root is required, but it can be placed on an unrooted tree arbitrarily, for example, any of the internal nodes can be taken as a root. The resulting optimal trees are of course unrooted, and in order to place roots on them one needs additional information about the OTUs in question. Since the number of unrooted topologies is smaller than the number of rooted ones (see Sect. 5.1), passing to unrooted trees somewhat reduces the computational complexity of the problem. However, for many datasets of interest the number of OTUs exceeds 20; in such cases even the space of unrooted topologies is far too large to be examined comprehensively, and only a minuscule part of it can be explored by various existing heuristic search algorithms.

We will now show how the maximum likelihood method can be utilized to produce a pseudodistance function on a set $\mathcal{M} = \{x^1, \ldots, x^N\}$ of OTUs. As we have seen in the preceding section, pseudodistance functions can be used by any distance method for inferring a phylogenetic tree. Let, as before, $D = \{\hat{x}^1, \ldots, \hat{x}^N\}$ be the collection of the sequences that form the reduced multiple alignment. A pseudodistance function on $\mathcal{M}$ comes from the pairwise alignments induced by the reduced multiple alignment, namely, d_{ij} is obtained by considering the alignment between $\hat{x}^i$ and $\hat{x}^j$. Specifically, the maximal likelihood method is applied to the dataset $D^{ij} = \{\hat{x}^i, \hat{x}^j\}$ of two aligned sequences. Any unrooted tree relating $\hat{x}^i$ and $\hat{x}^j$ is a segment as shown in Fig. 5.31.

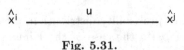

Fig. 5.31.

Maximizing $L(D^{ij}|T)$ over u and (in the case of the Kimura and HKY models) β, we obtain some optimal values of the parameters. Then d_{ij} is set to be equal to the optimal value of u (assuming that it is unique). Of course, the result depends on the evolutionary model chosen in advance.

As we will see in Example 5.16 below, it may happen that $d_{ij} = \infty$ for some i, j, and therefore the resulting "function" d may not be a pseudodistance function. In such cases further adjustments are required. For example, d_{ij} can be set to a very large number instead.

The above procedure is based on maximizing the likelihood of a dataset consisting of two DNA sequences of equal length put in a (unique) ungapped alignment. We will now give an example of such likelihood maximization (and hence pseudodistance calculation) for the case of the Jukes-Cantor model (see also Exercise 5.26).

Example 5.16. Consider two sequences $x = x_1, \ldots, x_n$ and $y = y_1, \ldots, y_n$, $x \neq y$, with $x_i, y_i \in \mathcal{Q} = \{A, C, G, T\}$ for $i = 1, \ldots, n$, aligned as follows

$$x : x_1 \ldots x_n$$
$$y : y_1 \ldots y_n.$$

We will find the distance $d(x, y)$ between x and y. Consider the two-sequence dataset $D = \{x, y\}$ and for any tree $\mathcal{T}$ relating x and y as shown in Fig. 5.32, calculate the likelihood $L(D|\mathcal{T})$.

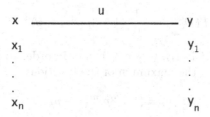

Fig. 5.32.

First of all, we need to find the site-specific likelihoods $L_i(D|\mathcal{T})$, $i = 1, \ldots, n$. We place a root in the middle of the branch and use formulas (5.10) with $\alpha = 1$. For each $1 \leq i \leq n$ we will consider two cases.

Let first $x_i = y_i = z$. Then we have

$$L_i(D|\mathcal{T}) = \frac{1}{4} \sum_{j \in \mathcal{Q}} p_{jz} \left(\frac{u}{2}\right)^2 = \frac{1}{4} \left(p_{zz} \left(\frac{u}{2}\right)^2 \right.$$

$$+ \sum_{j \in \mathcal{Q}, j \neq z} p_{jz} \left(\frac{u}{2}\right)^2 \right) = \frac{1}{4} \left(\left(\frac{1}{4} + \frac{3}{4} \exp\left(-\frac{u}{2}\right) \right)^2 \right. \qquad (5.12)$$

$$\left. +3 \left(\frac{1}{4} - \frac{1}{4} \exp\left(-\frac{u}{2}\right) \right)^2 \right) = \frac{1}{16} \left(1 + 3 \exp(-u) \right).$$

Suppose now that $x_i \neq y_i$. Then we obtain

$$L_i(D|T) = \frac{1}{4} \sum_{j \in Q} p_{jx_i}\left(\frac{u}{2}\right) p_{jy_i}\left(\frac{u}{2}\right) = \frac{1}{4}\left(p_{x_i x_i}\left(\frac{u}{2}\right) p_{x_i y_i}\left(\frac{u}{2}\right)\right.$$

$$+ p_{y_i x_i}\left(\frac{u}{2}\right) p_{y_i y_i}\left(\frac{u}{2}\right) + \left. \sum_{j \in Q, j \neq x_i, j \neq y_i} p_{jx_i}\left(\frac{u}{2}\right) p_{jy_i}\left(\frac{u}{2}\right)\right)$$

$$(5.13)$$

$$= \frac{1}{4}\left(2\left(\frac{1}{4} + \frac{3}{4}\exp\left(-\frac{u}{2}\right)\right)\left(\frac{1}{4} - \frac{1}{4}\exp\left(-\frac{u}{2}\right)\right)\right.$$

$$+ 2\left(\frac{1}{4} - \frac{1}{4}\exp\left(-\frac{u}{2}\right)\right)^2\right) = \frac{1}{16}\left(1 - \exp(-u)\right).$$

Let $0 \leq m < n$ be the number of indices $1 \leq i \leq n$ for which $x_i = y_i$. It then follows from the above formulas that the likelihood of D is given by

$$L(D|T) = \frac{1}{16^n}(1 + 3p)^m (1 - p)^{n-m},$$

where $p = \exp(-u)$. Clearly, $0 \leq p \leq 1$, and in order to maximize the likelihood, we need to find the maximum of the function

$$\psi(p) = (1 + 3p)^m (1 - p)^{n-m}$$

on the segment $[0, 1]$.

Let first $m = 0$. Then $\psi(p) = (1 - p)^n$ attains its maximum at $p = 0$ which gives $u = \infty$ and consequently $d(x, y) = \infty$.

Suppose now that $m > 0$. We will initially consider ψ on the whole real line and find its points of maximum. Differentiating we obtain

$$\psi'(p) = (1 + 3p)^{m-1}(1 - p)^{n-m-1}(4m - n - 3pn).$$

Clearly, $p = -1/3$ and $p = 1$ are not points of maximum since there the value of ψ is 0. The only remaining critical point is

$$p = p_0 = \frac{4m - n}{3n}.$$

Clearly, $-1/3 < p_0 < 1$ and

$$\psi''(p_0) = -3n(1 + 3p_0)^{m-1}(1 - p_0)^{n-m-1}$$

is negative. Therefore, p_0 is the unique point of maximum of ψ on $\mathbb{R}$.

If $4m > n$, then $p_0 > 0$ and ψ attains its maximum on $[0, 1]$ at p_0 which yields

$$d(x, y) = u = -\ln p_0 = -\ln \frac{4m - n}{3n} > 0.$$

If $4m \leq n$, then $p_0 \leq 0$, and ψ attains its maximum on $[0, 1]$ at 0, which, as for the case $m = 0$, gives $u = \infty$ and therefore $d(x, y) = \infty$.

We will now show a simpler way to compute the site-specific likelihoods from Example 5.16. We will do it for an arbitrary regular time-reversible model. As before, denote by φ the stationary probability distribution. Place a root on the tree in Fig. 5.32, and let the lengths of the branches leading from the root to x and y be u_1 and u_2 respectively, with $u_1 + u_2 = u$. In accordance with (vii) of the evolutionary assumption, the parameters of the model (such as β in the case of the Kimura and HKY models) are assumed to be constant on $\mathcal{T}$. Fix $1 \leq i \leq n$. Using (5.4) and (5.8) we obtain

$$L_i(D|\mathcal{T}) = \sum_{j \in \mathcal{Q}} \varphi_j p_{jx_i}(u_1) p_{jy_i}(u_2) = \varphi_{x_i} \sum_{j \in \mathcal{Q}} p_{x_i j}(u_1) p_{jy_i}(u_2)$$
$$= \varphi_{x_i} p_{x_i y_i}(u) = \varphi_{y_i} p_{y_i x_i}(u),$$

which in the case of the Jukes-Cantor model gives formulas (5.12) and (5.13).

This last calculation in fact proves the pulley principle for an arbitrary regular time-reversible model and $N = 2$. It is easy to generalize this proof to any number of OTUs, but one has to impose a condition analogous to (vii) of the evolutionary assumption: each model parameter (apart from α that is set to 1 everywhere) is branch-specific subject to the constraint that it has equal values on the two branches descending from the root (see Exercise 5.27).

5.6 Model Comparison

It is important to choose a correct evolutionary model to assess a particular dataset. Indeed, as we have seen in the preceding section, the analysis of a dataset by the maximum likelihood method depends on the model selected in advance. Choosing a model from a range of available ones is a difficult task. In this section we will address a somewhat easier question: given a pair of models, how can one select a model "most suitable" for the dataset in question? Specifically, we will describe a *statistical hypothesis testing* procedure that allows to compare the two models.

Suppose we are given a dataset D as in the preceding section and two regular reversible evolutionary models M_1 and M_2. Assume further that M_1 is a special case of M_2. For example, M_1 can be the Jukes-Cantor model and M_2 the Kimura model. We wish to know whether or not passing to the more general model M_2 gives us a *significantly better* understanding of the dataset compared to the understanding that we gained by applying model M_1 to it.

Since M_1 is a special case of M_2, using the latter model explains the data at least as well as using the former one. But if the more general model does not lead to a significantly better understanding of the dataset, we may as well be content with using model M_1 for it, which has its advantages, since M_1 is simpler.

Statistically, this setup is formalized as follows: consider the *null hypothesis*

$$H_0 : \text{the evolutionary assumption holds for } D \text{ with model } M_1,$$

and the *alternative hypothesis*

$$H_A : \text{the evolutionary assumption holds for } D \text{ with model } M_2.$$

The null hypothesis is a special case of the alternative one, that is, the hypotheses are *nested*. We wish to know whether the null hypothesis should be *accepted* or *rejected in favor of the alternative hypothesis*. Below we will briefly outline one procedure used in statistics for making a decision to either accept or reject the null hypothesis; more detail on this procedure will be given in Sect. 8.3, where a general approach to statistical hypothesis testing will be discussed.

The procedure is a special case of the *likelihood ratio test*. First of all, we find the maxima of the likelihoods of D under H_0 and H_A (assuming that points of maximum exist). We denote the maximal likelihoods by $L_0^{\max}(D)$ and $L_A^{\max}(D)$ respectively. Assume that $L_A^{\max}(D) \neq 0$ and form the *likelihood ratio*

$$\Delta(D) = \frac{L_0^{\max}(D)}{L_A^{\max}(D)}.$$

Clearly, $\Delta(D) \leq 1$. Intuitively, if $\Delta(D)$ is very close to 1, then H_0 probably explains the data almost as well as H_A, and should be accepted; on the other hand, if $\Delta(D)$ is much less than 1, then H_0 probably should be rejected in favor of H_A. However, we do not know what "$\Delta(D)$ is very close to 1" and "$\Delta(D)$ is much less than 1" exactly mean. It is in fact impossible to make a decision to either accept or reject the null hypothesis on the basis of $\Delta(D)$ alone. What we need to know is how close the value $\Delta(D)$ is to the "typical" values of the likelihood ratio calculated for datasets that genuinely obey H_0.

To determine such typical values we need to produce a large number of artificial datasets D_j that obey H_0. This can be done by *data simulation* under the null hypothesis. To simulate data at a single nucleotide site, we choose a rooted (not necessarily molecular clock) tree $\mathcal{T}$, assign one of the four nucleotide states to the root in accordance with the stationary probability distribution of model M_1 and let the site evolve down the tree to the tips according to M_1. Doing so for many nucleotide sites produces a simulated dataset that obeys H_0. This data simulation process is implemented in the software package Seq-Gen [RG].

Of course, before we can attempt such data simulation, we need to select the tree $\mathcal{T}$ as well as values of the parameters of model M_1 for each branch of T (note that the evolutionary rates are always set to 1). We obtain T by arbitrarily placing a root on a maximum likelihood tree found for D under the null hypothesis, and set the parameters of model M_1 to be equal to their optimal values found under the null hypothesis. With T and model parameters so selected we can simulate a number of datasets D_j, $j = 1, \ldots, L$, that obey H_0. These datasets can be assessed under each of H_0, H_A, and the values $\Delta(D_j)$ can be calculated for them (here we assume that $L_A^{\max}(D_j) \neq 0$ for all j). If L is large, we hope that $\Delta(D_j)$, $j = 1, \ldots, L$, give us a good idea about what a "typical" value of the likelihood ratio calculated for a dataset that obeys H_0 is.

On the basis of the values $\Delta(D_j)$, $j = 1, \ldots, L$, we will now make a decision whether or not H_0 should be accepted. We find the *significance point of the test*, which is the largest number K satisfying the condition that the ratio of the number of elements in the set $\{j : \Delta(D_j) \leq K\}$ and L does not exceed 0.05. Then H_0 is rejected if $\Delta(D) \leq K$. Thus, loosely speaking, H_0 is accepted if $\Delta(D)$ lies within the highest 95% of the values $\Delta(D_j)$, $j = 1, \ldots, L$, and rejected otherwise.

The hypothesis testing procedure outlined above is called the *parametric bootstrap method* and is widely used not just in phylogenetic reconstruction, but in statistics in general.

Exercises

5.1. Prove the formulas from Sect. 5.1 for the numbers of topologies of rooted and unrooted trees relating $N \geq 2$ OTUs.

5.2. Let $\mathcal{Q}$ be the three letter-alphabet $\{A, B, C\}$ and suppose that we are given the following reduced multiple alignment of three sequences

$$x^1 : A \ B \ C$$
$$x^2 : A \ A \ C$$
$$x^3 : B \ C \ C.$$

Find all the most parsimonious rooted topologies for x^1, x^2, x^3 together with all optimal sequence assignments.

5.3. Let T be a phylogenetic tree relating OTUs from a family $\mathcal{M}$, where branch lengths are allowed to be any non-negative numbers. Show that the tree-generated function d^T is a distance function on $\mathcal{M}$, if condition (i) from Definition 5.2 is satisfied.

5.4. Suppose that d is an additive distance function on a set of OTUs. Show by induction that there exists precisely one tree T relating the OTUs and having non-negative branch lengths, that generates d.

5.5. Show that any additive distance function on a set $\mathcal{M}$ of N OTUs with $N \geq 4$ satisfies the four-point condition.

5.6. Consider the following distance matrix

$$
\begin{array}{c|ccccccc}
M_d & x^1 & x^2 & x^3 & x^4 & x^5 & x^6 & x^7 \\
\hline
x^1 & 0 & 13 & 9 & 6 & 13 & 13 & 17 \\
x^2 & 13 & 0 & 4 & 15 & 16 & 16 & 20 \\
x^3 & 9 & 4 & 0 & 11 & 12 & 12 & 16 \\
x^4 & 6 & 15 & 11 & 0 & 15 & 15 & 19 \\
x^5 & 13 & 16 & 12 & 15 & 0 & 16 & 6 \\
x^6 & 13 & 16 & 12 & 15 & 16 & 0 & 20 \\
x^7 & 17 & 20 & 16 & 19 & 6 & 20 & 0
\end{array}
$$

Show that d is indeed a distance function and that it satisfies the four-point condition. Applying the neighbor-joining algorithm, find the tree that generates d.

5.7. Consider the following pseudodistance matrix

$$
\begin{array}{c|cccc}
M_d & x^1 & x^2 & x^3 & x^4 \\
\hline
x^1 & 0 & 3 & 2 & 7 \\
x^2 & 3 & 0 & 3 & 4 \\
x^3 & 2 & 3 & 0 & 3 \\
x^4 & 7 & 4 & 3 & 0.
\end{array}
$$

Show that d does not satisfy either the triangle inequality or the four-point condition, apply the neighbor-joining algorithm to d and find all trees that it can produce.

5.8. Let d be an ultrameric distance function on a set $\mathcal{M}$ of OTUs. Prove that d satisfies the four-point condition.

5.9. Prove that the distance function from Example 5.7 is ultrameric.

5.10. Show that if $\mathcal{T}$ is a molecular clock tree, then $d^{\mathcal{T}}$ is an ultrameric distance function, provided $d^{\mathcal{T}}$ satisfies condition (i) of Definition 5.2.

5.11. Consider the distance matrix

$$
\begin{array}{c|ccccccc}
M_d & x^1 & x^2 & x^3 & x^4 & x^5 & x^6 & x^7 \\
\hline
x^1 & 0 & 24 & 24 & 8 & 24 & 8 & 8 \\
x^2 & 24 & 0 & 6 & 24 & 2 & 24 & 24 \\
x^3 & 24 & 6 & 0 & 24 & 6 & 24 & 24 \\
x^4 & 8 & 24 & 24 & 0 & 24 & 6 & 4 \\
x^5 & 24 & 2 & 6 & 24 & 0 & 24 & 24 \\
x^6 & 8 & 24 & 24 & 6 & 24 & 0 & 6 \\
x^7 & 8 & 24 & 24 & 4 & 24 & 6 & 0
\end{array}
$$

Prove that d is indeed a distance function and that it is ultrameric. Find the corresponding molecular clock tree by applying the UPGMA algorithm.

5.12. Show that for the distance matrix from Example 5.9, d is indeed a distance function, that it satisfies the four-point condition, and that it is not ultrameric.

5.13. Consider the following distance matrix

$$
\begin{array}{c|cccccc}
M_d & x^1 & x^2 & x^3 & x^4 & x^5 & x^6 \\
\hline
x^1 & 0 & 7 & 3 & 8 & 8 & 11 \\
x^2 & 7 & 0 & 8 & 13 & 13 & 16 \\
x^3 & 3 & 8 & 0 & 7 & 7 & 10 \\
x^4 & 8 & 13 & 7 & 0 & 10 & 13 \\
x^5 & 8 & 13 & 7 & 10 & 0 & 11 \\
x^6 & 11 & 16 & 10 & 13 & 11 & 0 \\
\end{array}
$$

Prove that d is indeed a distance function, that it satisfies the four-point condition, and that it is not ultrameric. Apply the neighbor-joining algorithm and the UPGMA algorithm to d. How different are the two resulting trees?

5.14. Give an example of a pseudodistance function d on a set of four OTUs with the property that none of the trees produced from d by UPGMA is optimal in the sense of the least squares method.

5.15. Prove that in Example 5.10, the function ψ is minimized by the tree T_2 from Fig. 5.16 found in Example 5.6, that a point of minimum is unique and that the minimal value of ψ is $4(3/2 - a/4)^2$.

5.16. Show that each of the m^2 scalar series in the right-hand side of formula (5.6) converges for any matrix A.

5.17. Let $f(t)$ be a function defined on $[0, \infty)$, differentiable at any $t > 0$ and having a one-sided derivative at 0. Suppose that for all $t \geq 0$ f satisfies the differential equation

$$f'(t) = \alpha f(t),$$

where $\alpha \in \mathbb{R}$, with the initial condition $f(0) = 1$. Prove directly (without resorting to the general theory of differential equations) that $f(t) = \exp(\alpha t)$.

5.18. Prove that the elements in each row of the matrix of instantaneous change of a regular Markov chain sum up to 0.

5.19. Consider a discrete-time Markov chain with two states and the following matrix of transition probabilities

$$P = \begin{pmatrix} 1 - a & a \\ b & 1 - b \end{pmatrix},$$

where $0 \leq a, b \leq 1$ and $a + b > 0$. Show that this Markov chain has a unique stationary probability distribution and find it.

5.20. For the Markov chain from Exercise 5.19 find the limit of P^n, as $n \to \infty$, directly.

5.21. Show that $(1/4, 1/4, 1/4, 1/4)$ is the only stationary probability distribution of the Jukes-Cantor model and that the model is time-reversible.

5.22. Prove formulas (5.11).

5.23. Show that $(1/4, 1/4, 1/4, 1/4)$ is the only stationary probability distribution of the Kimura model and that the model is reversible.

5.24. Show that $(\pi_A, \pi_C, \pi_G, \pi_T)$ is the only stationary probability distribution of the Felsenstein model and that the model is reversible.

5.25. Show that $(\pi_A, \pi_C, \pi_G, \pi_T)$ is the only stationary probability distribution of the HKY model and that the model is reversible.

5.26. Generalize the result of Example 5.16 to the case of the Kimura model.

5.27. Prove the pulley principle for an arbitrary regular time-reversible evolutionary model, assuming that each model parameter (apart from α that is set to 1 everywhere) is branch-specific subject to the constraint that it has equal values on the two branches descending from the root.

5.28. Let M_1 and M_2 be the Jukes-Cantor and Kimura models respectively, and for a dataset D consider the corresponding null and alternative hypotheses. Let T be a maximum likelihood tree for D found using model M_1, and suppose that we have generated 60 artificial datasets D_j obeying H_0 by simulating data along the tree T. Let the likelihood ratios $\Delta(D_j)$ for $j = 1, \ldots, 60$ be as follows

$$\Big\{0.2, 0.007, 0.18, 0.15, 0.014, 0.35, 0.19, 0.17, 0.3, 0.013, 0.75, 0.2, 0.132, 0.08, 0.6,$$
$$0.6, 0.84, 0.013, 0.15, 0.6, 0.58, 0.63, 0.19, 0.2, 0.082, 0.12, 0.1, 0.11, 0.5, 0.9, 0.2,$$
$$0.4, 0.16, 0.165, 0.188, 0.154, 0.121, 0.132, 0.3, 0.14, 0.87, 0.9, 0.5, 0.15, 0.19, 0.19,$$
$$0.35, 0.014, 0.14, 0.178, 0.189, 0.19, 0.12, 0.14, 0.17, 0.567, 0.145, 0.17, 0.32, 0.2\Big\},$$

and suppose that $\Delta(D) = 0.0131$. Will the null hypothesis be accepted or rejected in favor of the alternative one by the parametric bootstrap method used with the above values $\Delta(D_j)$?

Mathematical Background for Sequence
Analysis

Mathematical Background for Sequence
Analysis

6
Elements of Probability Theory

In this chapter we will give a brief introduction to the theory of probability. Our exposition gives main constructions and illustrates them by many examples, but largely avoids detailed proofs. The interested reader can find all proofs in [KF], [W].

6.1 Sample Spaces and Events

We denote by S the set of all possible outcomes of a trial, experiment or operation that we are interested in and call S the *sample space*.

Example 6.1.

1. Tossing a coin once produces two possible outcomes: a head H or a tail T. Hence the sample space for this experiment is $S = \{H, T\}$.

2. Analogously, tossing a coin twice gives the following sample space $S = \{HH, HT, TH, TT\}$.

3. Suppose we are performing the experiment of shooting a bullet into the segment $[0, 1]$. In this case the outcome is the point of $[0, 1]$ hit by the bullet. Hence $S = [0, 1]$ and thus is an infinite set.

4. Suppose we are given a Markov chain without an end state. If we let the model run freely, it will generate all possible sequences of infinite length. Any sequence has the form $0\,x_1 x_2 \ldots$, where 0 denotes the begin state and all x_j belong to the set X of all non-zero states of the chain. The sample space S consists of all such sequences. Similarly, the sample space associated with a general continuous-time Markov chain is a collection of all sequences of the form $0\,x_{t_1}, x_{t_2}, \ldots$, where $\{t_j\}$ is a strictly increasing sequence of positive numbers and $x_{t_j} \in X$ for all j.

5. Suppose we are given a Markov chain with an end state (both the begin and end states are denoted by 0 below). Suppose that the model is non-trivially connected (see Sect. 3.1). If we let the model run freely, it will generate all possible sequences of either finite or infinite length. Any finite sequence has the form $0\,x_1x_2\ldots x_L\,0$ and any infinite sequence has the form $0\,x_1x_2\ldots$, where the x_js belong to the set X of all non-zero states of the chain (we will often omit the zeroes at the beginning and at the end). The sample space S consists of all such sequences.

6. Suppose we are given an HMM whose underlying Markov chain has an end state and is non-trivially connected as in Part 5 above. Suppose that the states of the HMM (apart from the begin and end states) emit symbols from an alphabet Q. There are two natural sample spaces that one can associate with the HMM.

   a. If we let the HMM run freely, it will generate all possible pairs of sequences (x, π), where x is a sequence of letters from Q and π is a sequence of elements from X called the *path for* x (as before, we will call sequences of elements from X *paths through the Markov chain* or simply *paths*). The lengths of x and π are equal and can be either finite or infinite. We will write x and π respectively as $x_1x_2\ldots x_L$ and $\pi = \pi_1\pi_2\ldots\pi_L$ in the finite case and $x_1x_2\ldots$ and $\pi_1\pi_2\ldots$ in the infinite case, where x_j is the element of Q emitted at the state π_j, for $j = 1, \ldots, L$. We denote the sample space that consists of all such pairs (x, π) by $S^{\mathbf{a}}$.

   b. Sometimes it will be useful for us to ignore paths and consider an HMM as a process that generates only sequences x of letters from Q. As above, such sequences can be either finite or infinite. We denote the sample space that consists of such sequences by $S^{\mathbf{b}}$.

7. Suppose that we are given a rooted phylogenetic tree with N labeled leaves. Fix a regular evolutionary model with particular branch-specific parameter values, where the values corresponding to the two branches descending from the root are equal, as stated in condition (vii) from Sect. 5.5, and the evolutionary rate is set to 1 on each branch. If we let the process of data simulation run freely starting at the root with the stationary probability distribution of the model, it will generate all possible nucleotide values at each OTU. Hence in this case the sample space S is the collection of all such N-tuples of nucleotides (corresponding to columns in ungapped multiple alignments of N sequences).

Subsets of a sample space S are called *events* and points from S are sometimes called *elementary events*. For example, in Part 2 of Example 6.1,

$E = \{HH, TH\}$ is an event that consists of two elementary events. We say that an event E *occurs* at a particular time, if the outcome of the experiment in question performed at that time is contained in E. We therefore say that the event S *always occurs* and the *empty event* $\emptyset$ *never occurs*.

We will use the ordinary set theory for dealing with events. If e is an element of an event E, we will write $e \in E$. Further, if for every $e \in E_1$ we have $e \in E_2$, we write $E_1 \subset E_2$ and say that *the occurrence of E_1 implies the occurrence of E_2* or that E_1 *is a subset of E_2*. Two events E_1 and E_2 are equal if $E_1 \subset E_2$ and $E_2 \subset E_1$. The empty event $\emptyset$ is a subset of any other event.

The *intersection* or *product* of two events E_1 and E_2 is the collection of all elementary events e such that $e \in E_1$ *and* $e \in E_2$, and is denoted by $E_1 \cap E_2$. If the event $E_1 \cap E_2$ occurs, we say that that we have an instance of *joint occurrence of E_1 and E_2*. In a similar way one can define the intersection of any finite number of events $\cap_{j=1}^n E_j$, of a countable infinite number of events $\cap_{j=1}^\infty E_j$ and in fact of any number of events $\cap_{\alpha \in A} E_\alpha$, where A is an index set.

The *union* or *sum* of two events E_1 and E_2 is the collection of all elementary events e such that $e \in E_1$ *or* $e \in E_2$, and is denoted by $E_1 \cup E_2$. If the event $E_1 \cup E_2$ occurs, we say that that we have an instance of *occurrence of at least one of E_1 and E_2*. As above, one can define the union of any finite number of events $\cup_{j=1}^n E_j$, of a countable infinite number of events $\cup_{j=1}^\infty E_j$ and of any number of events $\cup_{\alpha \in A} E_\alpha$.

The *difference* between events E_1 and E_2 is the collection of all elementary events e such that $e \in E_1$, but $e \notin E_2$ and is denoted by $E_1 \setminus E_2$. It translates into the *occurrence of E_1, but not E_2*. Note that the difference is not a symmetric operation: $E_1 \setminus E_2$ is generally different from $E_2 \setminus E_1$. The special case when $E_1 = S$ corresponds to taking the *complement* of an event and is denoted by $E^c = S \setminus E$.

Here are some useful rules that involve the operations with events defined above

$$(E_1 \cup E_2) \cup E_3 = E_1 \cup (E_2 \cup E_3),$$
$$(E_1 \cap E_2) \cap E_3 = E_1 \cap (E_2 \cap E_3),$$
$$E_1 \cap (E_2 \cup E_3) = (E_1 \cap E_2) \cup (E_1 \cap E_3),$$
$$E_1 \cup E_2 = E_1 \cup (E_2 \setminus (E_1 \cap E_2)),$$
$$E_1 \cap E_2 = (E_1^c \cup E_2^c)^c,$$
$$E_1 \cup E_2 = (E_1^c \cap E_2^c)^c.$$

The more general forms of the last two identities are

$$\bigcap_{\alpha \in A} E_\alpha = \left(\bigcup_{\alpha \in A} E_\alpha^c \right)^c, \tag{6.1}$$

$$\bigcup_{\alpha \in A} E_\alpha = \left(\bigcap_{\alpha \in A} E_\alpha^c \right)^c. \tag{6.2}$$

One more operation is the *symmetric difference*: $E_1 \triangle E_2 = (E_1 \backslash E_2) \cup (E_2 \backslash E_1)$. It translates into saying that *the event E_1 occurs, but E_2 does not, or the event E_2 occurs, but E_1 does not.*

Apart from individual events we will be also interested in specific families of events.

Definition 6.2. *A family $\mathcal{F}$ of events is called an algebra of events if the following holds*

(i) $S \in \mathcal{F}$,

(ii) for every $E \in \mathcal{F}$ we have $E^c \in \mathcal{F}$,

(iii) for all $E_1, E_2 \in \mathcal{F}$ we have $E_1 \cup E_2 \in \mathcal{F}$.

We note that (i) and (ii) imply that $\emptyset \in \mathcal{F}$ and that it follows from (ii), (iii) and identity (6.1) that for all $E_1, E_2 \in \mathcal{F}$ we have $E_1 \cap E_2 \in \mathcal{F}$. Of course, the sum and product of any *finite* number of events from $\mathcal{F}$ is contained in $\mathcal{F}$. We will be interested in the following special class of algebras that are closed with respect to taking any general *countable* sums and products.

Definition 6.3. *A family $\mathcal{F}$ of events is called a σ-algebra of events if the following holds*

(i) $S \in \mathcal{F}$,

(ii) for every $E \in \mathcal{F}$ we have $E^c \in \mathcal{F}$,

(iii) for any sequence of events $\{E_j\}$ with $E_j \in \mathcal{F}$ we have $\cup_{j=1}^{\infty} E_j \in \mathcal{F}$.

As before, (i) and (ii) imply that $\emptyset \in \mathcal{F}$ and it follows from (ii), (iii) and (6.1) that for or any sequence of events $\{E_j\}$ with $E_j \in \mathcal{F}$ we have $\cap_{j=1}^{\infty} E_j \in \mathcal{F}$. We also note that every σ-algebra is an algebra; this is proved by setting in (iii) of Definition 6.3 all but finitely many events to be equal to $\emptyset$. If an algebra contains only finitely many events, then it is clearly a σ-algebra as well. However, in general an algebra may not be a σ-algebra as shown in Part 3 of Example 6.4 below.

We will now give examples of algebras and σ-algebras of events.

Example 6.4.

1. For any S the family $\mathcal{F} = \{S, \emptyset\}$ is an algebra. Since it contains only finitely many events, it is also a σ-algebra.

2. For any S let $\mathcal{F}$ be the family of *all* events in S. Clearly, $\mathcal{F}$ is a σ-algebra.

3. Let $S = [0,1]$ and $\mathcal{F}$ be the family that consists of finite unions of intervals in S, where an interval is either $[a,b]$, or $[a,b)$, or $(a,b]$, or (a,b) for some $0 \leq a \leq b \leq 1$. First, we will show that $\mathcal{F}$ is an algebra by verifying that it satisfies the conditions of Definition 6.2. Indeed, (i) holds since S itself is an interval. Next, (ii) holds since the complement of a finite union of intervals is again a finite union of intervals. Finally, the sum of two finite unions of intervals is a finite union of intervals, and thus (iii) holds as well. Hence $\mathcal{F}$ is an algebra. It is also clear that $\mathcal{F}$ is not a σ-algebra. For example, the union of infinitely many disjoint intervals in S does not belong to $\mathcal{F}$.

For future considerations we will need the following theorem.

Theorem 6.5.

(i) The intersection of any number of algebras is an algebra.

(ii) The intersection of any number of σ-algebras is a σ-algebra.

(iii) If $\mathcal{F}$ is a non-empty family of events in S, then there exists a unique algebra $\mathcal{A}(\mathcal{F})$ that contains $\mathcal{F}$ and that is contained in every algebra containing $\mathcal{F}$.

(iv) If $\mathcal{F}$ is a non-empty family of events in S, then there exists a unique σ-algebra $\mathcal{B}(\mathcal{F})$ that contains $\mathcal{F}$ and that is contained in every σ-algebra containing $\mathcal{F}$.

(v) $\mathcal{A}(\mathcal{F}) \subset \mathcal{B}(\mathcal{F})$.

Proof: Parts (i) and (ii) are obvious and follow directly from Definitions 6.2 and 6.3.

We will now prove (iii). Let $\mathcal{M}$ be the algebra of all events in S (see Part 2 of Example 6.4) and let Σ be the collection of all algebras that contain $\mathcal{F}$. Σ is non-empty since $\mathcal{M} \in \Sigma$. Denote by $\mathcal{A}(\mathcal{F})$ the intersection of all the algebras in Σ. $\mathcal{A}(\mathcal{F})$ is an algebra by (i) and clearly contains $\mathcal{F}$. Suppose now that $\mathcal{R}$ is another algebra containing $\mathcal{F}$. Hence $\mathcal{R} \in \Sigma$ and therefore $\mathcal{A}(\mathcal{F}) \subset \mathcal{R}$. Thus $\mathcal{A}(\mathcal{F})$ satisfies the requirements of (iii). It is also clear that such an algebra is unique, and (iii) is proved. The proof of Part (iv) is analogous to the above.

Finally, (v) holds since $\mathcal{B}(\mathcal{F})$ is an algebra that contains $\mathcal{F}$. ∎

Definition 6.6. *The algebra $\mathcal{A}(\mathcal{F})$ and σ-algebra $\mathcal{B}(\mathcal{F})$ are called the algebra generated by $\mathcal{F}$ and σ-algebra generated by $\mathcal{F}$ respectively.*

We will now introduce another useful type of families of events.

Definition 6.7. *A family $\mathcal{F}$ of events is called a semi-algebra of events if the following holds*

(i) $S \in \mathcal{F}$,

(ii) $\emptyset \in \mathcal{F}$,

(iii) for all $E_1, E_2 \in \mathcal{F}$ we have $E_1 \cap E_2 \in \mathcal{F}$,

(iv) for all $E, E' \in \mathcal{F}$ such that $E' \subset E$, there exist finitely many pairwise disjoint events $E_1, \ldots, E_n \in \mathcal{F}$ such that $E = \cup_{j=1}^n E_j$ and $E_1 = E'$.

One can show by induction that (iv) of Definition 6.7 can be replaced by a stronger requirement: let events $E_1', \ldots, E_k' \in \mathcal{F}$ be pairwise disjoint and all contained in $E \in \mathcal{F}$, then there exist finitely many pairwise disjoint events $E_1, \ldots, E_n \in \mathcal{F}$ with $n \geq k$ such that $E = \cup_{j=1}^n E_j$ and $E_j = E_j'$ for $j = 1, \ldots, k$.

Any algebra of events is an example of a semi-algebra. Indeed, we only need to prove property (iv) in Definition 6.7. If $\mathcal{F}$ is an algebra we set $E_1 = E'$ and $E_2 = E \setminus E'$. The event E_2 belongs to $\mathcal{F}$ since $E_2 = E \cap E'^c$, and algebras are closed with respect to taking complements and finite intersections. We will now give a less trivial example of a semi-algebra.

Example 6.8. Let $S = [0,1]$ and $\mathcal{F}$ be the collection of all intervals in S, that is, $\mathcal{F} = \{[a,b], [a,b), (a,b], (a,b), 0 \leq a \leq b \leq 1\}$. To show that $\mathcal{F}$ is a semi-algebra we have to verify (i)-(iv) of Definition 6.7. Clearly, $S \in \mathcal{F}$ since S is an interval, and $\emptyset \in \mathcal{F}$ since $\emptyset$ is the interval (a,a), so (i) and (ii) hold. Further, the intersection of two intervals is an interval, so (iii) holds as well. To prove (iv) we notice that the difference between an interval and a subinterval is either one or two intervals. Thus $\mathcal{F}$ is a semi-algebra. It is also clear that $\mathcal{F}$ is not an algebra since the union of two disjoint intervals is not an interval.

If $\mathcal{F}$ is a semi-algebra, then the algebra $\mathcal{A}(\mathcal{F})$ introduced in Theorem 6.5 is easy to find, as stated in the following theorem. This fact will be of some importance for us later on.

Theorem 6.9. *If $\mathcal{F}$ is a semi-algebra of events then $\mathcal{A}(\mathcal{F})$ coincides with the collection $\mathcal{Z}(\mathcal{F})$ of all finite unions of pairwise disjoint events from $\mathcal{F}$.*

Proof: First, we will show that $\mathcal{Z}(\mathcal{F})$ is an algebra of events. Since $\mathcal{F}$ contains S, so does $\mathcal{Z}(\mathcal{F})$, and (i) of Definition 6.2 holds.

To prove (ii) of Definition 6.2, consider $E \in \mathcal{Z}(\mathcal{F})$, $E = \cup_{j=1}^n E_j$, where $E_j \in \mathcal{F}$ are pairwise disjoint. Then there exist events $D_1, \ldots, D_k \in \mathcal{F}$ such that

$$S = \left(\bigcup_{j=1}^n E_j \right) \cup \left(\bigcup_{i=1}^k D_i \right),$$

and all the events $E_1, \ldots, E_n, D_1, \ldots, D_k$ are pairwise disjoint. Hence $E^c = \cup_{i=1}^{k} D_i \in \mathcal{Z}(\mathcal{F})$, and (ii) of Definition 6.2 holds.

To show (iii) consider two events $E, D \in \mathcal{Z}(\mathcal{F})$, $E = \cup_{j=1}^{n} E_j$, $D = \cup_{i=1}^{k} D_i$, where $E_1, \ldots, E_n$ are pairwise disjoint and $D_1, \ldots, D_k$ are pairwise disjoint. Let $C_{ji} = E_j \cap D_i$ for all j, i. Then for every j there exist events $Q_{1j}, \ldots, Q_{p_j j} \in \mathcal{F}$ such that

$$E_j = \left(\bigcup_{i=1}^{k} C_{ji} \right) \cup \left(\bigcup_{q=1}^{p_j} Q_{qj} \right),$$

and the events $C_{j1}, \ldots, C_{jk}, Q_{1j}, \ldots, Q_{p_j j}$ are pairwise disjoint. Similarly, for every i there exist events $R_{1i}, \ldots, R_{s_i i} \in \mathcal{F}$ such that

$$D_i = \left(\bigcup_{j=1}^{k} C_{ji} \right) \cup \left(\bigcup_{r=1}^{s_i} R_{ri} \right),$$

and the events $C_{1i}, \ldots, C_{ni}, R_{1i}, \ldots, R_{s_i i}$ are pairwise disjoint. We note that each Q_{qj} is disjoint from $Q_{q'j'}$ for $j' \neq j$, each R_{ri} is disjoint from $R_{r'i'}$ for $i' \neq i$ and each Q_{qj} is disjoint from each R_{ri}. Hence

$$E \cup D = \left(\bigcup_{j=1}^{n} \bigcup_{i=1}^{k} C_{ji} \right) \cup \left(\bigcup_{j=1}^{n} \bigcup_{q=1}^{p_j} Q_{qj} \right) \cup \left(\bigcup_{i=1}^{k} \bigcup_{r=1}^{s_i} R_{ri} \right),$$

and all the sets in the right-hand side are pairwise disjoint. Thus $E \cup D \in \mathcal{Z}(\mathcal{F})$, and we have shown that $\mathcal{Z}(\mathcal{F})$ is an algebra of events. Clearly, $\mathcal{Z}(\mathcal{F})$ is contained in any algebra that contains $\mathcal{F}$ and therefore $\mathcal{A}(\mathcal{F}) = \mathcal{Z}(\mathcal{F})$. ∎

For a semi-algebra $\mathcal{F}$ the minimal σ-algebra $\mathcal{B}(\mathcal{F})$ is harder to describe than the minimal algebra $\mathcal{A}(\mathcal{F})$, and we do not do it here. We note that for $\mathcal{F}$ from Example 6.8, $\mathcal{B}(\mathcal{F})$ is the σ-algebra of *Borel sets in* $[0, 1]$ that play an important role in analysis. Borel sets can be much more complicated than sets from $\mathcal{A}(\mathcal{F})$. For example, all open sets and all closed sets in $[0, 1]$ are Borel. One famous example of a Borel set is the so-called *Cantor set* which is the countable intersection of sets $E_n \subset [0, 1]$, $n = \mathbb{N}$, where E_n is obtained from $[0, 1]$ by removing 3^{n-1} intervals of the form

$$\left(\frac{3k - 2}{3^n}, \frac{3k - 1}{3^n} \right), \qquad k = 1, \ldots, 3^{n-1}.$$

6.2 Probability Measure

We will now start assigning probabilities to events. The common notion of the probability of an event E is an abstraction of the idea of the relative frequency with which the event occurs in a sequence of trials of an experiment, that is

m_E/m, where m is the total number of trials and m_E is the number of trials for which the outcomes belong to E. In dealing with any events of interest, numbers between 0 and 1 can be assigned as the probabilities of events in some initial class of relatively simple events, and those will produce the probabilities of more complex events. In the actual assignment of probabilities to events in the initial class, one is usually guided by a hypothesis based on what one expects the relative frequencies of events to be in a large series of trials. The mathematical theory begins once an assignment of probabilities to events in the initial class has been made. In order to build a good theory, the initial class of events is always assumed to be a semi-algebra, and the probabilities of events in the initial class are assumed to satisfy certain conditions as indicated in the following definition.

Definition 6.10. *Let $\mathcal{F}$ be a semi-algebra of events in a sample space S. A probability measure or simply probability on $\mathcal{F}$ is a set function P defined for all events in $\mathcal{F}$ and having the following three properties*

(i) for every $E \in \mathcal{F}$ we have $P(E) \geq 0$,

(ii) $P(S) = 1$,

(iii) for any sequence $\{E_j\}$ of pairwise disjoint events in $\mathcal{F}$ such that $\cup_{j=1}^{\infty} E_j \in \mathcal{F}$ we have

$$P\left(\bigcup_{j=1}^{\infty} E_j \right) = \sum_{j=1}^{\infty} P(E_j).$$

Note that if $\mathcal{F}$ is in fact a σ-algebra, then the condition $\cup_{j=1}^{\infty} E_j \in \mathcal{F}$ holds automatically. Property (iii) in Definition 6.10 is called the σ-*additivity property* of the probability measure. It implies that $P(\emptyset) = 0$. Indeed, let $P(\emptyset) = c \geq 0$ and set $E_j = \emptyset$ for all j. Then $\cup_{j=1}^{\infty} E_j = \emptyset \in \mathcal{F}$ and the σ-additivity property gives

$$\sum_{j=1}^{\infty} c = c,$$

which is only possible if $c = 0$.

The σ-additivity property is stronger than the *finite additivity property* which is obtained from (iii) by replacing the sequence of events by a finite collection of pairwise disjoint events. Indeed, setting in the sequence $\{E_j\}$ all but finitely many events to be $\emptyset$ and taking into account that $P(\emptyset) = 0$, we obtain the finite additivity property.

Example 6.11. Consider the experiments from Example 6.1.

1. It is natural to choose $\mathcal{F}$ to be the collection of all events in S: $\mathcal{F} = \{S, \emptyset, H, T\}$ and, if the coin is fair, set $P(S) = 1$, $P(\emptyset) = 0$, $P(H) = 1/2$,

$P(T) = 1/2.$

2. Let $\mathcal{F} = \{S, \emptyset, HH, HT, TH, TT\}$. Clearly, $\mathcal{F}$ is a semi-algebra. Assuming again that the coin is fair, we set $P(S) = 1$, $P(\emptyset) = 0$, $P(\{HH\}) = 1/4$, $P(\{HT\}) = 1/4$, $P(\{TH\}) = 1/4$, $P(\{TT\}) = 1/4$.

3. Let $\mathcal{F}$ be as in Example 6.8. It is not hard to check that setting

$$P([a,b]) = P([a,b)) = P((a,b]) = P((a,b)) = b - a,$$

we obtain a probability measure on $\mathcal{F}$ (see Exercise 6.3).

4. We consider the following family $\mathcal{F}$ of events in S: $\mathcal{F}$ includes S, $\emptyset$ and all *cylinder events*, that is, events of the form

$$E_{x_{i_1}^0, \ldots, x_{i_m}^0} = \left\{ e = 0\, x_1 x_2 \ldots \in S : x_{i_1} = x_{i_1}^0, \ldots, x_{i_m} = x_{i_m}^0 \right\},$$

for all finite subsets of indices $i_1 < \ldots < i_m$ and all possible $x_{i_1}^0, \ldots, x_{i_m}^0 \in X$. It is easy to check that $\mathcal{F}$ is a semi-algebra. We define a probability measure on $\mathcal{F}$ as follows. We set $P(S) = 1$, $P(\emptyset) = 0$ and

$$P\left(E_{x_{i_1}^0, \ldots, x_{i_m}^0}\right) = \sum_{\substack{x_1, \ldots, x_{i_m} \in X: \\ x_{i_1} = x_{i_1}^0, \ldots, x_{i_m} = x_{i_m}^0}} \prod_{j=0}^{i_m - 1} p_{x_j\, x_{j+1}}, \qquad (6.3)$$

where $x_0 = 0$. It is not hard to check that P is a probability measure on $\mathcal{F}$ (see, in particular, Exercise 3.1).

The assignment of probabilities by formula (6.3) is natural in the following sense. Fix $L \in \mathbb{N}$, and for every sequence in S consider its first L elements (not counting the initial 0); we denote the collection of such truncated sequences by S_L. It can be shown that an arbitrary finite sequence $x = 0\, x_1 \ldots x_L$ is generated by the Markov chain among other elements of S_L with frequency that tends to $\prod_{j=0}^{L-1} p_{x_j\, x_{j+1}}$, as the number of model runs increases (see (3.1)). Setting $L = i_m$ and $E_{x_{i_1}^0, \ldots, x_{i_m}^0}^{i_m}$ to be the collection of all sequences $0\, x_1 \ldots x_{i_m}$ for which $x_{i_1} = x_{i_1}^0, \ldots, x_{i_m} = x_{i_m}^0$, it is then natural to define

$$P\left(E_{x_{i_1}^0, \ldots, x_{i_m}^0}^{i_m}\right) = \sum_{\substack{x_1, \ldots, x_{i_m} \in X: \\ x_{i_1} = x_{i_1}^0, \ldots, x_{i_m} = x_{i_m}^0}} \prod_{j=0}^{i_m - 1} p_{x_j\, x_{j+1}}.$$

This formula gives a probability measure on the σ-algebra of all events in S_L and is analogous to formula (6.3).

Formula (6.3) clarifies the need for introducing an end state for modeling finite sequences (see Sect. 3.1). Indeed, formula (3.1) gives the probability of a whole cylinder event, not a single finite sequence as required.

We note that it is possible to give a similar construction of a probability measure in the case of continuous-time Markov chains where the right-hand side in (6.3) is replaced with $p_{0\,x_{t_1}} p_{x_{t_1}\,x_{t_2}}(t_2-t_1) \times \ldots \times p_{x_{t_{m-1}}\,x_{t_m}}(t_m-t_{m-1})$.

5. In this case we choose $\mathcal{F}$ to be the collection of all events in S, and define a probability measure P on $\mathcal{F}$ as follows. If e is a finite sequence $0\,x_1 x_2 \ldots x_L\,0$, set

$$P(\{e\}) = p_{0\,x_1} \times \prod_{j=1}^{L-1} p_{x_j\,x_{j+1}} \times p_{x_L\,0}, \qquad (6.4)$$

a formula familiar from (3.4). If $E \in \mathcal{F}$ is an arbitrary event, we define $P(E)$ as the sum of the probabilities of all finite sequences contained in E (note that the number of finite sequences is countable). In particular, if E contains only infinite sequences, $P(E) = 0$. To check that P is indeed a probability measure on $\mathcal{F}$, we only need to check that $P(S) = 1$, that is

$$\sum_{L=1}^{\infty} \sum_{x_1,\ldots,x_L \in X} p_{0\,x_1} \times \prod_{j=1}^{L-1} p_{x_j\,x_{j+1}} \times p_{x_L\,0} = 1. \qquad (6.5)$$

One can show that identity (6.5) holds for all non-trivially connected Markov chains (see Exercise 3.2). Here we will only prove identity (6.5) for a Markov chain with $p_{a\,0} = \tau \neq 0$ for all $a \in X$ (note that, in general, if $p_{a\,0} \neq 0$ for all $a \in X$, then the model is non-trivially connected). Indeed, in this case the left-hand side in formula (6.5) becomes $\sum_{L=1}^{\infty} \tau(1-\tau)^{L-1}$ which is clearly equal to 1.

Hence when in Sect. 3.1 we calculated the probability of a sequence $y = y_1 \ldots y_L$ derived from real data by using formula (6.4), we in fact calculated the probability of y in the sense of the sample space S, that is, as if y was the outcome of a model run. In Sect. 3.1 we called this probability "the probability with which y arises from the model".

The above definition of probability measure can be justified as follows. As in Part 4 above, for any $L \geq 2$ consider the collection S_L of truncated sequences. Clearly, S_L consists of all finite sequences of length not exceeding L generated by the model (note, however, that one cannot distinguish between a finite sequence of length L, a truncation of a finite sequence of length $> L$ and a truncation of an infinite sequence). As we noted in Part 4, it is possible to prove that any finite sequence $x_1 \ldots x_m$ with $m \leq L - 1$ is generated by the Markov chain among other elements of S_L with frequency that tends to $p_{0\,x_1} \times \prod_{j=1}^{m-1} p_{x_j\,x_{j+1}} \times p_{x_m\,0}$, as the number of model runs increases.

6. We will consider two cases.

a. We choose $\mathcal{F}$ to be the collection of all events in $S^{\mathbf{a}}$ and for a pair of finite sequences $e = (x_1 \ldots x_L, \pi_1 \ldots \pi_L)$ define

$$P^{\mathbf{a}}(\{e\}) = p_{0\,\pi_1} \times \prod_{j=1}^{L-1} q_{\pi_j}(x_j) p_{\pi_j\,\pi_{j+1}} \times q_{\pi_L}(x_L) \times p_{\pi_L\,0}. \qquad (6.6)$$

Formula (6.6) is familiar from (3.8). The probability of any event $E \in \mathcal{F}$ is then defined to be the sum of the probabilities of all pairs of finite sequences contained in E. Verification that $P^{\mathbf{a}}(S) = 1$ is done as in Part 5 above, assuming that the underlying Markov chain is non-trivially connected (see Exercise 3.5). This assignment of probabilities can be justified as in Part 5.

Hence when in Sect. 3.2 we calculated the probability of a pair of finite sequences (y, ν) derived from real data, by using formula (6.6), we in fact calculated the probability of (y, ν) in the sense of the sample space $S^{\mathbf{a}}$, that is, as if y was emitted along the path ν during a model run. In Sect. 3.2 we also called this probability "the probability with which y arises from the model along the path ν".

b. We choose $\mathcal{F}$ to be the collection of all events in $S^{\mathbf{b}}$ and for a finite sequence $e = x_1 \ldots x_L$, define

$$P^{\mathbf{b}}(\{e\}) = \sum_{\pi_1, \ldots, \pi_L \in X} p_{0\,\pi_1} \times \prod_{j=1}^{L-1} q_{\pi_j}(x_j) p_{\pi_j\,\pi_{j+1}} \times q_{\pi_L}(x_L) \times p_{\pi_L\,0}. \qquad (6.7)$$

Formula (6.7) is familiar from (3.9). The probability of any event $E \in \mathcal{F}$ is then defined to be the sum of the probabilities of all finite sequences contained in E. This assignment of probabilities can be justified as in Part 5 as well.

Hence when in Sect. 3.2 we calculated the probability of a finite sequence y derived from real data, by using formula (6.7), we in fact calculated the probability of y in the sense of the sample space $S^{\mathbf{b}}$, that is, as if y was the outcome of a model run. In Sect. 3.2 we also called this probability "the probability with which y arises from the model".

7. We again choose $\mathcal{F}$ to be the collection of all events in S. We treat every N-tuple of nucleotides $e \in S$ as a nucleotide site and define $P(\{e\})$ to be the likelihood of the site, as introduced in Sect. 5.5. This assignment of probabilities is reasonable since it can be shown that the frequency of every particular N-tuple of nucleotides tends to the value specified above as the number of simulation runs increases. The probability of any event $E \subset S$ is then the sum of the probabilities of the elementary events in E (note that S is finite). It is straightforward to prove that $P(S) = 1$, and hence P is a probability measure on $\mathcal{F}$.

We will be usually interested in extending a probability measure P from the initial class of events (which is some semi-algebra $\mathcal{F}$) to more complex events. It is easy to extend P from $\mathcal{F}$ to $\mathcal{A}(\mathcal{F})$. Indeed, by Theorem 6.9, every element $E \in \mathcal{A}(\mathcal{F})$ can be represented as a finite union of pairwise disjoint events from $\mathcal{F}$: $E = \cup_{j=1}^{n} E_j$, and we set $P(E) = \sum_{j=1}^{n} P(E_j)$. It is easy to show that this definition does not depend on the representation of E as a union of events in $\mathcal{F}$ and that the extended function P is a probability measure on $\mathcal{A}(\mathcal{F})$, that is, satisfies the conditions in Definition 6.10 on $\mathcal{A}(\mathcal{F})$.

However, the σ-additivity property of P allows it to be extended past $\mathcal{A}(\mathcal{F})$, at least as far as $\mathcal{B}(\mathcal{F})$. An extension of P to $\mathcal{B}(\mathcal{F})$ can be constructed as follows.

Definition 6.12. *For any event $E \subset S$ define its outer measure $P^*(E)$ as follows*

$$P^*(E) = \inf \sum_{j=1}^{\infty} P(E_j),$$

where the infimum is taken over all countable coverings of E by elements from $\mathcal{F}$: $E \subset \cup_{j=1} E_j$, $E_j \in \mathcal{F}$.

Clearly, $P^*(E) = P(E)$ for all $E \in \mathcal{F}$. Further, the following important statement holds.

Theorem 6.13. P^* *is a probability measure on $\mathcal{B}(\mathcal{F})$.*

Theorem 6.13 shows that the probability measure P defined on $\mathcal{F}$ can be extended to a probability measure defined on $\mathcal{B}(\mathcal{F})$. In the future we will drop the superscript * when considering P^* on $\mathcal{B}(\mathcal{F})$ and call the triple $(S, \mathcal{B}(\mathcal{F}), P)$ the *probability space arising from S, $\mathcal{F}$ and P*.

Example 6.14. Consider the probability measures defined in Example 6.11.

1,5,6,7. In these cases $\mathcal{A}(\mathcal{F}) = \mathcal{B}(\mathcal{F}) = \mathcal{F}$, hence the extension procedure for P is trivial.

2. In this case $\mathcal{A}(\mathcal{F}) = \mathcal{B}(\mathcal{F})$ is the collection of all events in S, hence the extension of P to $\mathcal{A}(\mathcal{F})$ is the only non-trivial one. For example, for $E = \{HT, TH\}$ we have $P(E) = 1/4 + 1/4 = 1/2$.

3. In this case we get an extension of the probability measure P to all Borel sets in $[0, 1]$. In particular, the probability measure of the Cantor set is defined as well (and in fact is equal to zero). We note here that one can construct certain Borel sets similar to the Cantor set for which the probability measure is non-zero. Not every subset of $[0, 1]$ is Borel, and we will show in Example 6.16 that P cannot be reasonably extended to all subsets of $[0, 1]$.

4. In this case we obtain an extension of the probability measure P to the σ-algebra $\mathcal{B}(\mathcal{F})$ that, as in Part 3 above, one can show to be strictly smaller than the σ-algebra of all events in S.

We remark here that the probability space from Part 4 of Example 6.14 can be taken as a definition of a Markov chain. Similarly, the probability spaces from Parts 5 and 6 can be taken respectively as definitions of a non-trivially connected Markov chain with an end state and an HMM for which the underlying Markov chain has an end state and is non-trivially connected.

While for the purposes of the probability theory it will be generally sufficient to consider the extension of the probability measure only to $\mathcal{B}(\mathcal{F})$, it is interesting to remark that it can in fact be extended beyond events in $\mathcal{B}(\mathcal{F})$, to include so-called *Lebesgue measurable events*.

Definition 6.15. *An event $E \subset S$ is called Lebesgue measurable if for any $\varepsilon > 0$ there exists $E_0 \in \mathcal{A}(\mathcal{F})$ such that $P^*(E \Delta E_0) < \varepsilon$. The collection of all Lebesgue measurable events is denoted by $\mathcal{L}(\mathcal{F})$.*

When considering P^* on $\mathcal{L}(\mathcal{F})$ we will drop the superscript $*$. One can show that $\mathcal{B}(\mathcal{F}) \subset \mathcal{L}(\mathcal{F})$, $\mathcal{L}(\mathcal{F})$ is a σ-algebra, and P is a probability measure on $\mathcal{L}(\mathcal{F})$ which is called the *Lebesgue extension* of the original probability measure. Of course, for the probability measures defined in Parts 1 and 2 of Example 6.11, $\mathcal{B}(\mathcal{F})$ and $\mathcal{L}(\mathcal{F})$ coincide, but often $\mathcal{L}(\mathcal{F})$ is much larger than $\mathcal{B}(\mathcal{F})$. For instance, in Part 3 of Example 6.11, $\mathcal{L}(\mathcal{F})$ is substantially larger than $\mathcal{B}(\mathcal{F})$. In particular, one can show that not every subset of the Cantor set belongs to $\mathcal{B}(\mathcal{F})$, but every such subset belongs to $\mathcal{L}(\mathcal{F})$.

Despite the fact that one can extend every probability measure to the large σ-algebra $\mathcal{L}(\mathcal{F})$, one cannot hope that this extension defines the probability of every event $E \subset S$, as the following example shows.

Example 6.16. Consider the probability measure defined in Part 3 of Example 6.11 and extend it to the corresponding σ-algebra $\mathcal{L}(\mathcal{F})$. Define on $S = [0,1]$ the equivalence relation: $x \sim y$, if $x - y \in \mathbb{Q}$. Let $E \subset (0,1]$ be the event that contains one element from each equivalence class. For $r \in (0,1]$ define

$$E_r = \Big((r + E) \cup (r - 1 + E) \Big) \cap (0,1]. \tag{6.8}$$

We have

$$(0,1] = \bigcup_{r \in \mathbb{Q} \cap (0,1]} E_r,$$

and the events E_r are pairwise disjoint for $r \in \mathbb{Q} \cap (0,1]$.

If we suppose that $E \in \mathcal{L}(\mathcal{F})$, then $E_r \in \mathcal{L}(\mathcal{F})$ for all $r \in \mathbb{Q} \cap (0,1]$, and $P(E_r) = P(E)$. But then the σ-additivity of P on $\mathcal{L}(\mathcal{F})$ gives

$$1 = \sum_{r \in \mathbb{Q} \cap (0,1]} P(E),$$

which is impossible. Hence $E \notin \mathcal{L}(\mathcal{F})$, that is, E is a *non-measurable event*; one cannot set the value $P(E)$ in a reasonable way.

Example 6.16 shows that it is impossible to define a non-negative set function P for all events in $S = [0, 1]$ in such a way that: (i) for the translates E_r of $E \subset [0, 1]$ defined in (6.8) $P(E_r) = P(E)$; (ii) P is σ-additive.

We remark that the Lebesgue extension of the probability measure defined in Part 4 of Example 6.11 is also defined on a σ-algebra that generally does not include all possible events in S. In fact, if $p_{0j} \neq 0$ and $p_{ij} \neq 0$ for all $i, j \in X$, one can establish a natural almost one-to-one correspondence between the elements of the σ-algebras $\mathcal{L}(\mathcal{F})$ arising from Parts 3 and 4 of Example 6.11. Thus, in this situation Example 6.16 can be used to produce a Lebesgue non-measurable event for the probability measure defined in Part 4 of Example 6.11 as well.

Originally we assumed that the probability measure P is defined only on a semi-algebra $\mathcal{F}$. As we have seen, P can always be extended to the minimal σ-algebra $\mathcal{B}(\mathcal{F})$. For the purposes of the probability theory it is usually not necessary to extend P beyond $\mathcal{B}(\mathcal{F})$, and therefore we will assume from now on that any probability measure that we will consider is defined on a σ-algebra and ignore any possible extensions of the probability measure beyond the σ-algebra. Thus, we will be dealing with general *probability spaces*, that is, triples of the form $(S, \mathcal{B}, P)$, where S is a sample space, $\mathcal{B}$ is a σ-algebra of events in S and P is a probability measure defined for all events in $\mathcal{B}$. Sometimes for simplicity we say that P is a probability measure on the sample space S, meaning that P is in fact defined on a certain σ-algebra $\mathcal{B}$ of events in S. Any event outside of $\mathcal{B}$ will be considered *non-measurable* (even if a value of probability can be assigned to it by performing the Lebesgue extension), and events in $\mathcal{B}$ will be called *measurable*.

We will now list some useful properties of probability measures:

(i) for any $E \in \mathcal{B}$ we have $P(E) + P(E^c) = 1$,

(ii) for any $E_1, E_2 \in \mathcal{B}$ such that $E_1 \subset E_2$ we have $P(E_1) \leq P(E_2)$,

(iii) for any $E_1, E_2 \in \mathcal{B}$ we have $P(E_1 \cup E_2) = P(E_1) + P(E_2) - P(E_1 \cap E_2)$,

(iv) if $\{E_j\}$ is a sequence of non-decreasing events in $\mathcal{B}$, that is, $E_j \subset E_{j+1}$ for all j, then we have $\lim_{j \to \infty} P(E_j) = P\left(\cup_{j=1}^{\infty} E_j\right)$,

(v) if $\{E_j\}$ is a sequence of non-increasing events in $\mathcal{B}$, that is, $E_{j+1} \subset E_j$ for all j, then we have $\lim_{j \to \infty} P(E_j) = P\left(\cap_{j=1}^{\infty} E_j\right)$,

(vi) if $\{E_j\}$ is a sequence of events in $\mathcal{B}$ and $E \in \mathcal{B}$ is such that $E \subset \cup_{j=1}^{\infty} E_j$, then we have $P(E) \leq \sum_{j=1}^{\infty} P(E_j)$.

We will now mention one useful way of constructing new probability spaces from existing ones. Let $(S_j, \mathcal{B}_j, P_j)$, for $j = 1, \ldots, n$, be a collection of probability spaces. Let the new sample space be $S = S_1 \times \ldots \times S_n$. For every event $E \subset S$ of the form $E = E_1 \times \ldots \times E_n$, with $E_j \in \mathcal{B}_j$, define $P(E) = P_1(E_1) \times \ldots \times P_n(E_n)$. All such events E form a semi-algebra $\mathcal{F}$ of events in S, and it can be checked directly that P is a probability measure on $\mathcal{F}$. It can be extended to the σ-algebra $\mathcal{B} = \mathcal{B}(\mathcal{F})$, and we obtain the probability space $(S, \mathcal{B}, P)$. It is called the *Cartesian product of the probability spaces* $(S_j, \mathcal{B}_j, P_j)$, $j = 1, \ldots, n$ and is denoted by $\prod_{j=1}^{n}(S_j, \mathcal{B}_j, P_j)$. If the probability spaces $(S_j, \mathcal{B}_j, P_j)$ are all the same for $j = 1, \ldots, n$, and coincide with say a space $(S', \mathcal{B}', P')$, then we will call the Cartesian product $(S, \mathcal{B}, P)$ the nth *Cartesian power of* $(S', \mathcal{B}', P')$ and denote it by $(S', \mathcal{B}', P')^n$. In this case we will also call S the nth *Cartesian power of S'* and denote it by S'^n.

Suppose now that we have a sequence $\{(S_j, \mathcal{B}_j, P_j)\}$ of probability spaces. We will now construct the Cartesian product in this situation. Set $S = \prod_{j=1}^{\infty} S_j$, that is, let S consist of sequences $e = \{e_j\}$ with $e_j \in S_j$. For any m and any set of indices $i_1 < i_2 < \ldots < i_m$ define the projection $h_{i_1, \ldots, i_m} : S \to \prod_{j=1}^{m} S_{i_j}$ as follows

$$h_{i_1, \ldots, i_m}(e) = (e_{i_1}, \ldots, e_{i_m}).$$

If $E' \subset \prod_{j=1}^{m} S_{i_j}$ is a measurable event in the sense of the Cartesian product $\prod_{j=1}^{m}(S_{i_j}, \mathcal{B}_{i_j}, P_{i_j})$, then we call the event $E = h_{i_1, \ldots, i_m}^{-1}(E')$ a *cylinder event*. We will now set $P(E) = P(E')$, where $P(E')$ is computed in the sense of the Cartesian product $\prod_{j=1}^{m}(S_{i_j}, \mathcal{B}_{i_j}, P_{i_j})$. We also set $P(S) = 1$. The collection $\mathcal{F}$ of events in S that consists of all cylinder events and S itself, can be shown to be an algebra. One can prove that P is a probability measure on $\mathcal{F}$, and therefore it can be extended to the σ-algebra $\mathcal{B} = \mathcal{B}(\mathcal{F})$. Thus we obtain a probability space $(S, \mathcal{B}, P)$. It is called the *Cartesian product of the sequence of probability spaces* $\{(S_j, \mathcal{B}_j, P_j)\}$ and is denoted by $\prod_{j=1}^{\infty}(S_j, \mathcal{B}_j, P_j)$. If the probability spaces $(S_j, \mathcal{B}_j, P_j)$ are all the same for all j and coincide with a space $(S', \mathcal{B}', P')$, then we will call the Cartesian product $(S, \mathcal{B}, P)$ the *infinite Cartesian power of* $(S', \mathcal{B}', P')$ and denote it by $(S', \mathcal{B}', P')^\infty$. In this case we will also call S the *infinite Cartesian power of S'* and denote it by S'^∞.

6.3 Conditional Probability

Let $(S, \mathcal{B}, P)$ be a probability space and E an event in $\mathcal{B}$ such that $P(E) > 0$. Consider the triple $(E, \mathcal{B}_E, P_E)$, where $\mathcal{B}_E$ consists of those elements of $\mathcal{B}$ that are contained in E, and for $E_0 \in \mathcal{B}_E$ we define $P_E(E_0) = P(E_0)/P(E)$. One can check directly that $\mathcal{B}_E$ is a σ-algebra and that P_E is a probability measure on $\mathcal{B}_E$, hence the triple $(E, \mathcal{B}_E, P_E)$ is a probability space. In this reduced probability space E represents the event that always occurs.

By using the reduced probability space $(E, \mathcal{B}_E, P_E)$, for any event $E' \in \mathcal{B}$ one can define the probability of E' under the condition that the event E occurs, as the value of P_E calculated for the portion of E' contained in E.

Definition 6.17. *The* conditional probability *of the event* $E' \in \mathcal{B}$ *given that the event* $E \in \mathcal{B}$ *occurs is the following number*

$$P(E'|E) = P_E(E' \cap E) = \frac{P(E' \cap E)}{P(E)}.$$

Conditional probability leads to the important concept of independence of two events which takes place when the conditional probability $P(E'|E)$ coincides with $P(E')$, that is, when the value of the probability of E' does not depend on whether or not E occurs.

Definition 6.18. *Two events* $E, E' \in \mathcal{B}$ *with* $P(E) > 0$ *and* $P(E') > 0$ *are called* independent, *if* $P(E'|E) = P(E')$, *or, equivalently* $P(E|E') = P(E)$.

To incorporate events with zero probabilities a broader definition of independence is usually used.

Definition 6.19. *Two events* $E, E' \in \mathcal{B}$ *are called* independent, *if* $P(E' \cap E) = P(E')P(E)$.

Clearly, if either $P(E) = 0$ or $P(E') = 0$, then E and E' are independent. Note that independence of events depends on a particular probability measure and that two events that are independent for one choice of probability measure may not be independent for another.

Example 6.20.

1. Consider the probability space from Part 2 of Example 6.14 and choose $E_1 = \{HH, HT\}$, $E_2 = \{HT, TH\}$. We have $P(E_1 \cap E_2) = P(\{HT\}) = 1/4$ and $P(E_1)P(E_2) = (1/4 + 1/4) \times (1/4 + 1/4) = 1/2 \times 1/2 = 1/4$, hence E_1 and E_2 are independent.

Suppose now that the coin used to conduct the experiments is not fair and that it is reasonable to define a probability measure on $\mathcal{F}$ as follows: $P(S) = 1$, $P(\emptyset) = 0$, $P(\{HH\}) = 9/16$, $P(\{HT\}) = 3/16$, $P(\{TH\}) = 3/16$, $P(\{TT\}) = 1/16$. In this case $P(E_1 \cap E_2) = P(\{HT\}) = 3/16$ whereas $P(E_1)P(E_2) = (9/16 + 3/16) \times (3/16 + 3/16) = 3/4 \times 3/8 = 9/32$, hence E_1 and E_2 are not independent for this probability measure.

2. Let the sample space S to be the closed rectangle in the (x, y)-plane with vertices $(0, 0)$, $(0, 1)$, $(3/4, 0)$ and $(3/4, 1)$. Let $\mathcal{F}$ be the semi-algebra that consists of all rectangles in S with sides parallel to the x- and y-axes

$$\mathcal{F} = \Big\{ (a,b) \times (c,d); [a,b) \times (c,d); (a,b] \times (c,d); [a,b] \times (c,d);$$
$$(a,b) \times [c,d); [a,b) \times [c,d); (a,b] \times [c,d); [a,b] \times [c,d);$$
$$(a,b) \times (c,d]; [a,b) \times (c,d]; (a,b] \times (c,d]; [a,b] \times (c,d];$$
$$(a,b) \times [c,d]; [a,b) \times [c,d]; (a,b] \times [c,d]; [a,b] \times [c,d],$$
$$0 \le a \le b \le \frac{3}{4}, \, 0 \le c \le d \le 1 \Big\}.$$

For a rectangle E in $\mathcal{F}$ define $P(E)$ to be equal to the sum of the area of the intersection $E \cap \{[0,1/2] \times [0,1]\}$ and twice the area of the intersection $E \cap \{[1/2,3/4] \times [0,1]\}$. One can check directly that P is a probability measure on $\mathcal{F}$, and therefore one can extend it to the minimal σ-algebra $\mathcal{B}(\mathcal{F})$.

Consider two events: $E_1 = \{(x,y) \in S : x + y \le 1/2\}$ and $E_2 = \{(x,y) \in S : 1/8 \le x \le 19/32, 0 \le y \le 1\}$. It is not hard to see that $E_1, E_2 \in \mathcal{B}(\mathcal{F})$, and that $P(E_1 \cap E_2) = 9/128$, $P(E_1)P(E_2) = 1/8 \times 9/16 = 9/128$, which shows that E_1 and E_2 are independent.

Consider another pair of events: $E_3 = \{(x,y) \in S : x + 2y > 1\}$ and $E_4 = \{(x,y) \in S : y \ge x\}$. Again, $E_3, E_4 \in \mathcal{B}(\mathcal{F})$, and we have $P(E_3 \cap E_4) = 69/144$ and $P(E_3)P(E_4) = 23/32 \times 9/16 = 207/512$, which shows that E_3 and E_4 are not independent.

The concept of independence in Definition 6.19 can be extended to any finite number of events.

Definition 6.21. *Events $E_1, \ldots, E_n \in \mathcal{B}$ are called independent if for any collection of indices $1 \le i_1 < i_2 < \ldots < i_k \le n$ we have*

$$P\left(\bigcap_{j=1}^{k} E_{i_j} \right) = \prod_{j=1}^{k} P(E_{i_j}).$$

As the following example shows, it is not sufficient for independence of $E_1, \ldots, E_n$ that all the events are pairwise independent (that is, for every $1 \le i < j \le n$ the events E_i and E_j are independent).

Example 6.22. Suppose a fair die is thrown twice. In this case the sample space is

$$S = \{(1,1),(1,2),(1,3),(1,4),(1,5),(1,6),(2,1),(2,2),(2,3),(2,4),(2,5),$$
$$(2,6),(3,1),(3,2),(3,3),(3,4),(3,5),(3,6),(4,1),(4,2),(4,3),(4,4),$$
$$(4,5),(4,6),(5,1),(5,2),(5,3),(5,4),(5,5),(5,6),(6,1),(6,2),(6,3),$$
$$(6,4),(6,5),(6,6)\}.$$

Since the die is fair we define the probability of each elementary event to be $1/36$ and extend this probability measure to the algebra of all events in S. Let

$$E_1 = \{(1,1),(1,2),(1,3),(1,4),(1,5),(1,6),(3,1),(3,2),(3,3),(3,4),(3,5),$$
$$(3,6),(5,1),(5,2),(5,3),(5,4),(5,5),(5,6)\},$$
$$E_2 = \{(1,1),(1,3),(1,5),(2,1),(2,3),(2,5),(3,1),(3,3),(3,5),(4,1),(4,3),$$
$$(4,5),(5,1),(5,3),(5,5),(6,1),(6,3),(6,5)\},$$
$$E_3 = \{(1,2),(1,4),(1,6),(2,1),(2,3),(2,5),(3,2),(3,4),(3,6),(4,1),(4,3),$$
$$(4,5),(5,2),(5,4),(5,6),(6,1),(6,3),(6,5)\}.$$

Clearly, $P(E_1) = P(E_2) = P(E_3) = 1/2$, hence $P(E_i)P(E_j) = 1/4$ for all $i \neq j$. Also, $P(E_1 \cap E_2) = P(E_1 \cap E_3) = P(E_2 \cap E_3) = 1/4$ and hence the events E_1, E_2, E_3 are pairwise independent. However, $P(E_1 \cap E_2 \cap E_3) = 0$ since $E_1 \cap E_2 \cap E_3 = \emptyset$ whereas $P(E_1)P(E_2)P(E_3) = 1/8$. Hence E_1, E_2, E_3 are not independent.

6.4 Random Variables

For a given probability space $(S, \mathcal{B}, P)$ we will often consider real-valued functions on S. We will be interested in a particular class of functions as defined below.

Definition 6.23. *A function $f : S \to \mathbb{R}$ is called a random variable on the probability space $(S, \mathcal{B}, P)$, if for every $b \in \mathbb{R}$, the event $E_b(f) = \{e \in S : f(e) \leq b\}$ is measurable (that is, belongs to $\mathcal{B}$).*

It is not hard to show that f is a random variable if and only if $f^{-1}(E) \in \mathcal{B}$ for every Borel set E in $\mathbb{R}$ (Borel sets in $\mathbb{R}$ are defined analogously to Borel sets in $[0,1]$ – see Sect. 6.7). The property for a function to be a random variable depends on the choice of $\mathcal{B}$. The same function can be a random variable for one choice of $\mathcal{B}$ and fail to be a random variable for another. We will now give examples of random variables as well as functions that are not random variables.

Example 6.24.

1. Let S consist of all possible hands of 7 cards (it contains $\binom{25}{7}$ elementary events). Let $\mathcal{B}$ be the σ-algebra of all events in S and, assuming that all hands are equiprobable, for an event $E \in \mathcal{B}$ define $P(E)$ to be the number of elementary events in E divided by $\binom{25}{7}$. For each hand $e \in S$ define $f(e)$ to be the number of aces contained in e. Clearly, f is a random variable on $(S, \mathcal{B}, P)$ with integer values between 0 and 4.

2. Let $(S, \mathcal{B}(\mathcal{F}), P)$ be the probability space from Part 3 of Example 6.14 and let f be any real-valued continuous function on $S = [0, 1]$. Then for every $b \in \mathbb{R}$ the event $E_b(f)$ is a closed subset of $[0, 1]$ and hence a Borel set (that is, a set in $\mathcal{B}(\mathcal{F})$). Therefore f is a random variable on $(S, \mathcal{B}(\mathcal{F}), P)$.

Consider now the event E constructed in Example 6.16. Let $f : [0,1] \to \mathbb{R}$ be the function that takes value 0 on E and value 1 on $[0,1] \setminus E$. Clearly, in this case $E_{1/2}(f) = E$ and hence does not belong to $\mathcal{B}(\mathcal{F})$. Therefore f is not a random variable on $(S, \mathcal{B}(\mathcal{F}), P)$. In fact, it is not a random variable even on $(S, \mathcal{L}(\mathcal{F}), P)$.

3. Let the sample space S be as in Part 2 of Example 6.1 and let $\mathcal{B}$ be the following σ-algebra

$$\mathcal{B} = \{S, \emptyset, \{HH, HT\}, \{TH, TT\}\}.$$

Define a probability measure P on $\mathcal{B}$ as follows

$$P(S) = 1, \quad P(\emptyset) = 0, \quad P(\{HH, HT\}) = P(\{TH, TT\}) = \frac{1}{2}.$$

For each $e \in S$ define $f(e)$ to be the number of H's in e. Then we have

$$E_1(f) = \{HT, TH, TT\}.$$

Since $E_1(f) \notin \mathcal{B}$, f is not a random variable on $(S, \mathcal{B}, P)$. However, f becomes a random variable if we choose $\mathcal{B} = \mathcal{B}(\mathcal{F})$, where $\mathcal{F}$ is as in Part 2 of Example 6.11.

6.5 Integration of Random Variables

We will now discuss integration of random variables over an event. We will start with random variables that take only countably many values.

Definition 6.25. *A function* $f : S \to \mathbb{R}$ *is called* simple *if the number of values that it takes is countable.*

The functions from Parts 1 and 3 of Example 6.24 take only finitely many values and hence are simple.

Let f be a simple function and let $r_1, r_2, \ldots$ be its values, $r_i \neq r_j$ for $i \neq j$. It is easy to show that f is a random variable if and only if each of the events

$$E'_{r_j}(f) = \{e \in S : f(e) = r_j\}$$

is measurable.

Definition 6.26. *A random variable that is at the same time a simple function is called* discrete.

The function from Part 1 of Example 6.24 and the function from Part 3 of Example 6.24 with $\mathcal{B} = \mathcal{B}(\mathcal{F})$ are discrete random variables.

Definition 6.27. *Suppose that a discrete random variable f on a probability space $(S, \mathcal{B}, P)$ takes values $r_1, r_2, \ldots$, with $r_i \neq r_j$ for $i \neq j$, and let E be an event in $\mathcal{B}$. Then f is called integrable on E with respect to P, if the series*

$$\sum_{j=1}^{\infty} r_j P(E \cap E'_{r_j}(f)) \tag{6.9}$$

converges absolutely, that is, if the series

$$\sum_{j=1}^{\infty} \left| r_j P(E \cap E'_{r_j}(f)) \right|$$

converges. If f is integrable over E, the sum of series (6.9) is called the integral of f over E with respect to P and is denoted by $\int_E f(e)\, dP(e)$, or simply $\int_E f\, dP$.

Note that if a discrete random variable takes only finitely many values on S, then it is integrable on any $E \in \mathcal{B}$.

We will now define integrability for not necessarily discrete random variables. This will be done by means of approximating a random variable by discrete random variables. The sort of approximation we require is described in the following definition. Note that for the purposes of the definition the functions involved do not need to be random variables and do not need to be defined on all of S.

Definition 6.28. *Let f and $\{f_j\}$ be real-valued functions on $E \subset S$. We say that the sequence $\{f_j\}$ converges to f uniformly on E, if for every $\varepsilon > 0$ there exists a number $J \in \mathbb{N}$ such that $|f_j(e) - f(e)| \leq \varepsilon$ for all $j \geq J$ and all $e \in E$.*

We will now give a general definition of integrability.

Definition 6.29. *If f is a random variable on a probability space $(S, \mathcal{B}, P)$ and $E \in \mathcal{B}$, then f is called integrable on E with respect to P if there exists a sequence of discrete random variables $\{f_j\}$ integrable on E with respect to P that converges to f uniformly on E.*

We remark that if a sequence of random variables converges to a function f uniformly on S, then f is itself a random variable. Further, for any random variable f one can find a sequence of discrete random variables that uniformly converges to f on all of S (and hence on any $E \subset S$). Indeed, for $j \in \mathbb{N}$ and $k \in \mathbb{Z}$ define

$$E_{kj}(f) = \left\{ e \in S : \frac{k}{j} \leq f(e) < \frac{k+1}{j} \right\} \tag{6.10}$$

and set

$$f_j(e) = \frac{k}{j}, \quad \text{if} \quad e \in E_{kj}(f). \tag{6.11}$$

Clearly, f_j are discrete random variables. Further, $|f_j(e) - f(e)| \leq 1/j$ for all $e \in S$ and hence $\{f_j\}$ converges to f uniformly on S. However, the random variables f_j may not be integrable on E, and it is the existence of a sequence of *integrable* random variables that is required in Definition 6.29.

Suppose that $\{f_j\}$ is a sequence of discrete random variables integrable on E that converges uniformly on E to a random variable f as in Definition 6.29. We will show that the sequence of integrals $x_j = \int_E f_j \, dP$ also has a limit. If $P(E) = 0$, then $x_j = 0$ for all j and $\{x_j\}$ obviously converges. Assume that $P(E) > 0$. It is sufficient to prove that $\{x_j\}$ is a fundamental sequence, that is, for every $\delta > 0$ there exists $J \in \mathbb{N}$ such that $|x_k - x_l| \leq \delta$ for all $k, l \geq J$. We have

$$|x_k - x_l| = \left| \int_E f_k \, dP - \int_E f_l \, dP \right| = \left| \int_E (f_k - f_l) \, dP \right|$$

$$\leq \sup_{e \in E} |f_k(e) - f_l(e)| \, P(E)$$
(6.12)

$$= \sup_{e \in E} |(f_k(e) - f(e)) - (f_l(e) - f(e))| \, P(E)$$

$$\leq \left(\sup_{e \in E} |f_k(e) - f(e)| + \sup_{e \in E} |f_l(e) - f(e)| \right) P(E).$$

If in Definition 6.28 we choose J corresponding to $\varepsilon = \delta/(2P(E))$, then from (6.12) we obtain that for $k, l \geq J$ the following holds

$$|x_k - x_l| \leq 2\varepsilon P(E) = \delta,$$

which shows that $\{x_j\}$ is indeed a fundamental sequence.

Let $x = \lim_{j \to \infty} x_j$. Suppose now that $\{f'_j\}$ is another sequence of discrete random variables integrable on E that converges uniformly on E to f. Set $x'_j = \int_E f'_j \, dP$ and let $x' = \lim_{j \to \infty} x'_j$. We will show that $x' = x$. Indeed, suppose that $x' \neq x$. Consider the sequence $\{f''_j\}$ obtained by taking the union of the sequences $\{f_j\}$ and $\{f'_j\}$. Clearly, $\{f''_j\}$ converges uniformly on E to f and therefore the sequence of integrals $x''_j = \int_E f''_j \, dP$ also has a limit. On the other hand, it cannot have a limit since the subsequences $\{x_j\}$ and $\{x'_j\}$ converge to different numbers. This contradiction shows that in fact $x' = x$.

Hence, we have just shown that for any sequence of discrete random variables integrable on E that converges uniformly on E to a random variable f, the sequence of integrals over E also converges, and the limit of the sequence of integrals does not depend on the choice of approximating sequence of discrete random variables. We therefore will call the limit of the sequence of integrals the *integral of f over E with respect to P* and denote it by $\int_E f(e) \, dP(e)$, or simply $\int_E f \, dP$.

One very useful sufficient condition for the integrability of a random variable f on $E \in \mathcal{B}$ is the *boundedness of f on E*, that is, the existence of a number M such that $|f(e)| \leq M$ for all $e \in E$. Indeed, suppose that f is bounded on E. Consider the sequence $\{f_j\}$ of discrete random variables constructed in (6.10), (6.11). Since f is bounded, each f_j takes only finitely many values on E and thus is integrable on E. Thus we have found a sequence of discrete random variables integrable on E that converges uniformly on E to f. Therefore f is integrable on E. As we will see below, boundedness is not a necessary condition for integrability, there are integrable and non-integrable unbounded random variables.

We will often say that a condition holds *almost certainly*, or *a.c.* on S, if it holds for all elementary events in S except possibly for elementary events in an event of probability 0. For example, we say that two random variables f and g *coincide almost certainly* if $P\left(\{e \in S : f(e) \neq g(e)\}\right) = 0$. Clearly, if two such random variables are integrable, then the corresponding integrals coincide. In fact, in order to speak about the integrability of a random variable on an event E it is sufficient to have the random variable defined almost certainly on E. One can define the random variable on the rest of E arbitrarily, and doing so does not affect either the integrability of the random variable or the value of the integral. Note, however, that we always assume for simplicity that all random variables are defined everywhere on S.

Clearly, if a random variable is integrable on S, it is integrable on any $E \in \mathcal{B}$. The integral of a random variable over the whole sample space S has a special name and will be of great importance later on.

Definition 6.30. *Let f be a random variable on a probability space $(S, \mathcal{B}, P)$. Suppose that f is integrable on S. The expected value of f or the mean value of f or simply the mean of f is the integral*

$$\mathcal{E}(f) = \int_S f \, dP.$$

We will also introduce two more important characteristics of random variables.

Definition 6.31. *Let f be a random variable on a probability space $(S, \mathcal{B}, P)$. Suppose that both f and f^2 are integrable on S. The variance of f is defined as*

$$Var(f) = \mathcal{E}\left((f - \mathcal{E}(f))^2\right) = \mathcal{E}\left(f^2\right) - \mathcal{E}(f)^2,$$

and the standard deviation of f is defined as

$$\sigma(f) = \sqrt{Var(f)}.$$

We will now list without proof some important properties of integrals. In the formulas below E is a measurable event.

(i) If $c \in \mathbb{R}$, then $\int_E c \, dP = cP(E)$.

(ii) If f is integrable on E and $c \in \mathbb{R}$, then cf is integrable on E and $\int_E cf \, dP = c \int_E f \, dP$.

(iii) If f and g are integrable on E, then $f + g$ are integrable on E and $\int_E (f + g) \, dP = \int_E f \, dP + \int_E g \, dP$.

(iv) If f and g are integrable on E and $f(e) \leq g(e)$ for all $e \in E$, then $\int_E f \, dP \leq \int_E g \, dP$. In particular, if $m \leq f(e) \leq M$ for all $e \in E$, then $mP(E) \leq \int_E f \, dP \leq MP(E)$.

(v) A random variable f is integrable on E if and only if $|f|$ is integrable on E. In this case $\left| \int_E f \, dP \right| \leq \int_E |f| \, dP$.

(vi) If $|f(e)| \leq \varphi(e)$ for all $e \in E$ and φ is integrable on E, then f is integrable on E.

(vii) **The Cauchy-Bunyakowski Inequality.** If f^2 and g^2 are integrable on E, then fg is also integrable on E and

$$\left(\int_E fg \, dP \right)^2 \leq \int_E f^2 \, dP \int_E g^2 \, dP.$$

The above inequality becomes an equality if and only if f and g are almost certainly proportional on E. Note that the Cauchy-Bunyakowski inequality with $g \equiv 1$ implies that f is integrable on E if f^2 is integrable on E. This observation shows that if the variance of a random variable exists, then the mean exists as well. In particular, in Definition 6.31 above it is sufficient to assume only that f^2 is integrable on S.

(viii) **The Triangle Inequality.** If f^2 and g^2 are integrable on E, then $(f + g)^2$ is also integrable on E and

$$\sqrt{\int_E (f + g)^2 \, dP} \leq \sqrt{\int_E f^2 \, dP} + \sqrt{\int_E g^2 \, dP}.$$

We will now give examples of integrable and non-integrable random variables and calculate some integrals.

Example 6.32.

1. Consider the probability space from Part 2 of Example 6.14, and let f be the random variable defined in Part 3 of Example 6.24. Since f takes only finitely many values, f^2 is integrable on any $E \in \mathcal{B}$. We will find $\mathcal{E}(f)$, $Var(f)$ and $\sigma(f)$.

The values of f are 0, 1, 2 and we have

$$E_0'(f) = \{TT\}, \quad E_1'(f) = \{HT, TH\}, \quad E_2'(f) = \{HH\}.$$

Therefore

$$\mathcal{E}(f) = \int_S f \, dP = 0 \times P(E_0') + 1 \times P(E_1') + 2 \times P(E_2') = 1 \times \frac{1}{2} + 2 \times \frac{1}{4} = 1.$$

Further

$$Var(f) = \mathcal{E}(f^2) - 1 = 0 \times P(E_0') + 1 \times P(E_1') + 4 \times P(E_2') - 1$$

$$= 1 \times \frac{1}{2} + 4 \times \frac{1}{4} - 1 = \frac{1}{2},$$

and therefore $\sigma(f) = 1/\sqrt{2}$.

We will now compute the integral of f over a subset $E \subset S$. Let $E = \{HT, TH, TT\}$. Then $E \cap E_0' = E_0'$, $E \cap E_1' = E_1'$, $E \cap E_2' = \emptyset$. Therefore

$$\int_E f \, dP = 0 \times P(E_0') + 1 \times P(E_1') + 2 \times P(\emptyset) = 1 \times P(E_1') = 1 \times \frac{1}{2} = \frac{1}{2}.$$

2. Let $(S, \mathcal{B}(\mathcal{F}), P)$ be the probability space from Part 3 of Example 6.14. Consider the following function

$$f(e) = \begin{cases} 0, & \text{if } e = 0, \\[2mm] \dfrac{n^2 + n}{2^n}, & \text{if } \dfrac{1}{n+1} < e \leq \dfrac{1}{n}, \text{ for } n \in \mathbb{N}. \end{cases}$$

Clearly, f is a simple function that takes infinitely many values: 0 and $(n^2 + n)/2^n$ for $n \in \mathbb{N}$. We have

$$E_0'(f) = \{0\}, \quad E_{(n^2+n)/2^n}'(f) = \left(\frac{1}{n+1}, \frac{1}{n} \right].$$

All these sets clearly belong to $\mathcal{B}(\mathcal{F})$ and therefore f is a discrete random variable. We will now show that f is integrable on S. Indeed, consider the series from Definition 6.27

$$0 \times P(\{0\}) + \sum_{n=1}^{\infty} \frac{n^2 + n}{2^n} \times P\left(\left(\frac{1}{n+1}, \frac{1}{n} \right] \right)$$

$$= \sum_{n=1}^{\infty} \frac{n^2 + n}{2^n} \times \frac{1}{n^2 + n} = \sum_{n=1}^{\infty} \frac{1}{2^n}.$$

The series $\sum_{n=1}^{\infty} 1/2^n$ clearly converges absolutely and its sum is equal to 1. Therefore f is integrable on S, and $\mathcal{E}(f) = 1$.

We will now find the integral of f over $E = [0, 3/4]$. We have

$$E \cap E_0'(f) = \{0\}, \quad E \cap E_1'(f) = \left(\frac{1}{2}, \frac{3}{4}\right],$$

$$E \cap E_{(n^2+n)/2^n}'(f) = \left(\frac{1}{n+1}, \frac{1}{n}\right], \quad \text{for } n \geq 2.$$

Therefore by Definition 6.27 we have

$$\int_E f \, dP = 0 \times P(\{0\}) + 1 \times P\left(\left(\frac{1}{2}, \frac{3}{4}\right]\right) + \sum_{n=2}^{\infty} \frac{n^2+n}{2^n} \times P\left(\left(\frac{1}{n+1}, \frac{1}{n}\right]\right)$$

$$= \frac{1}{4} + \sum_{n=2}^{\infty} \frac{1}{2^n} = \frac{1}{4} + \frac{1}{2} = \frac{3}{4}.$$

3. Let $(S, \mathcal{B}(\mathcal{F}), P)$ be the probability space as in Part 2 above and let f be a continuous function on $S = [0, 1]$. Since f is continuous, it is a random variable. Further, since f is continuous, it is bounded on S and hence integrable on S. Next, it is not hard to prove that $\int_S f \, dP$ is equal to $\int_0^1 f(e) \, de$, the usual Riemann integral of f. In fact, for any $E = [a, b]$ with $0 \leq a < b \leq 1$ we have $\int_E f \, dP = \int_a^b f(e) \, de$.

Let, for example, $f(e) = e^2$. Then we have

$$\mathcal{E}(f) = \int_0^1 e^2 \, de = \left.\frac{e^3}{3}\right|_0^1 = \frac{1}{3},$$

$$Var(f) = \mathcal{E}\left(\left(e^2 - \frac{1}{3}\right)^2\right) = \mathcal{E}\left(e^4 - \frac{2}{3}e^2 + \frac{1}{9}\right) = \int_0^1 \left(e^4 - \frac{2}{3}e^2 + \frac{1}{9}\right) de$$

$$= \left.\left(\frac{e^5}{5} - \frac{2e^3}{9} + \frac{e}{9}\right)\right|_0^1 = \frac{1}{5} - \frac{2}{9} + \frac{1}{9} = \frac{4}{45},$$

$$\sigma(f) = \frac{2}{3\sqrt{5}},$$

and for $E = [1/4, 1/2]$ we have

$$\int_E f \, dP = \int_{1/4}^{1/2} e^2 \, de = \left.\frac{e^3}{3}\right|_{1/4}^{1/2} = \frac{7}{192}.$$

4. Let $(S, \mathcal{B}(\mathcal{F}), P)$ be the probability space as in Part 2 above and let

$$f(e) = \begin{cases} 0, & \text{if } e = 0, \\[2mm] \dfrac{1}{\sqrt{e}}, & \text{if } 0 < e \le 1. \end{cases}$$

The function f is not continuous on S, but is nevertheless a random variable. Indeed, we have

$$E_b(f) = \{0\} \cup \left[\frac{1}{b^2}, 1\right], \text{ if } b > 0,$$

$$E_0(f) = \{0\},$$

$$E_b(f) = \emptyset, \qquad\qquad \text{if } b < 0,$$

and hence $E_b(f) \in \mathcal{B}(\mathcal{F})$ for all $b \in \mathbb{R}$, that is, f is a random variable. We approximate f by the functions constructed in (6.10), (6.11). For the function f_j we have

$$E'_{k/j}(f_j) = \left(\frac{j^2}{(k+1)^2}, \frac{j^2}{k^2}\right], \text{ if } k \ge j,$$

$$E'_0(f_j) = \{0\},$$

$$E'_{k/j}(f_j) = \emptyset, \qquad\qquad \text{if } 0 < k < j \text{ or } k < 0.$$

Calculating series (6.9) for f_j we obtain the following series

$$\sum_{k=j}^{\infty} \frac{k}{j} \times P\left(\left(\frac{j^2}{(k+1)^2}, \frac{j^2}{k^2}\right]\right) = j\sum_{k=j}^{\infty} \frac{2k+1}{k(k+1)^2},$$

which converges absolutely. Hence f_j is integrable on S, and we have found a sequence of discrete random variables integrable on S that uniformly on S converges to f. Therefore f is integrable on S.

We also have

$$\int_S f_j \, dP = j\sum_{k=j}^{\infty} \frac{2k+1}{k(k+1)^2}. \tag{6.13}$$

By definition, $\mathcal{E}(f) = \lim_{j\to\infty} \int_S f_j \, dP$ and hence to calculate $\mathcal{E}(f)$ we need to find the limit of the expressions in the right-hand side of formula (6.13) as $j \to \infty$. We have

$$j \sum_{k=j}^{\infty} \frac{2k+1}{k(k+1)^2} = j \left(2 \sum_{k=j}^{\infty} \frac{1}{(k+1)^2} + \sum_{k=j}^{\infty} \frac{1}{k(k+1)^2} \right).$$

Since

$$0 \le j \sum_{k=j}^{\infty} \frac{1}{k(k+1)^2} \le \sum_{k=j}^{\infty} \frac{1}{(k+1)^2}$$

and the series $\sum_{k=1}^{\infty} 1/(k+1)^2$ converges, we have

$$\lim_{j \to \infty} j \sum_{k=j}^{\infty} \frac{1}{k(k+1)^2} = 0,$$

and hence we only need to determine the limit of $2j \sum_{k=j}^{\infty} 1/(k+1)^2$ as $j \to \infty$. The following estimates hold

$$\sum_{k=j}^{\infty} \frac{1}{(k+1)^2} \ge \int_{j+1}^{\infty} \frac{dx}{x^2} = -\frac{1}{x} \Big|_{j+1}^{\infty} = \frac{1}{j+1},$$

$$\sum_{k=j}^{\infty} \frac{1}{(k+1)^2} \le \int_{j}^{\infty} \frac{dx}{x^2} = -\frac{1}{x} \Big|_{j}^{\infty} = \frac{1}{j},$$

where the integrals involved are understood as Riemann improper integrals. Hence

$$\frac{2j}{j+1} \le 2j \sum_{k=j}^{\infty} \frac{1}{(k+1)^2} \le 2,$$

and therefore $\lim_{j \to \infty} 2j \sum_{k=j}^{\infty} 1/(k+1)^2 = 2$. Thus $\mathcal{E}(f) = 2$.

5. Let $(S, \mathcal{B}(\mathcal{F}), P)$ be the probability space as in Part 2 above and let

$$f(e) = \begin{cases} 0, & \text{if } e = 0, \\ \dfrac{1}{e}, & \text{if } 0 < e \le 1. \end{cases}$$

The function f is not continuous on S, but is nevertheless a random variable. Indeed, we have

$$E_b(f) = \{0\} \cup \left[\frac{1}{b}, 1 \right], \text{ if } b > 0,$$

$$E_0(f) = \{0\},$$

$$E_b(f) = \emptyset, \qquad\qquad \text{if } b < 0,$$

and hence $E_b(f) \in \mathcal{B}(\mathcal{F})$ for all $b \in \mathbb{R}$, that is, f is a random variable.

We will show that f is not integrable on S. Consider the following discrete random variable

$$g(e) = \begin{cases} 0, & \text{if } e = 0, \\ n, & \text{if } \dfrac{1}{n+1} < e \leq \dfrac{1}{n}, \, n \in \mathbb{N}. \end{cases}$$

Clearly, g is a discrete random variable and $|g(e)| \leq f(e)$ for all $e \in S$. Therefore, by property (vi) of integrals, if f were integrable, g would be integrable as well. For g we find

$$E'_n(g) = \left(\frac{1}{n+1}, \frac{1}{n} \right], \quad n \in \mathbb{N},$$

$$E'_0(g) = \{0\}.$$

Hence series (6.9) for g is

$$\sum_{n=1}^{\infty} n \times P\left(\left(\frac{1}{n+1}, \frac{1}{n} \right] \right) = \sum_{n=2}^{\infty} \frac{1}{n},$$

which is a divergent series. Thus f is not integrable on S. In particular, for the random variable from Part 4 above, the variance does not exist.

6.6 Monotone Functions on the Real Line

In our study of random variables in the forthcoming sections we will rely on a number of properties of certain functions on $\mathbb{R}$. In this section we will list all the required facts, mostly without proofs. At the same time we will present several important constructions some of which resemble those already given for probability measures and random variables.

We will start with constructing the *Lebesgue measure on* $\mathbb{R}$. This measure serves to indicate the size of a subset of $\mathbb{R}$ in the same way as the probability measure is used for events. First, we will construct the Lebesgue measure on $[0, 1]$. In fact, this has already been done in Sect. 6.2. Recall Part 3 of Example 6.11 and extend the probability measure P to the σ-algebra $\mathcal{L}(\mathcal{F})$ of all Lebesgue measurable events in $[0, 1]$ (see Definition 6.15). The extended probability measure on $\mathcal{L}(\mathcal{F})$ is called the *Lebesgue measure on* $[0, 1]$. In the same way we can construct the Lebesgue measure on every segment $[n, n+1]$, with $n \in \mathbb{Z}$. Denote the Lebesgue measure on $[n, n+1]$ by μ_n.

Definition 6.33. *A set* $A \subset \mathbb{R}$ *is called Lebesgue measurable if for every* $n \in \mathbb{Z}$ *the intersection* $A \cap [n, n+1]$ *is Lebesgue measurable in* $[n, n+1]$. *In this case the Lebesgue measure of* A *is*

$$\mu(A) = \sum_{n=-\infty}^{\infty} \mu_n(A \cap [n, n+1]).$$

The collection of all Lebesgue measurable sets in $\mathbb{R}$ is denoted by $\mathcal{L}(\mathbb{R})$.

One can verify that $\mathcal{L}(\mathbb{R})$ is a σ-algebra where $\mathbb{R}$ plays the role of S and that μ is σ-additive on $\mathcal{L}(\mathbb{R})$. Note that μ can take infinite values: for example $\mu(\mathbb{R}) = \mu([1, \infty)) = \infty$.

Next, we will consider certain real-valued functions on $\mathbb{R}$.

Definition 6.34. *A function $f : \mathbb{R} \to \mathbb{R}$ is called Lebesgue measurable or simply measurable if for every $b \in \mathbb{R}$ the set $A_b(f) = \{x \in \mathbb{R} : f(x) \leq b\}$ is Lebesgue measurable (that is, belongs to $\mathcal{L}(\mathbb{R})$).*

It is not hard to show that f is Lebesgue measurable if and only if $f^{-1}(E)$ is Lebesgue measurable for every Borel set E in $\mathbb{R}$ (see Sect. 6.7). Clearly, measurable functions are analogues of random variables for the Lebesgue measure.

We will now discuss the integrability of measurable functions on Lebesgue measurable sets in $\mathbb{R}$. Let first $A \in \mathcal{L}(\mathbb{R})$ be a set of finite measure, $\mu(A) < \infty$. Then the definition of integrability is identical to that for random variables. We first define the integrability of a simple measurable function as in Definition 6.27 and then treat arbitrary measurable functions as in Definition 6.29. The Lebesgue integral of a function f over the set A is denoted by $\int_A f(x)\, d\mu(x)$, or simply $\int_A f\, d\mu$. All properties of integrals with respect to probability measures stated in Sect. 6.5 remain true for Lebesgue integrals over sets of finite measure.

We will now briefly compare the classes of Lebesgue and Riemann integrable functions on a segment. If f is Riemann integrable on $[a, b]$, then f is also Lebesgue integrable on $[a, b]$, and $\int_{[a,b]} f\, d\mu = \int_a^b f(x)\, dx$. However, there are many functions on $[a, b]$ that are Lebesgue integrable, but are not Riemann integrable. Suppose now that f is not Riemann integrable on $[a, b]$, but the improper Riemann integral of f exists over $[a, b]$. Then f may not be Lebesgue integrable on $[a, b]$, but if the improper Riemann integral of $|f|$ over $[a, b]$ exists, then f is Lebesgue integrable on $[a, b]$. For example, let f be unbounded near a, Riemann integrable on every segment $[a + \varepsilon, b]$, and suppose that the improper Riemann integral $\int_a^b |f(x)|\, dx = \lim_{\varepsilon \to 0} \int_{a+\varepsilon}^b |f(x)|\, dx$ exists. Then the improper Riemann integral $\int_a^b f(x)\, dx = \lim_{\varepsilon \to 0} \int_{a+\varepsilon}^b f(x)\, dx$ also exists, f is Lebesgue integrable on $[a, b]$, and $\int_{[a,b]} f\, d\mu = \int_a^b f(x)\, dx$.

Suppose now that $A \in \mathcal{L}(\mathbb{R})$ has infinite measure, $\mu(A) = \infty$. Let $A_n = A \cap [-n, n]$ for $n \in \mathbb{N}$. Then $A_n \in \mathcal{L}(\mathbb{R})$, $\mu(A_n) < \infty$, $A_n \subset A_{n+1}$ and $A = \cup_{n=1}^\infty A_n$.

Definition 6.35. *A sequence $\{X_n\}$ of subsets of $A \in \mathcal{L}(\mathbb{R})$ is called exhausting if $X_n \in \mathcal{L}(\mathbb{R})$, $\mu(X_n) < \infty$, $X_n \subset X_{n+1}$ for all $n \in \mathbb{N}$, and $A = \cup_{n=1}^\infty X_n$.*

As we have seen, any $A \in \mathcal{L}(\mathbb{R})$ has an exhausting sequence of subsets.

Now we can give a definition of integrability over subsets of infinite measure.

Definition 6.36. *Let $A \in \mathcal{L}(\mathbb{R})$, $\mu(A) = \infty$, and suppose that $f : \mathbb{R} \to \mathbb{R}$ is measurable. Then f is called Lebesgue integrable on A if f is Lebesgue integrable over any $B \subset A$ with $B \in \mathcal{L}(\mathbb{R})$, $\mu(B) < \infty$, and for any exhausting sequence of subsets $\{X_n\}$ of A the limit*

$$\lim_{n \to \infty} \int_{X_n} f \, d\mu$$

exists and is independent of the sequence. This limit is called the Lebesgue integral of f over A and is denoted by $\int_A f(x) \, d\mu(x)$, or simply $\int_A f \, d\mu$.

Any function f can be represented as a difference of two non-negative functions $f = f_+ - f_-$, where

$$f_+(x) = \begin{cases} f(x), & \text{if } f(x) \geq 0, \\ 0, & \text{if } f(x) < 0, \end{cases} \qquad f_-(x) = \begin{cases} -f(x), & \text{if } f(x) \leq 0, \\ 0, & \text{if } f(x) > 0. \end{cases}$$

One can show that f is Lebesgue integrable on A if and only if f_+ and f_- are Lebesgue integrable on A in which case $\int_A f \, d\mu = \int_A f_+ \, d\mu - \int_A f_- \, d\mu$. Hence it is in fact sufficient to understand Lebesgue integrability only for non-negative functions. For such functions a more constructive definition of integrability on a set of infinite measure is possible.

Theorem 6.37. *Let $A \in \mathcal{L}(\mathbb{R})$, $\mu(A) = \infty$, and suppose that $f : \mathbb{R} \to \mathbb{R}$ is measurable and $f(x) \geq 0$ for all $x \in A$. Then f is Lebesgue integrable on A if and only if the following two conditions hold*

(i) for every segment $[a, b] \subset \mathbb{R}$, f is Lebesgue integrable on $A \cap [a, b]$,

(ii) there exists a sequence of segments $[a_n, b_n]$ such that $[a_n, b_n] \subset [a_{n+1}, b_{n+1}]$ for all n, $a_n \to -\infty$, $b_n \to \infty$ as $n \to \infty$, for which the limit

$$\lim_{n \to \infty} \int_{A \cap [a_n, b_n]} f \, d\mu \tag{6.14}$$

exists.

In this case $\int_A f \, d\mu$ coincides with limit (6.14).

Most properties of integrals with respect to probability measures stated in Sect. 6.5 remain true for Lebesgue integrals over sets of infinite measure. One should note, however, that a simple measurable function that takes finitely many values or, more generally, a bounded measurable function may not be integrable on a set of infinite measure. In particular, constant non-zero functions are not integrable on such sets, hence one has to make corresponding adjustments in properties (i) and (iv).

We will now briefly compare Lebesgue integrals over the half-lines $(-\infty, b]$, $[c, \infty)$ and over the whole real line with the corresponding improper Riemann integrals. This comparison is similar to that for the case of improper Riemann integrals over a segment discussed above. Suppose, for example, that a function f is Riemann integrable on every segment $[a, b]$ and the improper Riemann integral $\int_{-\infty}^{b} f(x)\,dx = \lim_{a\to-\infty} \int_{a}^{b} f(x)\,dx$ exists. Then f may not be Lebesgue integrable on $(-\infty, b]$. However, if the above improper integral converges absolutely, that is, if the improper Riemann integral $\int_{-\infty}^{b} |f(x)|\,dx = \lim_{a\to-\infty} \int_{a}^{b} |f(x)|\,dx$ exists, then f is Lebesgue integrable on $(-\infty, b]$ and $\int_{(-\infty,b]} f\,d\mu = \int_{-\infty}^{b} f(x)\,dx$. An analogous statement holds for the half-line $[c, \infty)$. Further, the improper Riemann integral over the real line $\mathbb{R}$ is just the sum of two improper integrals of the type discussed above $\int_{-\infty}^{\infty} f(x)\,dx = \int_{-\infty}^{a} f(x)\,dx + \int_{a}^{\infty} f(x)\,dx$ for any a, and hence f is Lebesgue integrable if each of these integrals converges absolutely, in which case $\int_{\mathbb{R}} f\,d\mu = \int_{-\infty}^{\infty} f(x)\,dx$.

We will often say that a condition holds *almost everywhere*, or *a.e.* on $\mathbb{R}$, if it holds for all points in $\mathbb{R}$ except possibly for points in a set of Lebesgue measure 0. For example, we say that two measurable functions f and g *coincide almost everywhere* if $\mu\left(\{x \in \mathbb{R} : f(x) \neq g(x)\}\right) = 0$. Clearly, if two such functions are Lebesgue integrable, then the corresponding integrals coincide. In fact, in order to speak about the integrability of a function on a set A it is sufficient to have the function defined almost everywhere on A and measurable there. One can define the function on the rest of A arbitrarily, and doing so does not affect either the integrability of the function or the value of the integral.

We will now discuss functions of the following type.

Definition 6.38. *A function $f : \mathbb{R} \to \mathbb{R}$ is called monotone non-decreasing, if for all $x \leq y$ we have $f(x) \leq f(y)$.*

It is not hard to show that a monotone non-decreasing function can only have discontinuities of the *first kind* (that is, such that the limits $f(x + 0) = \lim_{h>0,h\to0} f(x + h)$ and $f(x - 0) = \lim_{h>0,h\to0} f(x - h)$ exist for all $x \in \mathbb{R}$), and the set of points of discontinuity is countable. In particular, monotone non-decreasing functions are continuous almost everywhere. It turns out that such functions also have good differentiability properties.

Theorem 6.39. *Let $f : \mathbb{R} \to \mathbb{R}$ be a monotone non-decreasing function. Then f is differentiable almost everywhere on $\mathbb{R}$. The derivative f' is a non-negative function that is Lebesgue integrable on any segment $[a, b] \subset \mathbb{R}$.*

We will be interested in monotone non-decreasing functions of a special kind.

Definition 6.40. *A monotone non-decreasing function f is called absolutely continuous on $\mathbb{R}$ if for all $a \le b$ we have*

$$f(b) - f(a) = \int_{[a,b]} f' \, d\mu.$$

Not every monotone non-decreasing function is absolutely continuous.

Example 6.41. Consider the function

$$f(x) = \begin{cases} 0, & \text{if } x < 0, \\ 1, & \text{if } 0 \le x < 2, \\ 2, & \text{if } x \ge 2. \end{cases}$$

Clearly, $f'(x)$ exists and is equal to 0 for all x except $x = 0, 2$. Therefore for all $a \le b$ we have $\int_{[a,b]} f' \, d\mu = 0$. However, the function f is not constant. Hence f is not absolutely continuous.

In the next section for any random variable we will introduce a special function called the *distribution function*. Distribution functions are real-valued non-decreasing functions on $\mathbb{R}$. The special case when distribution functions are absolutely continuous will be of particular importance.

6.7 Distribution Functions

Many properties of a random variable are captured by a special function associated with the random variable as defined below.

Definition 6.42. *Let f be a random variable on a probability space $(S, \mathcal{B}, P)$. The distribution function of f is the function $F_f : \mathbb{R} \to \mathbb{R}$ defined as follows*

$$F_f(x) = P(E_x(f)).$$

In the future it will be also convenient to consider the "complementary" function $F_f^*(x) = 1 - F_f(x) = P(E_x(f)^c)$. The distribution function F_f has the following properties

(i) F_f is monotone non-decreasing, that is, $F(x) \le F(y)$, if $x \le y$,

(ii) $F(-\infty) = \lim_{x \to -\infty} F_f(x) = 0$,

(iii) $F(\infty) = \lim_{x \to \infty} F_f(x) = 1,$

(iv) F_f is upper semi-continuous on $\mathbb{R}$. Since F_f is monotone non-decreasing, this condition is equivalent to the condition $F(x + 0) = F(x)$ for all $x \in \mathbb{R}$, which follows from property (v) of probability measures in Sect. 6.2.

The function F_f can be used to define a probability measure on a certain σ-algebra of subsets of $\mathbb{R}$ in such a way that the identity function x on the resulting probability space will have the same distribution function F_f. This can be done as follows. Let $\mathcal{F}_0 = \{[a, b], [a, b), (a, b], (a, b), (a, \infty), [a, \infty), (-\infty, b),$ $(-\infty, b], (-\infty, \infty), -\infty < a \le b < \infty\}$ be the collection of all intervals in $\mathbb{R}$. Clearly, $\mathcal{F}_0$ is a semi-algebra. Define

$$
\begin{aligned}
P_{F_f}([a, b]) &= F_f(b) - F_f(a - 0), \\
P_{F_f}([a, b)) &= F_f(b - 0) - F_f(a - 0), \\
P_{F_f}((a, b]) &= F_f(b) - F_f(a), \\
P_{F_f}((a, b)) &= F_f(b - 0) - F_f(a), \quad \text{for } a < b, \\
P_{F_f}((a, \infty)) &= 1 - F_f(a), \\
P_{F_f}([a, \infty)) &= 1 - F_f(a - 0), \\
P_{F_f}((-\infty, b)) &= F_f(b - 0), \\
P_{F_f}((-\infty, b]) &= F_f(b), \\
P_{F_f}((-\infty, \infty)) &= 1.
\end{aligned}
\tag{6.15}
$$

One can check directly that P_{F_f} is a probability measure on $\mathcal{F}_0$. It can be extended to a probability measure on the minimal σ-algebra $\mathcal{B}(\mathcal{F}_0)$, which is the σ-algebra of all *Borel sets in* $\mathbb{R}$. This way we obtain the probability space $(\mathbb{R}, \mathcal{B}(\mathcal{F}_0), P_{F_f})$ such that the identity function x is a random variable on this space and $F_x = F_f$. Indeed, for every $t \in \mathbb{R}$ the set $E_t(x) = \{y \in \mathbb{R} : y \le t\} = (-\infty, t]$ is a Borel set and $F_x(t) = P_{F_f}(E_t(x)) = P_{F_f}((-\infty, t]) = F_f(t)$. This construction works for any function satisfying conditions (i)-(iv) above.

In problems involving random variables distribution functions are sufficient for many purposes, and in the future we will sometimes specify only distribution functions without specifying random variables and probability spaces. We wish to stress, however, that, as we have seen above, for a given distribution function, one can always construct a probability space and a random variable on it whose distribution function coincides with the given one.

To some extent the notion of integrability of random variables can be reformulated in terms of the sample space $(\mathbb{R}, \mathcal{B}(\mathcal{F}_0), P_{F_f})$. Let $E \in \mathcal{B}(\mathcal{F}_0)$ and let $E' = \{e \in S : f(e) \in E\}$. Since E is a Borel set and the sets $E_b(f)$ belong to $\mathcal{B}$ for all b, we have $E' \in \mathcal{B}$. It can be shown that f is integrable on E' with respect to P if and only if x is integrable on E with respect to P_{F_f}, and that in this case

$$
\int_{E'} f \, dP = \int_E x \, dP_{F_f}.
$$

In particular, if x is integrable on $\mathbb{R}$, we have

$$\mathcal{E}(f) = \int_{\mathbb{R}} x \, dP_{F_f}. \tag{6.16}$$

Let g be a random variable on $(\mathbb{R}, \mathcal{B}(\mathcal{F}_0), P_{F_f})$. Then $g(f)$ is a random variable on $(S, \mathcal{B}, P)$. One can show that $g(f)$ is integrable on E' if and only if $g(x)$ is integrable on E, and that in this case

$$\int_{E'} g(f) \, dP = \int_{E} g(x) \, dP_{F_f}. \tag{6.17}$$

In particular, for $g \equiv 1$ we have

$$P\Big(\{e \in S : f(e) \in E\}\Big) = P(E') = \int_{E'} dP = \int_{E} dP_{F_f} = P_{F_f}(E). \tag{6.18}$$

We will now derive another useful consequence of formula (6.17). If x^2 is integrable on $\mathbb{R}$ with respect to P_{F_f}, then x is integrable on $\mathbb{R}$ with respect to P_{F_f} as well, and setting $g(x) = (x - \mathcal{E}(f))^2$, we obtain

$$Var(f) = \int_{\mathbb{R}} (x - \mathcal{E}(f))^2 \, dP_{F_f} = \int_{\mathbb{R}} x^2 \, dP_{F_f} - \mathcal{E}(f)^2. \tag{6.19}$$

We will now give some examples of distribution functions and applications of formulas (6.16) and (6.19).

Example 6.43.

1. Consider the probability space from Part 2 of Example 6.14, and let f be the random variable defined in Part 3 of Example 6.24. Then we have

$$F_f(x) = \begin{cases} 0, & \text{if } x < 0, \\[2mm] \dfrac{1}{4}, & \text{if } 0 \le x < 1, \\[2mm] \dfrac{3}{4}, & \text{if } 1 \le x < 2, \\[2mm] 1, & \text{if } x \ge 2. \end{cases}$$

Let us calculate $\mathcal{E}(f)$ from formula (6.16). We approximate the function x by the discrete random variables defined in (6.10) and (6.11). For the corresponding function f_j with $j \ge 2$ we have

$$\int_{\mathbb{R}} f_j \, dP_{F_f} = 1 \times \left(\frac{3}{4} - \frac{1}{4} \right) + 2 \times \left(1 - \frac{3}{4} \right) = 1.$$

Hence $\mathcal{E}(f) = 1$ which agrees with the result of the calculation of $\mathcal{E}(f)$ in Part 1 of Example 6.32.

To calculate $Var(f)$ from formula (6.19), we approximate the function $(x - \mathcal{E}(f))^2 = (x - 1)^2$ by the discrete random variables defined in (6.10) and (6.11). For the corresponding function f_j with $j \geq 2$ we have

$$\int_{\mathbb{R}} f_j \, dP_{F_f} = (-1)^2 \times \frac{1}{4} + 1^2 \times \left(1 - \frac{3}{4}\right) = \frac{1}{2}.$$

Therefore $Var(f) = 1/2$ and $\sigma(f) = 1/\sqrt{2}$ which agrees with the corresponding result in Part 1 of Example 6.32.

2. Let $(S, \mathcal{B}(\mathcal{F}), P)$ be the probability space from Part 3 of Example 6.14. Consider the random variable $f(e) = e^2$ as in Part 3 of Example 6.32. Then we have

$$F_f(x) = \begin{cases} 0, & \text{if } x < 0, \\ \sqrt{x}, & \text{if } 0 \leq x < 1, \\ 1, & \text{if } x \geq 1. \end{cases} \tag{6.20}$$

Let us calculate $\mathcal{E}(f)$ from formula (6.16). We approximate the function x by the discrete random variables defined in (6.10) and (6.11). For the corresponding function f_j we have

$$\int_{\mathbb{R}} f_j \, dP_{F_f} = \frac{1}{j} \times \left(\frac{\sqrt{2}}{\sqrt{j}} - \frac{1}{\sqrt{j}}\right) + \frac{2}{j} \times \left(\frac{\sqrt{3}}{\sqrt{j}} - \frac{2}{\sqrt{j}}\right) + \ldots + \frac{j-1}{j}$$

$$\times \left(1 - \frac{j-1}{\sqrt{j}}\right) = \sum_{k=1}^{j-1} \frac{k}{j} \left(\frac{\sqrt{k+1}}{\sqrt{j}} - \frac{\sqrt{k}}{\sqrt{j}}\right) = \frac{1}{j^{3/2}} \sum_{k=1}^{j-1} k \left(\sqrt{k+1} - \sqrt{k}\right)$$

$$= 1 - \frac{1}{j^{3/2}} \sum_{k=2}^{j} \sqrt{k} - \frac{1}{j^{3/2}}.$$

It is easy to see that

$$\sum_{k=2}^{j} \sqrt{k} \leq \int_{2}^{j+1} \sqrt{x} \, dx = \frac{2}{3} \left((j+1)^{3/2} - 2^{3/2}\right),$$

and

$$\sum_{k=2}^{j} \sqrt{k} \geq \int_{1}^{j} \sqrt{x} \, dx = \frac{2}{3} \left(j^{3/2} - 1\right).$$

Hence

$$\frac{1}{j^{3/2}} \sum_{k=2}^{j} \sqrt{k} \to \frac{2}{3}, \quad \text{as } j \to \infty, \tag{6.21}$$

and thus $\mathcal{E}(f) = 1/3$ which agrees with the result of the calculation of $\mathcal{E}(f)$ in Part 3 of Example 6.32.

To calculate $Var(f)$ from formula (6.19), we approximate the function $(x - \mathcal{E}(f))^2 = (x - 1/3)^2$ by the discrete random variables defined in (6.10) and (6.11). For the corresponding function f_j we have

$$\int_{\mathbb{R}} f_j \, dP_{F_f} = \left(\frac{1}{j} - \frac{1}{3}\right)^2 \times \left(\frac{\sqrt{2}}{\sqrt{j}} - \frac{1}{\sqrt{j}}\right) + \left(\frac{2}{j} - \frac{1}{3}\right)^2 \times \left(\frac{\sqrt{3}}{\sqrt{j}} - \frac{2}{\sqrt{j}}\right) + \dots$$

$$+ \left(\frac{j-1}{j} - \frac{1}{3}\right)^2 \times \left(1 - \frac{j-1}{\sqrt{j}}\right) = \sum_{k=1}^{j-1} \left(\frac{k}{j} - \frac{1}{3}\right)^2 \left(\frac{\sqrt{k+1}}{\sqrt{j}} - \frac{\sqrt{k}}{\sqrt{j}}\right)$$

$$= \frac{1}{\sqrt{j}} \sum_{k=1}^{j-1} \left(\frac{k^2}{j^2} - \frac{2k}{3j} + \frac{1}{9}\right) \left(\sqrt{k+1} - \sqrt{k}\right) = \frac{1}{j^{5/2}} \sum_{k=1}^{j-1} k^2 (\sqrt{k+1} - \sqrt{k})$$

$$- \frac{2}{3j^{3/2}} \sum_{k=1}^{j-1} k(\sqrt{k+1} - \sqrt{k}) + \frac{1}{9\sqrt{j}} \sum_{k=1}^{j-1} (\sqrt{k+1} - \sqrt{k}).$$

From the calculation of $\mathcal{E}(f)$ we know that

$$\frac{1}{j^{3/2}} \sum_{k=1}^{j-1} k(\sqrt{k+1} - \sqrt{k}) \to \frac{1}{3}, \quad \text{as } j \to \infty,$$

and therefore

$$\frac{2}{3j^{3/2}} \sum_{k=1}^{j-1} k(\sqrt{k+1} - \sqrt{k}) \to \frac{2}{9}, \quad \text{as } j \to \infty.$$

Further,

$$\frac{1}{9\sqrt{j}} \sum_{k=1}^{j-1} (\sqrt{k+1} - \sqrt{k}) = \frac{1}{9\sqrt{j}}(-1 + \sqrt{j}) \to \frac{1}{9}, \quad \text{as } j \to \infty.$$

We also have

$$\frac{1}{j^{5/2}} \sum_{k=1}^{j-1} k^2 (\sqrt{k+1} - \sqrt{k}) = 1 - \frac{2}{j^{5/2}} \sum_{k=2}^{j} k^{3/2} + \frac{1}{j^{5/2}} \sum_{k=2}^{j} \sqrt{k} - \frac{1}{j^{5/2}}.$$

It follows from (6.21) that

$$\frac{1}{j^{5/2}} \sum_{k=2}^{j} \sqrt{k} \to 0, \quad \text{as } j \to \infty.$$

Also, it is easy to see that

$$\sum_{k=2}^{j} k^{3/2} \le \int_{2}^{j+1} x^{3/2}\, dx = \frac{2}{5} \left((j+1)^{5/2} - 2^{5/2} \right),$$

and

$$\sum_{k=2}^{j} k^{3/2} \ge \int_{1}^{j} x^{3/2}\, dx = \frac{2}{5} \left(j^{5/2} - 1 \right).$$

Hence

$$\frac{1}{j^{5/2}} \sum_{k=2}^{j} k^{3/2} \to \frac{2}{5}, \quad \text{as } j \to \infty.$$

Thus $Var(f) = 1 - 4/5 - 2/9 + 1/9 = 4/45$ and $\sigma(f) = 2/(3\sqrt{5})$ which agrees with the corresponding result in Part 3 of Example 6.32.

6.8 Common Types of Random Variables

Most random variables that arise in statistical analysis and that will be of interest to us belong to one of two types: the *discrete* type and the *continuous* type.

6.8.1 The Discrete Type

We have encountered discrete random variables before. Let f be a discrete random variable on a probability space $(S, \mathcal{B}, P)$ that takes values $r_1, r_2, \ldots$, with $r_i \ne r_j$ for $i \ne j$ and recall that

$$E'_{r_j} = \{e \in S : f(e) = r_j\}.$$

Definition 6.44. *The collection of probabilities* $\{p_f(r_j) = P(E'_{r_j}(f))\}$ *is called the probability distribution of* f.

The distribution function F_f of f is fully determined by the probability distribution of f. Namely, F_f is a step function whose value changes only at the points r_j where jumps of the sizes $p_f(r_j)$ occur

$$F(x) = \sum_{r_j \le x} p_f(r_j).$$

Hence

$$F(r_j) - F(r_j - 0) = p_f(r_j).$$

Recall that if f is a discrete random variable, $\mathcal{E}(f)$ and $Var(f)$ are easy to find either directly by using Definition 6.27 or from formulas (6.16) and (6.19)

$$\mathcal{E}(f) = \sum_{j=1}^{\infty} r_j p_f(r_j), \qquad (6.22)$$

$$Var(f) = \sum_{j=1}^{\infty} (r_j - \mathcal{E}(f))^2 p_f(r_j) = \sum_{j=1}^{\infty} r_j^2 p_f(r_j) - \mathcal{E}(f)^2. \qquad (6.23)$$

6.8.2 The Continuous Type

Definition 6.45. *A random variable f on a probability space is called continuous, if its distribution function is absolutely continuous. The derivative F_f' is called the* density function of f, *or simply the* density of f, *and is denoted by ϱ_f.*

The concept of continuity defined above should not be confused with the usual continuity of functions. Indeed, consider the random variable x on a probability space $(\mathbb{R}, \mathcal{B}(\mathcal{F}_0), P_{F_f})$. Regarded as a function from $\mathbb{R}$ into $\mathbb{R}$, x is continuous at every point. However, the distribution function of x coincides with F_f and may not be absolutely continuous; in this case x is not continuous as a random variable on the given probability space.

Recall that strictly speaking the density function ϱ_f is only defined almost everywhere on $\mathbb{R}$ (see Theorem 6.39). However, since ϱ_f will be almost exclusively used for integration, we can set it to be equal to any non-negative number on the set of Lebesgue measure 0, where it was not originally defined. Hence in the future we will often assume that ϱ_f is a non-negative function defined everywhere on $\mathbb{R}$. For a continuous random variable f we have

$$F_f(x) = F_f(t) + \int_{[t,x]} \varrho_f \, d\mu, \qquad (6.24)$$

for all $t \le x$. Since $\lim_{t \to -\infty} F(t) = 0$ and $\lim_{x \to \infty} F(x) = 1$, the limit $\lim_{n \to \infty} \int_{[a_n,b_n]} \varrho_f \, d\mu$ exists and is equal to 1 for any expanding sequence of segments as in Theorem 6.37. Therefore this theorem implies that ϱ_f is integrable on $\mathbb{R}$ and hence on any Lebesgue measurable subset of $\mathbb{R}$.

Letting in formula (6.24) $t \to -\infty$ we obtain a formula for $F_f(x)$ in terms of ϱ_f

$$F_f(x) = \int_{(-\infty,x]} \varrho_f \, d\mu.$$

By the definition of F_f this means

$$P(\{e \in S : f(e) \le x\}) = \int_{(-\infty,x]} \varrho_f \, d\mu.$$

It turns out that this last fact can be generalized to an arbitrary Borel set in $\mathbb{R}$ as follows.

Theorem 6.46. *Let f be a continuous random variable on a probability space $(S, \mathcal{B}, P)$. Then for every Borel set $E \subset \mathbb{R}$ we have*

$$P(\{e \in S : f(e) \in E\}) = \int_E \varrho_f \, d\mu. \tag{6.25}$$

Note that formula (6.25) is a version of formula (6.18) for the special case of continuous random variables.

If ϱ_f is continuous on $\mathbb{R}$ then for any $a \le b$ the probability $P(\{e \in S : a \le f(e) \le b\})$ is equal to the area under the graph of ϱ_f on the segment $[a, b]$. If $b = a$ we obtain $P(\{e \in S : f(e) = a\}) = 0$ for all $a \in \mathbb{R}$ which is not the case for any discrete random variable. More generally, if $\mu(E) = 0$, then the probability in the left-hand side of formula (6.25) is equal to zero. This shows, in particular, that the classes of discrete and continuous random variables do not intersect (one can also make this observation by arguing as in Example 6.41). In fact, discrete and continuous random variables represent two extremes among all random variables.

We will now show that the random variable from Part 3 of Example 6.32 is continuous.

Example 6.47. Let $(S, \mathcal{B}(\mathcal{F}), P)$ be the probability space from Part 3 of Example 6.14. Consider the random variable $f(e) = e^2$ as in Part 3 of Example 6.32. In Part 2 of Example 6.43 we found the distribution function F_f of f (see formula (6.20)). Clearly, $F_f'(x)$ exists for all $x \in \mathbb{R}$ except $x = 0, 1$, and

$$F_f'(x) = \begin{cases} 0, & \text{if } x < 0 \text{ or } x > 1, \\ \dfrac{1}{2\sqrt{x}}, & \text{if } 0 < x < 1. \end{cases}$$

Let $[a, b]$ be a segment. If $b \le 0$ or $a \ge 1$ we clearly have

$$\int_{[a,b]} F_f' \, d\mu = 0 = F(b) - F(a).$$

Suppose that $0 < a < 1$ and let $c = \min\{1, b\}$. Then

$$\int_{[a,b]} F_f' \, d\mu = \sqrt{x} \Big|_a^c = \sqrt{c} - \sqrt{a} = F(b) - F(a).$$

Let now $a = 0$ and set again $c = \min\{1, b\}$. The improper Riemann integral $\int_0^b F_f'(x) \, dx$ exists and is equal to $\sqrt{c}$. Hence $\int_{[0,c]} F_f' \, d\mu$ exists and

$$\int_{[0,b]} F'_f \, d\mu = \sqrt{c} = F(b) - F(a).$$

Finally, for $a < 0$, $b > 0$ we argue as in the case $a = 0$ above and again obtain the required identity.

Thus we have proved that f is absolutely continuous.

For a continuous random variable f it is not hard to show that if E is a Borel set in $\mathbb{R}$, then x is integrable on E with respect to P_{F_f} if and only if $x\varrho_f(x)$ is Lebesgue integrable on E, and in this case

$$\int_E x \, dP_{F_f} = \int_E x\varrho_f(x) \, d\mu.$$

More generally, a random variable g on $(\mathbb{R}, \mathcal{B}(\mathcal{F}_0), P_{F_f})$ is integrable on E with respect to P_{F_f}, if and only if $g(x)\varrho_f(x)$ is Lebesgue integrable on E, and in this case

$$\int_E g(x) \, dP_{F_f} = \int_E g(x)\varrho_f(x) \, d\mu.$$

Therefore formulas (6.16) and (6.19) for continuous random variables take the forms

$$\mathcal{E}(f) = \int_{\mathbb{R}} x\varrho_f(x) \, d\mu, \tag{6.26}$$

and

$$Var(f) = \int_{\mathbb{R}} (x - \mathcal{E}(f))^2 \varrho_f(x) \, d\mu = \int_{\mathbb{R}} x^2 \varrho_f \, d\mu - \mathcal{E}(f)^2 \tag{6.27}$$

respectively.

In the future we will often speak about the *distribution of a random variable*. In the discrete case this will mean the corresponding probability distribution or the distribution function, in the continuous case the corresponding density function or the distribution function. In the general case this will mean the distribution function.

6.9 Common Discrete and Continuous Distributions

In this section we will list some standard discrete and continuous distributions that are of interest in bioinformatics and in statistics in general.

6.9.1 The Discrete Case

We will start with the simplest distribution.

The Uniform Distribution

A *uniformly distributed* discrete random variable takes finitely many (say N) values, and for each value r_j, $p_f(r_j) = 1/N$. If, for example, a random variable f takes values $a, a + 1, a + 2, \ldots, a + N - 1$, then formulas (6.22) and (6.23) give

$$\mathcal{E}(f) = a + \frac{N - 1}{2},$$

$$Var(f) = \frac{N^2 - 1}{12}.$$

In the above calculations we used the identities

$$\sum_{k=1}^{n} k = \frac{n(n + 1)}{2},$$

$$\sum_{k=1}^{n} k^2 = \frac{n(n + 1)(2n + 1)}{6}.$$

A Bernoulli Trial

A *Bernoulli trial* is a single experiment with two possible outcomes: success and failure. The probability of success is denoted by p and hence the probability of failure is $1 - p$. We have already seen a Bernoulli trial with $p = 1/2$ in Part 1 of Example 6.11. The Bernoulli random variable f is the number of successes obtained in the trial. Clearly, f takes only two values 0 and 1, and the probability distribution of f is $\{p_f(0) = 1 - p, \ p_f(1) = p\}$. Further, by applying formulas (6.22) and (6.23) we obtain

$$\mathcal{E}(f) = p,$$
$$Var(f) = p(1 - p).$$

The Binomial Distribution

The *binomial* random variable f is the number of successes in a fixed number n of independent Bernoulli trials. It takes values $0, 1, \ldots, n$, and the probability distribution of f is

$$\left\{ p_f(k) = \binom{n}{k} p^k (1 - p)^{n-k}, \quad k = 0, \ldots, n \right\},$$

where we set $0^0 = 1$.

We will now find $\mathcal{E}(f)$ and $Var(f)$ from formulas (6.22) and (6.23). Consider the function of two variables

$$\varphi(x,y) = \sum_{k=0}^{n} \binom{n}{k} x^k y^{n-k} = (x+y)^n.$$

Differentiating φ with respect to x for $x \neq 0$ we obtain

$$\frac{\partial \varphi}{\partial x} = \frac{1}{x} \sum_{k=1}^{n} k \binom{n}{k} x^k y^{n-k} = n(x+y)^{n-1}.$$

Setting in the above formula $x = p$, $y = 1 - p$ we obtain for $p \neq 0$

$$\mathcal{E}(f) = \sum_{k=1}^{n} k \binom{n}{k} p^k (1-p)^{n-k} = np.$$

For $p = 0$ it is obvious that $\mathcal{E}(f) = 0$, hence we always have $\mathcal{E}(f) = np$. In order to find $Var(f)$ we differentiate φ twice with respect to x for $x \neq 0$

$$\frac{\partial^2 \varphi}{\partial x^2} = \frac{1}{x^2} \sum_{k=1}^{n} k(k-1) \binom{n}{k} x^k y^{n-k} = n(n-1)(x+y)^{n-2}.$$

Therefore for $p \neq 0$

$$Var(f) = \sum_{k=1}^{n} k^2 \binom{n}{k} p^k(1-p)^{n-k} - n^2 p^2 = p^2 n(n-1) + np - n^2 p^2 = np(1-p).$$

For $p = 0$ it is obvious that $Var(f) = 0$, hence we always have $Var(f) = np(1-p)$.

The Geometric Distribution

In this case an infinite sequence of independent Bernoulli trials with $p \neq 1$ is conducted and a random f variable is defined as the number of trials before but not including the first failure. This random variable takes values $0, 1, 2 \ldots$ and its probability distribution is the *geometric distribution*

$$\{p_f(k) = (1-p)p^k, \quad k = 0, 1, 2, \ldots\},$$

where we set $0^0 = 1$.

From formulas (6.22) and (6.23) we have

$$\mathcal{E}(f) = \sum_{k=1}^{\infty} k(1-p)p^k = p(1-p) \left(\sum_{k=0}^{\infty} p^k \right)' = p(1-p) \left(\frac{1}{1-p} \right)' = \frac{p}{1-p},$$

and

$$Var(f) = \sum_{k=1}^{\infty} k^2 (1-p)p^k - \frac{p^2}{(1-p)^2} = p(1-p)\left(\sum_{k=1}^{\infty} kp^k\right)' - \frac{p^2}{(1-p)^2}$$

$$= p(1-p)\left(p\left(\sum_{k=0}^{\infty} p^k\right)'\right)' - \frac{p^2}{(1-p)^2} = p(1-p)\left(p\left(\frac{1}{1-p}\right)'\right)' - \frac{p^2}{(1-p)^2}$$

$$= \frac{p(1+p)}{(1-p)^2} - \frac{p^2}{(1-p)^2} = \frac{p}{(1-p)^2}.$$

Let us also find the distribution function $F_f(x)$. Clearly, $F_f(x) = 0$, if $x < 0$. Suppose that $k \leq x < k+1$ for some non-negative integer k. Then we have

$$F_f(x) = \sum_{j=0}^{k} p_f(j) = (1-p)\sum_{j=1}^{k} p^j = (1-p)\frac{1-p^{k+1}}{1-p} = 1 - p^{k+1}.$$

Hence $F_f^*(x) = p^{k+1}$. In particular, $F_f^*(k) = p^{k+1}$ for all $k = 0, 1, 2, \ldots$. In the future we will be interested in random variables whose distribution functions have a similar property.

Definition 6.48. *Suppose a discrete random variable f takes values $0, 1, 2, \ldots$. Then f is called geometric-like if*

$$F_f^*(k) \sim Cp^{k+1}, \quad \text{as } k \to \infty,$$

for some constants $C > 0$, $0 < p \leq 1$, where $k \in \mathbb{N}$.

The above condition means that there exists a number $0 < p \leq 1$ such that the limit

$$\lim_{k \to \infty, \, k \in \mathbb{N}} \frac{F_f^*(k)}{p^{k+1}}$$

exists and is positive. In this case $F_f^*(k)$ becomes arbitrarily close to Cp^{k+1} as $k \to \infty$, and for large k we can write

$$F_f^*(k) \approx Cp^{k+1}.$$

The Poisson Distribution

A *Poisson-distributed* random variable f also takes infinitely many values $0, 1, 2, \ldots$, and the corresponding probability distribution is

$$\left\{ p_f(k) = \frac{\exp(-\lambda)\lambda^k}{k!}, \quad k = 0, 1, 2, \ldots \right\},$$

where $\lambda \geq 0$, and we set $0^0 = 1$.

For $\mathcal{E}(f)$ from formula (6.22) we obtain

$$\mathcal{E}(f) = \sum_{k=1}^{\infty} k \frac{\exp(-\lambda)\lambda^k}{k!} = \exp(-\lambda)\lambda \sum_{k=0}^{\infty} \frac{\lambda^k}{k!} = \lambda.$$

For $Var(f)$ from formula (6.23) we obtain

$$Var(f) = \sum_{k=1}^{\infty} k^2 \frac{\exp(-\lambda)\lambda^k}{k!} - \lambda^2 = \exp(-\lambda)\left(\lambda^2 \sum_{k=0}^{\infty} \frac{\lambda^k}{k!} + \lambda \sum_{k=0}^{\infty} \frac{\lambda^k}{k!}\right) - \lambda^2 = \lambda.$$

6.9.2 The Continuous Case

Again, we will start with the simplest distribution.

The Uniform Distribution

A continuous random variable f is said to have a *uniform distribution* if for some $a < b$ its density function is

$$\varrho_f(x) = \begin{cases} 0, & \text{if } x \leq a \text{ or } x > b, \\[2mm] \dfrac{1}{b-a}, & \text{if } a < x \leq b. \end{cases}$$

The distribution function of f is easily found

$$F_f(x) = \begin{cases} 0, & \text{if } x < a, \\[2mm] \dfrac{x-a}{b-a}, & \text{if } a \leq x < b, \\[2mm] 1, & \text{if } x \geq b. \end{cases}$$

From formulas (6.26) and (6.27) it is straightforward to calculate $\mathcal{E}(f)$ and $Var(f)$.

$$\mathcal{E}(f) = \int_a^b \frac{x}{b-a}\,dx = \frac{x^2}{2(b-a)}\Big|_a^b = \frac{a+b}{2},$$

$$Var(f) = \int_a^b \frac{x^2}{b-a}\,dx - \frac{(a+b)^2}{4} = \frac{x^3}{3(b-a)}\Big|_a^b - \frac{(a+b)^2}{4} = \frac{(b-a)^2}{12}.$$

The Normal Distribution

The density function for the *normal distribution* is

$$\varrho_f(x) = \frac{1}{\sqrt{2\pi\sigma^2}} \exp\left(-\frac{(x-\mu)^2}{2\sigma^2}\right),$$

where $\mu \in \mathbb{R}$ and $\sigma^2 > 0$. We calculate $\mathcal{E}(f)$ from formula (6.26).

$$\mathcal{E}(f) = \frac{1}{\sqrt{2\pi\sigma^2}} \int_{-\infty}^{\infty} x \exp\left(-\frac{(x-\mu)^2}{2\sigma^2}\right) dx$$

$$= \mu + \frac{1}{\sqrt{2\pi\sigma^2}} \int_{-\infty}^{\infty} (x-\mu) \exp\left(-\frac{(x-\mu)^2}{2\sigma^2}\right) dx = \mu.$$

(6.28)

In the above calculation we used the well-known identity

$$\int_{-\infty}^{\infty} \exp\left(-\frac{(x-\mu)^2}{2\sigma^2}\right) dx = \sqrt{2\pi\sigma^2}$$

(6.29)

and the fact that the graph of $\varrho_f(x)$ is symmetric with respect to the line $x = \mu$ which gives that the last integral in formula (6.28) is equal to 0. Next, we calculate $Var(f)$ from formula (6.27) using integration by parts.

$$Var(f) = \frac{1}{\sqrt{2\pi\sigma^2}} \int_{-\infty}^{\infty} (x-\mu)^2 \exp\left(-\frac{(x-\mu)^2}{2\sigma^2}\right) dx$$

$$= \frac{1}{\sqrt{2\pi\sigma^2}} \int_{-\infty}^{\infty} t^2 \exp\left(-\frac{t^2}{2\sigma^2}\right) dt = -\frac{\sqrt{\sigma^2}}{\sqrt{2\pi}} \int_{-\infty}^{\infty} t\, d\left(\exp\left(-\frac{t^2}{2\sigma^2}\right)\right)$$

$$= -\frac{\sqrt{\sigma^2}}{\sqrt{2\pi}} \left(t \exp\left(-\frac{t^2}{2\sigma^2}\right)\Big|_{-\infty}^{\infty} - \int_{-\infty}^{\infty} \exp\left(-\frac{t^2}{2\sigma^2}\right) dt\right) = \sigma^2,$$

where we again used identity (6.29).

The *standard normal distribution* is the normal distribution with $\mu = 0$ and $\sigma^2 = 1$. If a random variable f has a normal distribution with parameters μ and σ^2, then it can be "standardized" by introducing the new random variable $g = (f - \mu)/\sqrt{\sigma^2}$ that has the standard normal distribution.

The Exponential Distribution

For an *exponentially distributed* random variable the density function is

$$\varrho_f(x) = \begin{cases} 0, & \text{if } x \leq 0, \\ \lambda \exp(-\lambda x), & \text{if } x > 0, \end{cases}$$

where $\lambda > 0$. We will now find $\mathcal{E}(f)$ from formula (6.26) using integration by parts.

$$\mathcal{E}(f) = \int_0^\infty x\lambda \exp(-\lambda x)\, dx = -\int_0^\infty x\, d(\exp(-\lambda x))$$

$$= -\left(x\exp(-\lambda x)\Big|_0^\infty - \int_0^\infty \exp(-\lambda x)\, dx \right) = -\frac{1}{\lambda}\exp(-\lambda x)\Big|_0^\infty = \frac{1}{\lambda}.$$

Similarly, for $Var(f)$ from formula (6.27) we obtain

$$Var(f) = \int_0^\infty x^2\lambda \exp(-\lambda x)\, dx - \frac{1}{\lambda^2} = -\int_0^\infty x^2\, d(\exp(-\lambda x)) - \frac{1}{\lambda^2}$$

$$= -\left(x^2\exp(-\lambda x)\Big|_0^\infty - 2\int_0^\infty x\exp(-\lambda x)\, dx \right) - \frac{1}{\lambda^2} = \frac{2}{\lambda}\mathcal{E}(f) - \frac{1}{\lambda^2} = \frac{1}{\lambda^2}.$$

The Gamma Distribution

For the *gamma distribution* the density function is

$$\varrho_f(x) = \begin{cases} 0, & \text{if } x \le 0, \\[2mm] \dfrac{\lambda^p x^{p-1} \exp(-\lambda x)}{\Gamma(p)}, & \text{if } x > 0, \end{cases}$$

where $\lambda, p > 0$ and $\Gamma(p)$ is the *gamma function*

$$\Gamma(p) = \int_0^\infty t^{p-1}\exp(-t)\, dt.$$

The gamma function extends the factorial function to all positive numbers since one can show that $\Gamma(n) = (n-1)!$ for all $n \in \mathbb{N}$.

We will now find $\mathcal{E}(f)$ from formula (6.26) using integration by parts as for the exponential distribution.

$$\mathcal{E}(f) = \frac{\lambda^p}{\Gamma(p)}\int_0^\infty x^p \exp(-\lambda x)\, dx = -\frac{\lambda^{p-1}}{\Gamma(p)}\int_0^\infty x^p\, d(\exp(-\lambda x))$$

$$= -\frac{\lambda^{p-1}}{\Gamma(p)}\left(x^p\exp(-\lambda x)\Big|_0^\infty - p\int_0^\infty x^{p-1}\exp(-\lambda x)\, dx \right) = \frac{p}{\lambda}.$$

Similarly, for $Var(f)$ from formula (6.27) we obtain

$$Var(f) = \frac{\lambda^p}{\Gamma(p)} \int_0^\infty x^{p+1} \exp(-\lambda x)\, dx - \frac{p^2}{\lambda^2} = -\frac{\lambda^{p-1}}{\Gamma(p)} \int_0^\infty x^{p+1}\, d(\exp(-\lambda x))$$

$$-\frac{p^2}{\lambda^2} = -\frac{\lambda^{p-1}}{\Gamma(p)} \left(x^{p+1} \exp(-\lambda x) \Big|_0^\infty - (p+1) \int_0^\infty x^p \exp(-\lambda x)\, dx \right)$$

$$-\frac{p^2}{\lambda^2} = \frac{p+1}{\lambda} \mathcal{E}(f) - \frac{p^2}{\lambda^2} = \frac{p}{\lambda^2}.$$

The exponential distribution is a special case of the gamma distribution for $p = 1$. Another special case of the gamma distribution is the *chi-square distribution with ν degrees of freedom* obtained for $\lambda = 1/2$ and $p = \nu/2$, where $\nu \in \mathbb{N}$. The chi-square distribution is important for statistical hypothesis testing (see Sect. secthyptes).

There is an interesting connection between the normal and chi-square distributions, as shown in the following proposition.

Proposition 6.49. *If a random variable f has the standard normal distribution, then f^2 has the chi-square distribution with 1 degree of freedom.*

Proof: Let f be defined on a probability space $(S, \mathcal{B}, P)$. Clearly, $F_{f^2}(x) = 0$, if $x < 0$. For $x \geq 0$ using formula (6.25) we obtain

$$F_{f^2}(x) = P\Big(\{e \in S : f^2(e) \leq x\}\Big) = P\Big(\{e \in S : -\sqrt{x} \leq f(e) \leq \sqrt{x}\}\Big)$$

$$= \frac{1}{\sqrt{2\pi}} \int_{-\sqrt{x}}^{\sqrt{x}} \exp\left(-\frac{x^2}{2}\right) dx.$$

Differentiation with respect to x for $x > 0$ yields

$$\varrho_{f^2}(x) = \frac{1}{\sqrt{2\pi}} \frac{1}{\sqrt{x}} \exp\left(-\frac{x}{2}\right),$$

which is the density function of the chi-square distribution with 1 degree of freedom (it follows from identity (6.29) that $\Gamma(1/2) = \sqrt{\pi}$). ∎

6.10 Vector-Valued Random Variables

A *vector-valued random variable* is a finite collection $f = (f_1, \ldots, f_n)$ of ordinary random variables all of which are defined on the same probability space $(S, \mathcal{B}, P)$. If $b_1, \ldots, b_n \in \mathbb{R}$, then the event $E_{b_1, \ldots, b_n}(f) = \{e \in S : f_1(e) \leq b_1, \ldots, f_n(e) \leq b_n)\}$ is measurable since $E_{b_1, \ldots, b_n}(f) = \cap_{k=1}^n E_{b_k}(f_k)$. Hence

$P(E_{b_1,\ldots,b_n}(f))$ is defined and we can consider the *distribution function of the vector-valued random variable* f

$$F_f(x_1,\ldots,x_n) = P(E_{x_1,\ldots,x_n}(f)).$$

Clearly, F_f is defined on all of $\mathbb{R}^n$.

One can also consider the distribution functions $F_{f_k}(x)$ of the components f_k. They are called the *marginal distribution functions of f* and obtained from F_f as follows

$$F_{f_k}(x) = F_f(\infty,\ldots,\infty,x,\infty,\ldots,\infty),$$

where in the right-hand side x occupies the kth position. Marginal distribution functions are used to introduce the following important concept.

Definition 6.50. *Let $f_1,\ldots,f_n$ be random variables defined on the same probability space. They are called independent if for the vector-valued random variable $f = (f_1,\ldots,f_n)$ we have*

$$F_f(x_1,\ldots,x_n) = F_{f_1}(x_1) \times \ldots \times F_{f_n}(x_n), \tag{6.30}$$

for all $(x_1,\ldots,x_n) \in \mathbb{R}^n$.

We note that if $f_1,\ldots,f_n$ are independent, then any smaller number of random variables are independent as well. Indeed, consider random variables $f_{k_1},\ldots,f_{k_m}$ with $1 \le k_1 < \ldots < k_m \le n$ and the corresponding vector-valued random variable $f' = (f_{k_1},\ldots,f_{k_m})$. Then $F_{f'}(y_1,\ldots,y_m)$ is obtained from $F_f(x_1,\ldots,x_n)$ by setting all values of the variables to be infinite except $x_{k_1},\ldots,x_{k_m}$ and replacing x_{k_j} with y_j for $j = 1,\ldots,m$. Making these changes in formula (6.30) turns it into

$$F_{f'}(y_1,\ldots,y_m) = F_{f_{k_1}}(y_1) \times \ldots \times F_{f_{k_m}}(y_m)$$

which means that $f_{k_1},\ldots,f_{k_m}$ are independent.

If $f = (f_1,\ldots,f_n)$ is a vector-valued random variable, then it is possible to generalize formula (6.15) to obtain a probability measure P_{F_f} on *Borel sets in $\mathbb{R}^n$* from the distribution function F_f (the σ-algebra $\mathcal{B}(\mathcal{F}_0^n)$ of Borel sets in $\mathbb{R}^n$ is generated by the semi-algebra $\mathcal{F}_0^n$ of all – possibly unbounded – parallelepipeds in $\mathbb{R}^n$ with faces parallel to the coordinate hyperplanes). One can show that the distribution function of the vector-valued random variable $(x_1,\ldots,x_n)$ on the probability space $(\mathbb{R}^n, \mathcal{B}(\mathcal{F}_0^n), P_{F_f})$ coincides with F_f. The condition of independence of $f_1,\ldots,f_n$ can be formulated in terms of P_{F_f}. Namely, $f_1,\ldots,f_n$ are independent if and only if for every Borel set $E \subset \mathbb{R}^n$ of the form $E = E_1 \times \ldots \times E_n$, with $E_j \in \mathcal{B}(\mathcal{F}_0)$ (that is, each E_j is a Borel set in $\mathbb{R}$), we have $P_{F_f}(E) = P_{F_{f_1}}(E_1) \times \ldots \times P_{F_{f_n}}(E_n)$.

We will now describe an important construction that leads to independent random variables. Let $f_1,\ldots,f_n$ be random variables defined on probability spaces $(S_j, \mathcal{B}_j, P_j)$, for $j = 1,\ldots,n$, respectively. Consider the Cartesian product $(S, \mathcal{B}, P)$ of these probability spaces. For an elementary event

$e = (e_1, \ldots, e_n) \in S$ set $\tilde{f}_j(e) = f_j(e_j)$, $j = 1, \ldots, n$. Then the random variables $\tilde{f}_1, \ldots, \tilde{f}_n$ are independent. If $(S_j, \mathcal{B}_j, P_j) = (S', \mathcal{B}', P')$ and $f_j = f$ for $j = 1, \ldots, n$, we denote the vector-valued random variable $(\tilde{f}_1, \ldots, \tilde{f}_n)$ by $f^{\times n}$ and call it the nth *Cartesian power of* f. Clearly, $f^{\times n}$ is defined on $(S', \mathcal{B}', P')^n$, and its components are independent.

One is often interested in discrete and continuous vector-valued random variables.

Definition 6.51. *A vector-valued random variable* $f = (f_1, \ldots, f_n)$ *is called discrete, if* f_k *is discrete for* $k = 1, \ldots, n$. *Let* $r_{1,k}, r_{2,k}, \ldots$, *with* $r_{i,k} \neq r_{j,k}$ *for* $i \neq j$, *be the values of* f_k. *Then the collection of numbers*

$$\left\{ p_f(r_{j_1,1}, \ldots r_{j_n,n}) = P\{e \in S : f_1(e) = r_{j_1,1}, \ldots, f_n(e) = r_{j_n,n}\} \right.$$
$$\left. = P\left(\bigcap_{k=1}^{n} E'_{r_{j_k,k}}(f_k) \right) \right\}$$

is called the probability distribution of the vector-valued random variable f. *The probability distributions* $\{p_{f_k}(r_{j,k})\}$ *are called the marginal probability distributions of* f.

The distribution function of a discrete vector-valued random variable is entirely determined by its probability distribution, namely

$$F_f(x_1, \ldots, x_n) = \sum_{r_{j_1,1} \leq x_1, \ldots r_{j_n,n} \leq x_n} p_f(r_{j_1,1}, \ldots r_{j_n,n}).$$

It is not hard to show that discrete random variables $f_1, \ldots, f_n$ defined on the same probability space are independent if and only if the probability distribution of $f = (f_1, \ldots, f_n)$ is related to its marginal probability distributions as follows

$$p_f(r_{j_1,1}, \ldots r_{j_n,n}) = p_{f_1}(r_{j_1,1}) \times \ldots \times p_{f_n}(r_{j_n,n}),$$

for all possible values $r_{j_k,k}$.

One can give a definition of continuous vector-valued random variable along the lines of Definition 6.45, but this approach requires some preliminary work on analogues of "monotone non-decreasing functions" on $\mathbb{R}^n$ (cf. Sect. 6.6). However, since in the future we will be dealing mainly with vector-valued random variables with independent components, we will only give a definition of continuous vector-valued random variable in this situation and quickly explain how the definition can be generalized to the case of not necessarily independent components.

Definition 6.52. *Let* $f = (f_1, \ldots, f_n)$ *be a vector-valued random variable whose components* $f_1, \ldots, f_n$ *are independent. Then* f *is called continuous if each* f_k *is continuous. In this case the function* $\varrho_f(x_1, \ldots, x_n) = \varrho_{f_1}(x_1) \times \ldots \times \varrho_{f_n}(x_n)$ *is called the density function of* f *or simply the density of* f.

It follows from formula (6.30) that if the components of $f = (f_1, \ldots, f_n)$ are independent and f is continuous in the sense of Definition 6.52, then

$$F(x_1, \ldots, x_n) = \int_{(-\infty, x_1]} \varrho_{f_1} \, d\mu \times \ldots \times \int_{(-\infty, x_n]} \varrho_{f_n} \, d\mu. \qquad (6.31)$$

Identity (6.31) can be written in a more compact form if we introduce a Lebesgue measure μ^n on $\mathbb{R}^n$. It is produced in the same way as the Lebesgue measure μ on $\mathbb{R}$. Namely, μ^n is first constructed on the unit cube $\{(x_1, \ldots, x_n) \in \mathbb{R}^n : 0 \le x_1 \le 1, \ldots, 0 \le x_n \le 1\}$ starting with the semi-algebra of parallelepipeds lying inside the cube with faces parallel to the co-ordinate hyperplanes, and afterwards extended to $\mathbb{R}^n$. Lebesgue integrability with respect to μ^n is defined analogously. Since ϱ_{f_k} is Lebesgue integrable on $\mathbb{R}$ for all k, the function ϱ_f is Lebesgue integrable on $\mathbb{R}^n$, and identity (6.31) can be written as

$$F(x_1, \ldots, x_n) = \int_{(-\infty, x_1] \times \ldots \times (-\infty, x_n]} \varrho_f \, d\mu^n. \qquad (6.32)$$

More generally, if $E \subset \mathbb{R}^n$ is a Borel set, we have

$$P(\{e \in S : f(e) \in E\}) = \int_E \varrho_f \, d\mu^n,$$

which is a version of formula (6.25) for the case of vector-valued random variables.

In fact, the existence of a non-negative Lebesgue integrable on $\mathbb{R}^n$ function ϱ_f such that (6.32) holds can be taken as a general definition of continuous $\mathbb{R}^n$-valued random variable in which case the corresponding function ϱ_f is called the *density function* or *density* of f. Moreover, in this case the nth order partial derivative $\partial^n F_f / \partial x_1 \ldots \partial x_n$ exists almost everywhere on $\mathbb{R}^n$ and coincides almost everywhere with ϱ_f.

As we noted above, for a vector-valued random variable $f = (f_1, \ldots, f_n)$ on a probability space $(S, \mathcal{B}, P)$ one can construct a probability measure P_{F_f} on the σ-algebra $\mathcal{B}(\mathcal{F}_0^n)$ of all Borel sets in $\mathbb{R}^n$ in such a way that the distribution function of the vector-valued random variable $(x_1, \ldots, x_n)$ on the probability space $(\mathbb{R}^n, \mathcal{B}(\mathcal{F}_0^n), P_{F_f})$ coincides with F_f. Let $E \in \mathcal{B}(\mathcal{F}_0^n)$ and let $E' = \{e \in S : f(e) \in E\}$. Since E is Borel, the event E' belongs to $\mathcal{B}$. Let g be a random variable on $(\mathbb{R}^n, \mathcal{B}(\mathcal{F}_0^n), P_{F_f})$. Then $g(f)$ is a random variable on $(S, \mathcal{B}, P)$. One can show that $g(f)$ is integrable on E' with respect to P if and only if $g(x_1, \ldots, x_n)$ is integrable on E with respect to P_{F_f}, and that in this case

$$\int_{E'} g(f) \, dP = \int_E g(x_1, \ldots, x_n) \, dP_{F_f}. \qquad (6.33)$$

Formula (6.33) is a generalization of formula (6.17). If we set $g \equiv 1$ in (6.33), we obtain a useful generalization of formula (6.18)

$$P\Big(\{e \in S : f(e) \in E\}\Big) = P(E') = \int_{E'} dP = \int_{E} dP_{F_f} = P_{F_f}(E). \quad (6.34)$$

If $f_1, \ldots, f_n$ are independent and $E = E_1 \times \ldots \times E_n$ with E_j being Borel sets in $\mathbb{R}$, formula (6.33) becomes

$$\int_{E'} g(f) \, dP = \int_{E_1} \cdots \int_{E_n} g(x_1, \ldots, x_n) \, dP_{F_{f_n}} \cdots dP_{F_{f_1}}.$$

In particular, if $g(x_1, \ldots, x_n) = g_1(x_1) \times \ldots \times g_n(x_n)$, we have

$$\int_{E'} g(f) \, dP = \int_{E_1} g_1(x_1) \, dP_{F_{f_1}} \times \ldots \times \int_{E_n} g_n(x_n) \, dP_{F_{f_n}}.$$

The last formula has an important application. Setting $E = \mathbb{R}^n = \mathbb{R} \times \ldots \times \mathbb{R}$ and $g(x_1, \ldots x_n) = x_1 \times \ldots \times x_n$ we notice that $E' = S$ and therefore obtain

$$\mathcal{E}(f_1 \times \ldots \times f_n) = \int_{S} f_1 \times \ldots \times f_n \, dP = \int_{\mathbb{R}} x_1 \, dP_{F_{f_1}} \times \ldots$$

$$(6.35)$$

$$\times \int_{\mathbb{R}} x_n \, dP_{F_{f_n}} = \mathcal{E}(f_1) \times \ldots \times \mathcal{E}(f_n).$$

Identity (6.35) is very useful when independent random variables are considered.

Further, for a pair of random variables the following quantity is frequently used.

Definition 6.53. Let f_1, f_2 be two random variables defined on the same probability space $(S, \mathcal{B}, P)$. The *covariance* of f_1 and f_2 is the following number

$$\mathrm{Cov}(f_1, f_2) = \mathcal{E}\Big((f_1 - \mathcal{E}(f_1))(f_2 - \mathcal{E}(f_2))\Big).$$

Of course, for the covariance to exist the random variables f_1, f_2 and $f_1 f_2$ must be integrable on S with respect to P. Expanding the formula in Definition 6.53, we obtain

$$\mathrm{Cov}(f_1, f_2) = \mathcal{E}(f_1 f_2) - \mathcal{E}(f_1)\mathcal{E}(f_2).$$

For a vector-valued random variable $f = (f_1, \ldots, f_n)$ define the *variance-covariance matrix of f* as

$$(\Sigma(f))_{ii} = Var(f_i),$$
$$(\Sigma(f))_{ij} = \mathrm{Cov}(f_i, f_j), \quad i \neq j,$$

for $i, j = 1, \ldots, n$. Clearly, $\Sigma(f)$ is a symmetric $n \times n$ matrix. Variance-covariance matrices are widely used in statistical analysis.

It follows from formula (6.35) that for a vector-valued random variable with independent components the corresponding variance-covariance matrix is diagonal.

6.11 Sequences of Random Variables

In this section we will consider sequences $\{f_k\}$ of random variables all of which are defined on the same probability space $(S, \mathcal{B}, P)$. One important example of such sequences is associated with Markov chains. Indeed, consider a discrete-time Markov chain without an end state and denote its begin state by 0. Let $\mathbf{P} = (p_{ij})$ be the $N \times N$-matrix of transition probabilities and $p_0 = (p_{01}, \ldots, p_{0N})$ the vector of initialization probabilities. As before, let X denote the set of all non-zero states of the chain, and we assume that X is realized as the set $\{1, \ldots, N\}$, so that G_j corresponds to j for $j = 1, \ldots, N$. Let $(S, \mathcal{B}, P)$ be the probability space arising from the Markov chain (see Part 4 of Example 6.14). For every $k \in \mathbb{N}$ define a function on S as follows: for $e = 0\, x_1 x_2 \ldots$ set $f_k(e) = x_k$. Loosely speaking, f_k represents the state reached by the Markov chain after k steps (counting the initialization step). Clearly, f_k is a discrete random variable on $(S, \mathcal{B}, P)$. Its probability distribution is easy to find. We have

$$P\Big(\{e \in S : f_k(e) = i\}\Big) = (p_0 \mathbf{P}^{k-1})_i,$$

where $(p_0 \mathbf{P}^{k-1})_i$ is the ith component of the vector $p_0 \mathbf{P}^{k-1}$. Further, for the joint distribution of f_k and f_{k+1} we have

$$P\Big(\{e \in S : f_{k+1}(e) = j, f_k(e) = i\}\Big) = (p_0 \mathbf{P}^{k-1})_i p_{ij}.$$

Therefore, for all $i, j \in X$ we have

$$p_{ij} = P\Big(\{e \in S : f_{k+1}(e) = j | f_k(e) = i\}\Big),$$

for every $k \in \mathbb{N}$ such that $P\Big(\{e \in S : f_k(e) = i\}\Big) \neq 0$. Hence in some cases the transition probabilities of a Markov chain can be treated as certain conditional probabilities. More generally, for all $i, j \in X$ and $n \in \mathbb{N}$ the following holds

$$P\Big(\{e \in S : f_{k+n}(e) = j, f_k(e) = i\}\Big) = (p_0 \mathbf{P}^{k-1})_i p(n)_{ij} \qquad (6.36)$$

and hence

$$p(n)_{ij} = P\Big(\{e \in S : f_{k+n}(e) = j | f_k(e) = i\}\Big),$$

for every $k \in \mathbb{N}$ such that $P\Big(\{e \in S : f_k(e) = i\}\Big) \neq 0$, where $p(n)_{ij}$ are the entries of the matrix $\mathbf{P}(n) = \mathbf{P}^n$. The matrix $\mathbf{P}(n)$ can be thought of as the matrix of transition probabilities after $n + 1$ steps of the Markov chain.

We will now turn to the general theory of sequences of random variables. We will be mainly interested in the *convergence properties* of such sequences. There are many types of convergence for random variables, and in this section we will explore some of them.

We will start with perhaps the most popular type of convergence.

Definition 6.54. *A sequence $\{f_k\}$ of random variables on a probability space $(S, \mathcal{B}, P)$ is said to converge in probability to a random variable f defined on the same probability space, if for every $\varepsilon > 0$ we have*

$$\lim_{k \to \infty} P\Big(\{e \in S : |f_k(e) - f(e)| \geq \varepsilon\}\Big) = 0. \tag{6.37}$$

In this case we write $f_k \xrightarrow{P} f$.

An important tool for verifying convergence in probability is the following inequality called *Chebyshev's inequality*.

Theorem 6.55. (Chebyshev's Inequality) *Let $f \geq 0$ be an integrable random variable on a probability space $(S, \mathcal{B}, P)$, and $c > 0$. Then*

$$P\Big(\{e \in S : f(e) \geq c\}\Big) \leq \frac{\mathcal{E}(f)}{c}. \tag{6.38}$$

Proof: Let $E_c = \{e \in S : f(e) \geq c\}$. Since f is a random variable, E_c is measurable. We have

$$\mathcal{E}(f) \geq \int_{E_c} f \, dP \geq cP(E_c),$$

which gives inequality (6.38).

The theorem is proved. ∎

If φ is a random variable with mean μ and variance σ^2, then setting in inequality (6.38) $f = (\varphi - \mu)^2$ and $c = \lambda^2 \sigma^2$, with $\lambda > 0$, we obtain the classical form of Chebyshev's inequality

$$P\Big(\{e \in S : |\varphi(e) - \mu| \geq \lambda\sqrt{\sigma^2}\}\Big) \leq \frac{1}{\lambda^2}.$$

We will now prove the following important theorem.

Theorem 6.56. (Weak Law of Large Numbers) *Let $\{f_k\}$ be a sequence of pairwise independent random variables on a probability space $(S, \mathcal{B}, P)$. Suppose that f_k^2 is integrable on S for each k. Let $\mathcal{E}(f_k) = \mu$ for all k. Also let $\sigma_k^2 = Var(f_k)$ and assume that*

$$\lim_{n \to \infty} \frac{1}{n^2} \sum_{k=1}^{n} \sigma_k^2 = 0.$$

Then $\overline{f_n} \xrightarrow{P} \mu$, where

$$\overline{f_n} = \frac{1}{n} \sum_{k=1}^{n} f_k.$$

Proof: Clearly, $\overline{f_n}$ is integrable on S and $\mathcal{E}(\overline{f_n}) = \mu$. To show that $Var(\overline{f_n})$ exists we need to prove that $\overline{f_n}^2$ is also integrable on S. We have

$$\overline{f_n}^2 = \frac{1}{n^2}\left(\sum_{k=1}^{n} f_k^2 + 2 \sum_{1 \leq i < j \leq n} f_i f_j\right).$$

The random variables f_k^2 are integrable on S by assumption. The products $f_i f_j$ with $i \neq j$ are integrable on S by the Cauchy-Bunyakowski inequality (see property (vii) of integrals in Sect. 6.5). One can also deduce the integrability of $\overline{f_n}^2$ from the triangle inequality (see property (viii) of integrals in Sect. 6.5).

We further have

$$Var(\overline{f_n}) = \mathcal{E}(\overline{f_n}^2) - \mu^2 = \frac{1}{n^2}\left(\sum_{k=1}^{n} \mathcal{E}(f_k^2) + 2 \sum_{1 \leq i < j \leq n} \mathcal{E}(f_i f_j)\right) - \mu^2$$

$$= \frac{1}{n^2}\left(\sum_{k=1}^{n} \sigma_k^2 + n\mu^2 + (n^2 - n)\mu^2\right) - \mu^2 = \frac{1}{n^2}\sum_{k=1}^{n} \sigma_k^2,$$

where in the above calculation we used $\mathcal{E}(f_i f_j) = \mathcal{E}(f_i)\mathcal{E}(f_j)$ for $i \neq j$ which follows from pairwise independence of the random variables f_k (see formula (6.35)).

Fix $\varepsilon > 0$ and apply Chebyshev's inequality (6.38) to $|\overline{f_n} - \mu|$.

$$P\Big(\{e \in S : |\overline{f_n}(e) - \mu| \geq \varepsilon\}\Big) = P\Big(\{e \in S : (\overline{f_n}(e) - \mu)^2 \geq \varepsilon^2\}\Big)$$

$$\leq \frac{Var(\overline{f_n})}{\varepsilon^2} = \frac{1}{\varepsilon^2}\frac{1}{n^2}\sum_{k=1}^{n} \sigma_k^2,$$

and hence $P\Big(\{e \in S : |\overline{f_n}(e) - \mu| \geq \varepsilon\}\Big) \to 0$ as $n \to \infty$. Thus we have shown that $\overline{f_n} \xrightarrow{P} \mu$.

The theorem is proved. ∎

We will often consider sequences of random variables satisfying the following condition.

Definition 6.57. Let $\{f_k\}$ be either a sequence or a finite collection of random variables. The random variables f_k are said to be independent identically distributed or simply iid if:

(i) for every $n \geq 2$ the random variables $f_1, \ldots, f_n$ are independent,

(ii) the distributions of all f_k are identical.

If $\{f_k\}$ is a sequence of random variables for which only (i) is known to hold, we say that f_k are *independent*.

If in Definition 6.57, $\{f_k\} = \{f_1, \ldots, f_m\}$ is a finite collection of random variables, then it is of course sufficient to require that (i) holds only for $n = m$. Examples of finite collections of iid random variables are the components of $f^{\times m}$, the mth Cartesian power of any random variable f, in which case each component has the same distribution as f.

It is also possible to give examples of sequences of independent and iid random variables. Let f_j be random variables defined on probability spaces $(S_j, \mathcal{B}_j, P_j)$, for $j = 1, 2, \ldots$, respectively. Consider the Cartesian product $(S, \mathcal{B}, P)$ of these probability spaces. For an elementary event $e = \{e_j\} = (e_1, e_2, \ldots) \in S$ set $\tilde{f}_j(e) = f_j(e_j)$, $j = 1, 2 \ldots$. Then the random variables $\tilde{f}_j$ are independent. If $(S_j, \mathcal{B}_j, P_j) = (S', \mathcal{B}', P')$ and $f_j = f$ for $j = 1, 2 \ldots$, then the random variables $\tilde{f}_j$ are iid. In this case we denote the sequence $\{\tilde{f}_j\}$ by $f^{\times \infty}$ and call it the *infinite Cartesian power of f*. We also denote the vector-valued random variable $(\tilde{f}_1, \ldots, \tilde{f}_n)$ by $f^{\times \infty, n}$ and call it the *sample of size n from the infinite Cartesian power of f*. Clearly, $f^{\times \infty, n}$ is defined on $(S', \mathcal{B}', P')^\infty$ for every $n \in \mathbb{N}$.

Remark 6.58. Obviously, Theorem 6.56 holds for a sequence of iid random variables for which the variance exists. In fact, for a sequence of iid random variables Theorem 6.56 can be proved only with the assumption of the existence of the mean, by an argument that does not use Chebyshev's inequality.

We will now introduce another type of convergence.

Definition 6.59. *Let $\{f_k\}$ be a sequence of random variables on a sample space $(S, \mathcal{B}, P)$ such that f_k^2 is integrable on S for every k. Let f be a random variable on the same sample space. Then the sequence $\{f_k\}$ is said to converge to f in the square mean or simply in the mean if $(f_k - f)^2$ is integrable on S for all k and*

$$\lim_{k \to \infty} \mathcal{E}((f_k - f)^2) = 0.$$

In this case we write $f_k \overset{M}{\to} f$.

We note that in Definition 6.59, f^2 is automatically integrable on S which follows from the triangle inequality (see property (viii) of integrals in Sect. 6.5).

It is not hard to show that if $f_k \overset{M}{\to} f$, then $f_k \overset{P}{\to} f$. Indeed, it follows from Chebyshev's inequality (6.38) that

$$P\left(\{e \in S : |f_k(e) - f(e)| \geq \varepsilon\}\right) = P\left(\{e \in S : (f_k(e) - f(e))^2 \geq \varepsilon^2\}\right)$$

$$\leq \frac{\mathcal{E}\left((f_k - f)^2\right)}{\varepsilon^2},$$

and hence convergence of $\{f_k\}$ to f in the mean implies convergence of $\{f_k\}$ to the same random variable f in probability. The converse is not true as the following example shows.

Example 6.60. Consider the probability space from Part 3 of Example 6.14 and the following sequence of random variables

$$f_k(e) = \begin{cases} \sqrt{k}, & \text{if } 0 \leq e \leq \dfrac{1}{k}, \\ 0, & \text{if } \dfrac{1}{k} < e \leq 1. \end{cases}$$

We have $f_k \overset{P}{\to} 0$ since for any $\varepsilon < 1$ the following holds

$$P\left(\{e \in [0,1] : |f_k(e)| \geq \varepsilon\}\right) = \frac{1}{k} \to 0, \quad \text{as } k \to \infty.$$

It is also clear that f_k^2 is integrable on $S = [0,1]$ for all k. However, $\mathcal{E}(f_k^2) = 1$ for all k, and thus $\{f_k\}$ does not converge to 0 in the mean. In fact, $\{f_k\}$ does not converge in the mean to any random variable. Indeed, suppose that there exists a random variable f such that $f_k \overset{M}{\to} f$. Then from the triangle inequality (see property (viii) of integrals in Sect. 6.5) we have

$$\mathcal{E}((f_k - f_n)^2) = \mathcal{E}\left(\left((f_k - f) + (f - f_n)\right)^2\right) \leq \left(\mathcal{E}((f_k - f)^2) + \mathcal{E}((f_n - f)^2)\right)^2,$$

and hence $\mathcal{E}((f_k - f_n)^2) \to 0$ as $k, n \to \infty$. However, we have

$$\mathcal{E}((f_k - f_{2k})^2) = 2 - \sqrt{2}, \quad \text{for all } k,$$

which shows that $\{f_k\}$ does not converge in the mean to any random variable.

We will now introduce one more type of convergence.

Definition 6.61. *A sequence $\{f_k\}$ is said to converge in distribution to a random variable f if for every point $x \in \mathbb{R}$ at which F_f is continuous, we have*

$$\lim_{k \to \infty} F_{f_k}(x) = F_f(x).$$

In this case we write $f_k \overset{D}{\to} f$.

Note that the limit of a sequence that converges in distribution is not unique. Indeed, if $f_k \xrightarrow{D} f$ and if $\tilde{f}$ is another random variable such that $F_{\tilde{f}} = F_f$, then we also have $f_k \xrightarrow{D} \tilde{f}$. Also note that in Definition 6.61 one can in fact allow each of the random variables f_k and f to be defined on its own probability space. Convergence in distribution is often understood in this broader sense.

It is not hard to show that if $f_k \xrightarrow{D} f$ and F_f is continuous on $\mathbb{R}$, then F_{f_k} converges to F_f *uniformly* on $\mathbb{R}$ (see Exercise 6.21).

We will now show that convergence in distribution is a weaker condition than convergence in probability.

Theorem 6.62. *If $f_k \xrightarrow{P} f$, then $f_k \xrightarrow{D} f$.*

Proof: Suppose F_f is continuous at $x_0 \in \mathbb{R}$ and let $x' < x_0$. Consider the following three events

$$E_1 = E_{x'}(f),$$
$$E_2^k = E_{x_0}(f_k),$$
$$E_3^k = \left(E_{x_0-x'}(|f_k - f|) \right)^c = \{ e \in S : |f_k(e) - f(e)| > x_0 - x' \}.$$

Clearly, $E_1 \subset E_2 \cup E_3$, and therefore we have

$$F_f(x') = P(E_1) \leq P(E_2^k) + P(E_3^k) = F_{f_k}(x_0) + P(E_3^k).$$

Since $f_k \xrightarrow{P} f$, we have $P(E_3^k) \to 0$ as $k \to \infty$, and therefore we obtain $F_f(x') \leq \liminf_{k\to\infty} F_{f_k}(x_0)$.

Similarly, by taking $x'' > x_0$ we obtain $F_f(x'') \geq \limsup_{k\to\infty} F_{f_k}(x_0)$. We therefore have

$$F_f(x') \leq \liminf_{k\to\infty} F_{f_k}(x_0) \leq \limsup_{k\to\infty} F_{f_k}(x_0) \leq F_f(x'').$$

Now let in the above inequality $x', x'' \to x_0$. Since F_f is continuous at x_0, $F_f(x') \to F_f(x_0)$ as $x' \to x_0$ and $F_f(x'') \to F_f(x_0)$ as $x'' \to x_0$. Hence $\lim_{k\to\infty} F_{f_k}(x_0)$ exists and is equal to $F_f(x_0)$. Thus $f_k \xrightarrow{D} f$.

The theorem is proved. ∎

Theorem 6.62 implies that if $f_k \xrightarrow{M} f$, then $f_k \xrightarrow{D} f$. The converse is not true as shown in the following example.

Example 6.63. Consider again the probability space from Part 3 of Example 6.14 and the following sequence of random variables

$$f_{2m-1}(e) = \begin{cases} 1, & \text{if } 0 \leq e \leq \frac{1}{2}, \\ 0, & \text{if } \frac{1}{2} < e \leq 1, \end{cases} \qquad f_{2m}(e) = \begin{cases} 0, & \text{if } 0 \leq e \leq \frac{1}{2}, \\ 1, & \text{if } \frac{1}{2} < e \leq 1, \end{cases}$$

for $m \in \mathbb{N}$. All f_k are discrete random variables with the same probability distribution $\{p_{f_k}(0) = 1/2,\ p_{f_k}(1) = 1/2\}$, and hence their distribution functions coincide. Therefore, $f_k \overset{D}{\to} f_1$ (and, in fact, $f_k \overset{D}{\to} f_2$). On the other hand, for $\varepsilon \leq 1$, we have

$$P(\{e \in [0,1] : |f_{2m} - f_1| \geq \varepsilon\}) = 1,$$

for all $m \in \mathbb{N}$, and hence $\{f_k\}$ does not converge to f_1 in probability. In fact, $\{f_k\}$ does not converge in probability to any random variable. Indeed, suppose that $f_k \overset{P}{\to} f$. Then we have

$$P(\{e \in [0,1] : |f_{2m+1} - f| \geq \varepsilon\}) = P(\{e \in [0,1] : |f_1 - f| \geq \varepsilon\}),$$
$$P(\{e \in [0,1] : |f_{2m} - f| \geq \varepsilon\}) \ \ = P(\{e \in [0,1] : |f_2 - f| \geq \varepsilon\}).$$

Therefore $P(\{e \in [0,1] : |f_1 - f| \geq \varepsilon\}) = P(\{e \in [0,1] : |f_2 - f| \geq \varepsilon\}) = 0$ for every $\varepsilon > 0$, which implies that f almost certainly coincides with each of f_1 and f_2. However, f_1 and f_2 do not coincide almost certainly. Therefore, $\{f_k\}$ does not converge in probability to any random variable.

The sequence of random variables from Example 6.63 is also an example of a sequence that converges in distribution but does not converge in the mean.

We will now state without proof a fundamental theorem of the probability theory and statistics. The theorem is formulated in terms of convergence in distribution.

Theorem 6.64. (Central Limit Theorem)_Let $\{f_k\}$ be a sequence of iid random variables for which the variance exists. Let μ and σ^2 be the mean and variance of the f_k's respectively. Assume that $\sigma^2 > 0$ and consider the random variables_

$$g_n = \frac{\displaystyle\sum_{k=1}^{n} f_k - n\mu}{\sqrt{n\sigma^2}}.$$

Then $g_n \overset{D}{\to} f$, where f is a continuous random variable whose distribution is the standard normal distribution, and, moreover, F_{g_n} converges to F_f uniformly on $\mathbb{R}$.

Theorem 6.64 shows that for a sequence of iid random variables, the distribution function of the sum $\sum_{k=1}^{n} f_k$ is uniformly on $\mathbb{R}$ close to the normal distribution with mean $n\mu$ and variance $n\sigma^2$, thus providing one indication of the importance of the normal distribution in probability and statistics.

Next, we will consider another frequently used type of convergence.

Definition 6.65. _A sequence $\{f_k\}$ of random variables is said to converge almost certainly or a.c. to a random variable f defined on the same probability space, if_

$$\lim_{k \to \infty} f_k(e) = f(e),$$

for all $e \in E$, where E is an event with $P(E) = 1$. In this case we write $f_k \overset{a.c.}{\to} f$ and call E the set of convergence of $\{f_k\}$.

We note that in the above definition it is not necessary to require that f is a random variable. One can show that the almost certain limit of a sequence of random variables is a random variable as well (see Exercise 6.22).

We will now show that almost certain convergence is stronger than convergence in probability.

Theorem 6.66. If $f_k \overset{a.c.}{\to} f$, then $f_k \overset{P}{\to} f$.

Proof: Let $(S, \mathcal{B}, P)$ be the probability space on which the random variables f_k are defined. Fix $\varepsilon > 0$ and consider the following sequence of events

$$E_k = \{e \in S : |f_n(e) - f(e)| < \varepsilon \text{ for all } n \geq k\}.$$

Clearly, $E_k \in \mathcal{B}$ and $E_k \subset E_{k+1}$ for all $k \in \mathbb{N}$. Let E be the set of convergence of $\{f_k\}$. We obviously have $E \subset \cup_{k=1}^{\infty} E_k$. Since $P(E) = 1$, we obtain $P(\cup_{k=1}^{\infty} E_k) = 1$ and hence by property (iv) of probability measures (see Sect. 6.2), $P(E_k) \to 1$ as $k \to \infty$.

We now have

$$P(\{e \in S : |f_k(e) - f(e)| < \varepsilon\}) \geq P(E_k) \to 1, \quad \text{as } k \to \infty,$$

and therefore

$$P(\{e \in S : |f_k(e) - f(e)| \geq \varepsilon\}) \to 0, \quad \text{as } k \to \infty,$$

which shows that $f_k \overset{P}{\to} f$.

The theorem is proved. ■

We will now formulate a version of Theorem 6.56 in which convergence in probability is replaced with almost certain convergence.

Theorem 6.67. (Strong Law of Large Numbers) Let $\{f_k\}$ be a sequence of independent random variables on a probability space $(S, \mathcal{B}, P)$. Suppose that f_k^2 is integrable on S for each k. Let $\mathcal{E}(f_k) = \mu$ for all k. Also let $\sigma_k^2 = Var(f_k)$ and assume that the series

$$\sum_{k=1}^{\infty} \frac{\sigma_k^2}{k^2}$$

converges. Then $\overline{f}_n \overset{a.c.}{\to} \mu$, where

$$\overline{f}_n = \frac{1}{n} \sum_{k=1}^{n} f_k.$$

Note that Theorem 6.67 holds for sequences of iid random variables.

Theorem 6.66 implies that if $f_k \xrightarrow{a.c.} f$, then $f_k \xrightarrow{D} f$. The following example shows that the converse to Theorem 6.66 does not hold.

Example 6.68. As before, consider the probability space from Part 3 of Example 6.14. We also consider a sequence of random variables $\{f_{k,n}\}$ indexed for convenience by two indices, $k \in \mathbb{N}$, $n = 1, \ldots, k$, defined as follows

$$
f_{k,n}(e) = \begin{cases} 1, & \text{if } \dfrac{n-1}{k} < e \leq \dfrac{n}{k}, \\[2mm] 0, & \text{otherwise.} \end{cases}
$$

The sequence $\{f_{k,n}\}$ converges in probability to 0. Indeed, for $\varepsilon < 1$ we have

$$
P(\{e \in [0,1] : |f_{k,n}(e)| \geq \varepsilon\}) = \frac{1}{k} \to 0, \quad \text{as } k \to \infty.
$$

However, $\{f_{k,n}\}$ does not converge to 0 almost certainly since the only point $e \in [0,1]$ for which $\lim_{k \to \infty} f_{k,n}(e) = 0$ is $e = 0$.

Nevertheless, one can show that if $f_k \xrightarrow{P} f$, then one can always find a subsequence $\{f_{k_m}\}$ of $\{f_k\}$ such that $f_{k_m} \xrightarrow{a.c.} f$. We do not prove this fact here but notice that in Example 6.68 for the subsequence $\{f_{k,1}\}$ we have $f_{k,1} \xrightarrow{a.c.} 0$.

Note that the sequence from Example 6.68 in fact converges to 0 in the mean since

$$
\mathcal{E}(f_{k,n}^2) = \frac{1}{k} \to 0, \quad \text{as } k \to \infty.
$$

Hence this example also shows that convergence in the mean does not imply almost certain convergence. Nevertheless, if $f_k \xrightarrow{M} f$, then one can always find a subsequence $\{f_{k_m}\}$ of $\{f_k\}$ such that $f_{k_m} \xrightarrow{a.c.} f$. Also note that the sequence from Example 6.60 shows that almost certain convergence does not imply convergence in the mean. Finally, Example 6.63 also shows that convergence in distribution does not imply almost certain convergence.

Recall that in Sect. 6.5 we defined yet another kind of convergence, namely uniform convergence (see Definition 6.28). Clearly, uniform convergence implies almost certain convergence (the set of convergence in this case is all of the sample space) and hence convergence in probability and convergence in distribution. A proof analogous to that in Sect. 6.5 for convergence of integrals of discrete random variables shows that uniform convergence also implies convergence in the mean. However, none of the above types of convergence implies uniform convergence. Indeed, the sequence from Example 6.60 converges in probability, but does not converge uniformly, the sequence from Example 6.63 converges in distribution, but does not converge uniformly, the sequence from Example 6.68 converges in the mean, but does not converge uniformly, its subsequence $\{f_{k,1}\}$ converges almost certainly, but not uniformly.

We summarize the relationships among the various types of convergence in Fig. 6.1. Note that uniform convergence is the strongest type of convergence and convergence in distribution is the weakest type of convergence among all convergence types considered in this section.

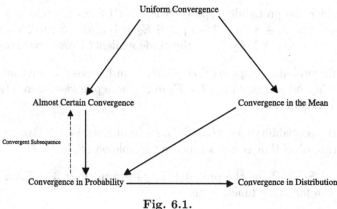

Fig. 6.1.

Exercises

6.1. Prove identities (6.1) and (6.2).

6.2. Let the sample space S be the closed square in the (x, y)-plane with vertices $(0, 0)$, $(0, 1)$, $(1, 0)$ and $(1, 1)$. Define $\mathcal{F}$ to be the collection of all rectangles in S with sides parallel to the x- and y-axes

$$
\mathcal{F} = \Big\{ (a, b) \times (c, d), [a, b) \times (c, d), (a, b] \times (c, d), [a, b] \times (c, d),
$$
$$
(a, b) \times [c, d), [a, b) \times [c, d), (a, b] \times [c, d), [a, b] \times [c, d),
$$
$$
(a, b) \times (c, d], [a, b) \times (c, d], (a, b] \times (c, d], [a, b] \times (c, d],
$$
$$
(a, b) \times [c, d], [a, b) \times [c, d], (a, b] \times [c, d], [a, b] \times [c, d],
$$
$$
0 \le a, b, c, d \le 1 \Big\}.
$$

Prove that $\mathcal{F}$ is a semi-algebra of events in S.

6.3. Prove that P defined in Part 3 of Example 6.11 is a probability measure on the semi-algebra $\mathcal{F}$ from Example 6.8.

6.4. Let S and $\mathcal{F}$ be as in Exercise 6.2. For a rectangle E in $\mathcal{F}$ define $P(E)$ to be equal to the area of E. Prove that P is a probability measure on $\mathcal{F}$.

6.5. Let S and $\mathcal{F}$ be as in Exercise 6.2 and P be the probability measure on $\mathcal{F}$ defined in Exercise 6.4. Consider the minimal σ-algebra $\mathcal{B}(\mathcal{F})$ generated by $\mathcal{F}$ (which is the σ-algebra of Borel sets in S) and extend the probability measure to events in $\mathcal{B}(\mathcal{F})$. Prove that every open or closed triangle in S with two sides parallel to the x- and y-axes is measurable with respect to this probability measure. Give an example of an non-measurable event in S.

6.6. Consider the probability space $(S, \mathcal{B}(\mathcal{F}), P)$ from Exercise 6.5 and two events $E_1 = \{(x, y) \in S : y \leq 1 - x\}$ and $E_2 = \{(x, y) \in S : y < x\}$ in S. Are E_1 and E_2 measurable? If yes, are they independent? Prove your conclusions.

6.7. For the probability space $(S, \mathcal{B}(\mathcal{F}), P)$ from Exercise 6.5 give an example of three independent events E_1, E_2, E_3 in S whose pairwise intersections have non-zero probabilities.

6.8. For the probability space $(S, \mathcal{B}(\mathcal{F}), P)$ from Exercise 6.5 give an example of a function on S that is not a random variable on $(S, \mathcal{B}(\mathcal{F}), P)$.

6.9. Let $(S, \mathcal{B}(\mathcal{F}), P)$ be the probability space from Part 3 of Example 6.14. Consider the following function on S

$$f(e) = \begin{cases} 0, & \text{if } e = 0, \\ (-1)^n(n+1), & \text{if } \dfrac{1}{n+1} < e \leq \dfrac{1}{n}, \text{ for } n \in \mathbb{N}. \end{cases}$$

Is f a random variable on $(S, \mathcal{B}(\mathcal{F}), P)$? If yes, is it integrable on either of the events $E_1 = [0, 1/2]$, $E_2 = [1/2, 1]$? Prove your conclusions.

6.10. Let $(S, \mathcal{B}(\mathcal{F}), P)$ be the probability space from Part 3 of Example 6.14. Consider the following function on S

$$f(e) = \begin{cases} 0, & \text{if } e = 0, \\ e^{-1/3}, & \text{if } 0 < e \leq 1. \end{cases}$$

Prove that f is a random variable on $(S, \mathcal{B}(\mathcal{F}), P)$, that f^2 is integrable on S and find the expected value $\mathcal{E}(f)$ and variance $Var(f)$ of f.

6.11. Give an example of an everywhere continuous, but not absolutely continuous monotone non-decreasing function on the real line. [Hint: use the Cantor set defined at the end of Sect. 6.1.]

6.12. Let $(S, \mathcal{B}(\mathcal{F}), P)$ be the probability space from Part 3 of Example 6.14. Find the distribution function of the random variable f defined as $f(e) = e^3$. Is f a continuous random variable? If yes, what is its density function? Prove your conclusions.

6.13. Consider the probability space $(S, \mathcal{B}(\mathcal{F}), P)$ from Exercise 6.5 and let $f : S \to \mathbb{R}$ be defined as follows

$$f(e) = x + 2y,$$

where $e = (x, y)$. Prove that f is a random variable on $(S, \mathcal{B}(\mathcal{F}), P)$, that f^2 is integrable on S and find the expected value $\mathcal{E}(f)$ and variance $Var(f)$ of f. In addition, find the distribution function F_f of f. Is f a continuous random variable? If yes, what is its density function? Prove your conclusions.

6.14. Let f be a random variable on a probability space $(S, \mathcal{B}, P)$ having a binomial distribution. Find $\int_S f^3 dP$.

6.15. Let f be a random variable on a probability space $(S, \mathcal{B}, P)$ having a geometric distribution. Find $\int_S f^4 dP$.

6.16. Let f be a random variable on a probability space $(S, \mathcal{B}, P)$ having a normal distribution. Find $\int_S f^3(f - 1) dP$.

6.17. Let f be a random variable on a probability space $(S, \mathcal{B}, P)$ having an exponential distribution. Find $\int_S f^5 dP$.

6.18. Consider the probability space $(S, \mathcal{B}(\mathcal{F}), P)$ from Exercise 6.5 and let $f : S \to \mathbb{R}^2$ be the following vector-valued function

$$f(e) = (x^2, \sin y),$$

where $e = (x, y)$. Prove that f is a vector-valued random variable on $(S, \mathcal{B}(\mathcal{F}), P)$, find the distribution function F_f of f and show that the components of f are independent random variables.

6.19. Let $(S, \mathcal{B}(\mathcal{F}), P)$ be the probability space from Part 3 of Example 6.14 and let $f : S \to \mathbb{R}^3$ be the following vector-valued random variable

$$f(e) = (e^2, e^3, \exp(e)).$$

Are the components of f independent random variables? Prove your conclusion.

6.20. Let $(S, \mathcal{B}(\mathcal{F}), P)$ be the probability space from Part 3 of Example 6.14. Consider the following sequence $\{f_k\}$ of random variables on it: $f_k(e) = e^k + e$. Prove that $f_k \xrightarrow{P} f$, where $f(e) = e$.

6.21. Let $f_k \xrightarrow{D} f$, and assume that F_f is continuous everywhere on $\mathbb{R}$. Prove that F_{f_k} converges to F_f uniformly on $\mathbb{R}$.

6.22. Let $\{f_k\}$ be a sequence of random variables on a probability space $(S, \mathcal{B}, P)$ that converges almost certainly to a function $f : S \to \mathbb{R}$. Prove that f is a random variable on $(S, \mathcal{B}, P)$.

6.23. Let $(S, \mathcal{B}(\mathcal{F}), P)$ be the probability space from Part 3 of Example 6.14. Define a sequence of functions $f_k : S \to \mathbb{R}$ as follows

$$f_k(e) = \begin{cases} -k^2 e + k, & \text{if } 0 \leq e \leq \dfrac{1}{k}, \\ \\ 0, & \text{if } \dfrac{1}{k} \leq e \leq 1. \end{cases}$$

Show that $\{f_k\}$ converges almost certainly, find its limit and the set of convergence.

6.24. Let $(S, \mathcal{B}(\mathcal{F}), P)$ be the probability space from Part 3 of Example 6.14. Define a sequence of random variables $f_k : S \to \mathbb{R}$ as follows

$$f_k(e) = \begin{cases} 2k\sqrt{e}, & \text{if } 0 \leq e \leq \dfrac{1}{2k}, \\ \\ \sqrt{-4k^2 e + 4k}, & \text{if } \dfrac{1}{2k} \leq e \leq \dfrac{1}{k}, \\ \\ 0, & \text{if } \dfrac{1}{k} \leq e \leq 1. \end{cases}$$

Prove that $\{f_k\}$ converges to $f \equiv 0$ almost certainly, but does not converge to f in the mean.

7

Significance of Sequence Alignment Scores

In this chapter we generally follow the line of argument in [EG], but our exposition is more focused and mathematically rigorous. In particular, we carefully describe the sample spaces on which each of the random variables involved is defined.

7.1 The Problem

In this section we return to the problem of aligning two biological sequences considered in Chap. 2. Suppose that for two given sequences we have been able to find optimal alignments (local or global, gapped or ungapped) for some scoring scheme, and assume that these alignments have a high score that we denote by s_0. Now we want to know if the alignments found are biologically meaningful and give evidence for homology, or they are just some of the best alignments between two unrelated sequences. There are two possible approaches to this problem: one is *classical*, the other is *Bayesian*. In this book we only deal with the more commonly used classical approach.

The classical approach will require applying the probability theory presented in Chap. 6 to the sequence alignment problem. We wish to calculate the probability of the event that the best alignments between two "randomly generated sequences" (as explained below) have score greater than or equal to s_0. If this probability is small (for example, does not exceed 0.05), then we say that the similarity observed between the original sequences is significant which indicates that the sequences are possibly homologous. Otherwise, we say that the similarity is not significant which indicates that the sequences are probably unrelated.

The problem is most interesting for local alignments. We will consider the problem only for the case of *ungapped* local alignments. It is complex enough even in this case. The corresponding problem for global ungapped alignments is much easier and will be dealt with in Sect. 8.4 within the general framework of statistical hypothesis testing.

Let x^0 be a sequence of length N_1, y^0 be a sequence of length N_2, and suppose that we have found some of the optimal ungapped alignments between segments in x^0 and segments in y^0. Denote by s_0 the score of these alignments. Let $\mathcal{Q}$ be the alphabet corresponding to x^0 and y^0 (recall that in our applications $\mathcal{Q}$ is either the DNA alphabet, or the RNA alphabet, or the amino acid alphabet). For ungapped alignments scores are calculated using a substitution matrix alone, and such a matrix whose elements we, as before, denote by $s(a, b)$, with $a, b \in \mathcal{Q}$, will be fixed from now on. We assume that $s(a, b)$ are integers for all $a, b \in \mathcal{Q}$ and hence s_0 is assumed to be a large integer. Let p_a be the frequency of $a \in \mathcal{Q}$ determined from x^0 and p'_b be the frequency of $b \in \mathcal{Q}$ determined from y^0, calculated for all $a, b \in \mathcal{Q}$ (in applications when a query sequence x^0 is compared to every sequence y^0 in a database, $\{p'_b\}$ are either calculated from the whole database, or taken from some published sources). Let S be the collection of all pairs (x, y) of sequences of letters from $\mathcal{Q}$, where x has length N_1 and y has length N_2. We will define a probability measure on S as follows: for $(x, y) \in S$ with $x = x_1 \ldots x_{N_1}$, $y = y_1 \ldots y_{N_2}$ set

$$P\Big(\{(x, y)\}\Big) = \prod_{i=1}^{N_1} p_{x_i} \times \prod_{j=1}^{N_2} p'_{y_j}.$$

Since S is a finite set and since the probability measure P is defined for every elementary event, it can be extended to the σ-algebra $\mathcal{B}$ of all events in S. One can think of the probability space $(S, \mathcal{B}, P)$ as the collection of pairs of sequences (x, y), where x has length N_1, y has length N_2, and x and y are generated by independent random processes from the frequencies $\{p_a\}$ and $\{p'_b\}$ respectively, with each site in x and y being generated independently of the others.

On the probability space $(S, \mathcal{B}, P)$ we define a random variable s as follows: for $(x, y) \in S$ we set $s((x, y))$ to be the score of any optimal local alignment between x and y. Under the classical approach we are interested in determining the probability $P\Big(\{(x, y) \in S : s((x, y)) \geq s_0\}\Big) = P\Big(\{(x, y) \in S : s((x, y)) > s_0 - 1\}\Big) = 1 - F_s(s_0 - 1) = F_s^*(s_0 - 1)$. If $F_s^*(s_0 - 1) < 0.05$, then we say that the similarity observed between x^0 and y^0 is *significant*; otherwise we say that it is *not significant*. The problem of calculating an approximation to $F_s(\alpha)$ for large values of α will be addressed in considerable detail in subsequent sections. In this section we will only present an argument that gives some indication of what the answer may look like.

Let $(x, y) \in S$ with $x = x_1 \ldots x_{N_1}$, $y = y_1 \ldots y_{N_2}$. A general local ungapped alignment between x and y is given as follows

$$x_i \ldots x_{i+N}$$
$$y_j \ldots y_{j+N},$$

for some $1 \leq i \leq N_1$, $1 \leq j \leq N_2$ and $N \leq \min\{N_1 - i, N_2 - j\}$. If we denote this alignment by $A_{i,j,N}(x, y)$, then for its score $\mathcal{S}\Big(A_{i,j,N}(x, y)\Big)$ we have

$$\mathcal{S}\Big(A_{i,j,N}(x,y)\Big) = s(x_i, y_j) + s(x_{i+i}, y_{j+1}) + \ldots + s(x_{i+N}, y_{j+N}).$$

The random variables $s(x_{i+k}, y_{i+k})$ on the probability space $(S, \mathcal{B}, P)$ are clearly iid. Hence by the Central Limit Theorem (Theorem 6.64), if N is large, the distribution function of $\mathcal{S}\Big(A_{i,j,N}(x,y)\Big)$ is approximated by the distribution function of a normally distributed random variable. Hence, the scores of long local alignments can be assumed to be normally distributed. Clearly, we have

$$s((x, y)) = \max_{i,j,N} \mathcal{S}\Big(A_{i,j,N}(x,y)\Big).$$

The random variables $\mathcal{S}\Big(A_{i,j,N}(x,y)\Big)$ are not independent, but for the purposes of our (not entirely rigorous) argument we assume that they are independent. We also assume that they are identically distributed. Hence, we are considering the following situation: we are given a collection $\{f_1, \ldots, f_M\}$ of iid normally distributed random variables on some probability space $(S', \mathcal{B}', P')$, and we need to find the distribution function of the random variable $f_{\max} = \max\{f_1, \ldots, f_M\}$. The distribution of $f_{\max}$ is called the *extreme value distribution for* $\{f_1, \ldots, f_M\}$. Let F be the distribution function of the f_k's. Then we have

$$F_{f_{\max}}(\alpha) = P\Big(\{e \in S' : f_{\max}(e) \leq \alpha\}\Big) = P\Big(\{e \in S' : f_1(e) \leq \alpha, \ldots,$$

$$f_M(e) \leq \alpha\}\Big) = \prod_{k=1}^{M} P\Big(\{e \in S' : f_k(e) \leq \alpha\}\Big) = F(\alpha)^M.$$

One can now show that as M and α become large, we have

$$F_{f_{\max}}(\alpha) \approx \exp(-KM \exp(-\lambda\alpha)),$$

or

$$F^*_{f_{\max}}(\alpha) \approx 1 - \exp(-KM \exp(-\lambda\alpha)), \tag{7.1}$$

for some constants $K > 0$ and $\lambda > 0$. In the following sections we will derive in a much more rigorous way a distribution of a form similar to that in (7.1) that gives an approximation to $F_s^*(\alpha)$, if N_1, N_2 and α are sufficiently large.

7.2 Random Walks

In this section we will introduce and study an important class of random processes.

Definition 7.1. *A random walk is a discrete-time process that starts at 0 and can move up or down by one of finitely many prescribed values (step sizes) with prescribed probabilities independently of previously made moves.*

We will always assume that the set of all possible step sizes in a random walk has the form $T = \{-c, -c+1, \ldots, 0, \ldots, d-1, d\}$ for some fixed $c, d \in \mathbb{N}$ and denote the respective probabilities by $p_{-c}, p_{-c+1}, \ldots, p_d$; these probabilities are required to sum up to 1. We will now introduce a random variable called the *step size* whose probability distribution is $\{p_j, j = -c, \ldots, d\}$. Define a probability measure on T by setting $P(\{j\}) = p_j$ for all $j \in T$ and let the step size on the resulting probability space $(T, \mathcal{B}_T, P)$, where $\mathcal{B}_T$ is the σ-algebra of all events in T, be the identity mapping from T into itself. The random walk can be identified with this random variable.

We assume three conditions throughout

(i) $p_{-c} > 0$ and $p_d > 0$,

(ii) the step size has a negative mean, that is, $\sum_{j=-c}^{d} j p_j < 0$,

(iii) the greatest common divisor of all positive elements $j \in T$ for which $p_j > 0$ is equal to 1.

We will now associate a random walk with the substitution matrix $(s(a, b))$ and frequencies $\{p_a\}$, $\{p_b'\}$. Let T be the collection of all integers between the minimal element $\tilde{s}$ and the maximal element $\hat{s}$ of the substitution matrix. If j is an element of the substitution matrix, we define the corresponding probability as $p_j = \sum_{(a,b):s(a,b)=j} p_a p_b'$. If $\tilde{s} < j < \hat{s}$ and $s(a, b) \neq j$ for all $a, b \in \mathcal{Q}$, we set $p_j = 0$. In accordance with the above requirements, in the future we will always assume that the elements of the substitution matrix satisfy the following conditions

(iv) $\hat{s} > 0$, $p_{\hat{s}} > 0$, and $p_{\tilde{s}} > 0$,

(v) $\sum_{a,b \in \mathcal{Q}} s(a, b) p_a p_b' < 0$,

(vi) the greatest common divisor of the positive elements of the substitution matrix is equal to 1.

Note that condition (v) implies that the minimal element $\tilde{s}$ in the substitution matrix is negative, as required.

In the remainder of this section we will consider random walks without reference to sequence comparison, to which we will return in the following section. The results presented in this section were first obtained in [KD].

Let S_w be the collection of all possible *trajectories* of a random walk, that is, the collection of all infinite sequences $t = t_1, t_2, \ldots$, where $t_j \in T$. Figure 7.1 shows a trajectory for a random walk with step sizes $-1, 0, 1$.

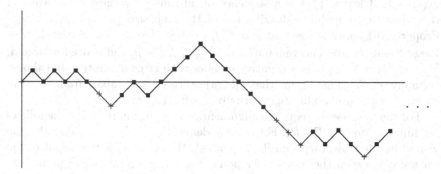

Fig. 7.1.

The crosses in this figure relate to *ladder points*, that is, to points in the trajectory lower than any previously reached point. The part of the trajectory from a ladder point until the highest point visited before the next ladder point, is called an *excursion*.

One can think of S_w as the sample space that consists of all outcomes of the random walk if we think of it as a sequence-generation process. We will now introduce a probability measure on S_w. Consider the following family $\mathcal{F}_w$ of events in S_w: $\mathcal{F}_w$ includes S_w, $\emptyset$ and all *cylinder events*, that is, events of the form

$$E_{t^0_{i_1},\ldots,t^0_{i_m}} = \left\{ t \in S_w : t_{i_1} = t^0_{i_1}, \ldots, t_{i_m} = t^0_{i_m} \right\}, \tag{7.2}$$

for all finite subsets of indices $i_1 < \ldots < i_m$ and all possible $t^0_{i_1}, \ldots, t^0_{i_m} \in T$. It is easy to check that $\mathcal{F}_w$ is a semi-algebra (see Exercise 7.1), hence we can consider the minimal σ-algebra $\mathcal{B}_w = \mathcal{B}(\mathcal{F}_w)$ generated by $\mathcal{F}_w$. We define a probability measure on $\mathcal{F}_w$ as follows. Set $P(S_w) = 1$, $P(\emptyset) = 0$ and

$$P\left(E_{t^0_{i_1},\ldots,t^0_{i_m}}\right) = \prod_{k=1}^{m} p_{t^0_{i_j}}. \tag{7.3}$$

It is not hard to check that P is a probability measure on $\mathcal{F}_w$ (see Exercise 7.2), and therefore it can be extended to $\mathcal{B}_w$. The resulting probability space $(S_w, \mathcal{B}_w, P)$ is the space we will work on while studying random walks. We remark that $(S_w, \mathcal{B}_w, P)$ is simply the infinite Cartesian power of the probability space $(T, \mathcal{B}_T, P)$ as defined in Sect. 6.2 (see Exercise 7.3). We also note that if $p_j \neq 0$ for all $j \in T$, then the σ-algebra to which P extends from $\mathcal{F}_w$ by the Lebesgue extension procedure can be essentially identified with the σ-algebra of Lebesgue measurable sets in $[0,1]$.

For every $k \in \mathbb{N}$ define a random variable f_k on $(S_w, \mathcal{B}_w, P)$ by setting $f_k(t) = t_k$. Clearly, $\{f_k\}$ is a sequence of iid random variables, and each of them has the probability distribution of the step size $\{p_{-c}, p_{-c+1}, \ldots, p_d\}$. Property (ii) above states that $\mu = \mathcal{E}(f_k)$ is negative. By the Strong Law of Large Numbers (see Theorem 6.67), $1/n \sum_{k=1}^{n} f_k \overset{a.c.}{\to} \mu$, and hence for large n, $\sum_{k=1}^{n} f_k(t) = \sum_{k=1}^{n} t_k$ is a negative number with large absolute value almost certainly on S_w. This means that all trajectories $t \in S_w$ apart from those in an event E_0 of probability 0, eventually "drift" to $-\infty$.

For the purposes of sequence alignment, we will be interested in the following function on $(S_w, \mathcal{B}_w, P)$. For $t \in S_w$, define $Y(t)$ to be the maximal value visited by the trajectory t until t reaches -1, that is, $Y(t)$ is the height of the excursion between the first ladder point 0 and the second ladder point (the first negative value visited by t). Clearly, Y is defined everywhere on E_0^c. Since we will only be interested in the distribution of Y, we define Y arbitrarily on E_0. It can be shown that Y is a random variable on $(S_w, \mathcal{B}_w, P)$ (see Exercise 7.4). Our goal is to determine the behavior of $F_Y(\alpha)$ or $F_Y^*(\alpha)$, where α is a positive integer, as $\alpha \to \infty$. Specifically, we will show that Y is geometric-like (see Definition 6.48). For convenience we introduce another function $\tilde{Y}$ defined as follows: for a trajectory $t \in E_0^c$ define $\tilde{Y}(t)$ as the maximal value visited by t and define $\tilde{Y}$ arbitrarily on E_0. It can be shown that $\tilde{Y}$ is also a random variable on $(S_w, \mathcal{B}_w, P)$ (see Exercise 7.4). We will study the behavior of $F_{\tilde{Y}}^*$ and $F_{\tilde{Y}}^*$.

Let α be a positive integer. For $j = 1, \ldots, c$ denote by R_{-j} the probability of the event that the first negative value that a trajectory visits is $-j$. More precisely, R_{-j} is the probability of the following event

$$E_{-j} = \Big\{ t \in S_w : \text{for some } n \in \mathbb{N} \text{ we have } \sum_{i=1}^{m} t_i \geq 0, \text{ for all } m < n$$

$$\text{and } \sum_{i=1}^{n} t_i = -j \Big\} \tag{7.4}$$

(see Exercise 7.5). Then we have

$$F_{\tilde{Y}}^*(\alpha) = F_Y^*(\alpha) + \sum_{j=1}^{c} R_{-j} F_{\tilde{Y}}^*(\alpha + j),$$

hence

$$F_Y^*(\alpha) = F_{\tilde{Y}}^*(\alpha) - \sum_{j=1}^{c} R_{-j} F_{\tilde{Y}}^*(\alpha + j). \tag{7.5}$$

Therefore, to study the behavior of $F_Y^*(\alpha)$ (or, equivalently, $F_Y(\alpha)$) we can study that of $F_{\tilde{Y}}^*(\alpha)$ (or, equivalently, $F_{\tilde{Y}}(\alpha)$).

We now need the following definition.

Definition 7.2. *Let f be a discrete random variable that takes finitely many values $r_1, r_2, \ldots, r_n$, with $r_i \neq r_j$ for $i \neq j$, and let $\{p_f(r_j),\ j = 1, \ldots, n\}$ be the probability distribution of f. Then the following function of a real variable*

$$\mathbb{M}_f(\theta) = \mathcal{E}(\exp(\theta f)) = \sum_{j=1}^{n} \exp(\theta r_j) p_f(r_j), \quad \theta \in \mathbb{R},$$

is called the moment-generating function of f.

Of course, a moment-generating function can be introduced for any random variable, but it may not be defined for all $\theta \in \mathbb{R}$ (see Exercise 7.6).

We will now prove the following theorem.

Theorem 7.3. *Let f be a discrete random variable that takes finitely many values and such that $\mathcal{E}(f) \neq 0$. Suppose that f takes a positive value a and a negative value b with positive probabilities $p_f(a)$ and $p_f(b)$ respectively. Then there exists a unique non-zero $\theta^* \in \mathbb{R}$ such that $\mathbb{M}_f(\theta^*) = 1$.*

Proof: We have

$$\mathbb{M}_f(\theta) > \exp(\theta a) p_f(a), \qquad \mathbb{M}_f(\theta) > \exp(\theta b) p_f(b),$$

for all $\theta \in \mathbb{R}$. Thus $\mathbb{M}_f(\theta) \to \infty$ as $\theta \to \pm\infty$. Further,

$$\mathbb{M}_f''(\theta) = \sum_{j=1}^{n} r_j^2 \exp(\theta r_j) p_f(r_j) > 0,$$

for all $\theta \in \mathbb{R}$, that is, $\mathbb{M}_f$ is a convex function. We also have $\mathbb{M}_f(0) = 1$ and $\mathbb{M}_f'(0) = \mathcal{E}(f) \neq 0$. Together with the convexity of f this implies that there exists a unique non-zero $\theta^* \in \mathbb{R}$ such that $\mathbb{M}_f(\theta^*) = 1$. ∎

It is clear from the proof of Theorem 7.3 that $\theta^* > 0$, if $\mathcal{E}(f) < 0$ and $\theta^* < 0$, if $\mathcal{E}(f) > 0$.

We will now apply Theorem 7.3 to the random walk we are considering. Let $\mathbb{M}$ be the moment-generating function of the step size

$$\mathbb{M}(\theta) = \sum_{j=-c}^{d} \exp(\theta j) p_j.$$

Clearly, the step size satisfies the conditions of Theorem 7.3, and therefore there exists a unique non-zero θ^* such that $\mathbb{M}(\theta^*) = 1$, that is

$$\sum_{j=-c}^{d} \exp(\theta^* j) p_j = 1. \tag{7.6}$$

Since the step size has a negative mean, we have $\theta^* > 0$.

For $k = 1, 2 \ldots$ denote by Q_k the probability of the event that a trajectory visits k before visiting any other positive value. More precisely, Q_k is the probability of the following event

$$E_k = \left\{ t \in S_w : \text{for some } n \in \mathbb{N} \text{ we have } \sum_{i=1}^{m} t_i \leq 0, \text{ for all } m < n \right.$$

$$\left. \text{and } \sum_{i=1}^{n} t_i = k \right\} \tag{7.7}$$

(see Exercise 7.5). Clearly, $Q_k = 0$ for all $k > d$, and we also set for convenience $Q_0 = 0$. Since the probability of the event that a trajectory never visits any positive values is non-zero, we have $\sum_{k=1}^{d} Q_k < 1$. To find the behavior of $F_{\tilde{Y}}(\alpha)$, where α is a positive integer, we note that

$$F_{\tilde{Y}}(\alpha) = \overline{Q} + \sum_{k=0}^{\alpha} Q_k F_{\tilde{Y}}(\alpha - k), \tag{7.8}$$

where $\overline{Q} = 1 - \sum_{k=1}^{d} Q_k$. Define

$$V(\alpha) = F_{\tilde{Y}}^*(\alpha) \exp(\theta^* \alpha). \tag{7.9}$$

We will now show that the limit $\lim_{\alpha \to \infty} V(\alpha)$ exists and determine its value. Equation (7.8) can be rewritten as

$$1 - V(\alpha) \exp(-\theta^* \alpha) = \overline{Q} + \sum_{k=0}^{\alpha} Q_k \left(1 - V(\alpha - k) \exp(-\theta^*(\alpha - k)) \right),$$

which gives

$$V(\alpha) = \exp(\theta^* \alpha) \sum_{k=\alpha+1}^{d} Q_k + \sum_{k=0}^{\alpha} (Q_k \exp(\theta^* k)) V(\alpha - k), \quad \text{if } \alpha < d, \tag{7.10}$$

and

$$V(\alpha) = \sum_{k=0}^{d} (Q_k \exp(\theta^* k)) V(\alpha - k), \quad \text{if } \alpha \geq d. \tag{7.11}$$

Next, we will use the following fact that we state without proof.

Theorem 7.4. (The Renewal Theorem) *Suppose three sequences* $\{a_0, a_1, \ldots\}$, $\{b_0, b_1, \ldots\}$ *and* $\{c_0, c_1, \ldots\}$ *of non-negative numbers satisfy the equation*

$$c_j = a_j + (c_j b_0 + c_{j-1} b_1 + \ldots + c_1 b_{j-1} + c_0 b_j), \tag{7.12}$$

for all $j \geq 0$. *Suppose further that the sequence* $\{c_j\}$ *is bounded,* $\sum_{j=0}^{\infty} b_j = 1$ *and that the series* $\sum_{j=0}^{\infty} a_j$ *and* $\sum_{j=0}^{\infty} j b_j$ *converge. Denote respectively by* A

and μ the sums of these series. Assume in addition that the greatest common divisor of the integers j for which $b_j > 0$ is equal to 1. Then the limit $\lim_{j \to \infty} c_j$ exists and

$$\lim_{j \to \infty} c_j = \frac{A}{\mu}.$$

A proof of Theorem 7.4 can be found in [Kar].

Theorem 7.4 can be used to find the limit $\lim_{\alpha \to \infty} V(\alpha)$. Indeed, set $a_j = \exp(\theta^* j) \sum_{k=j+1}^{d} Q_k$ for $j < d$ and $a_j = 0$ for $j \geq d$. Also set $b_j = Q_j \exp(\theta^* j)$, $c_j = V(j)$ for all $j \geq 0$. Clearly, the series $\sum_{j=0}^{\infty} a_j$ and $\sum_{j=0}^{\infty} j b_j$ converge. Identities (7.10) and (7.11) show that condition (7.12) is satisfied. Also, requirement (iii) from the beginning of this section guarantees that the greatest common divisor of the integers j for which $b_j > 0$ is equal to 1. In order to use Theorem 7.4 we further need to show that $\sum_{j=0}^{\infty} b_j = 1$, that is,

$$\sum_{k=1}^{d} Q_k \exp(\theta^* k) = 1. \tag{7.13}$$

To obtain (7.13), choose $L \in \mathbb{N}$, and denote by $Q_k(L)$, for $k = 1, \ldots, d$, the probability of the event that a trajectory visits k before visiting any other positive value and before reaching $-L$; also denote by $Q_k(L)$, for $k = -L - c + 1, \ldots, -L$, the probability of the event that a trajectory visits k before visiting any positive values and does not reach $-L$ before visiting k. Clearly, we have

$$\lim_{L \to \infty} Q_k(L) = Q_k, \quad \text{for } k = 1, \ldots, d.$$

We now need the following theorem that we also state without proof. A proof can be found in [KT].

Theorem 7.5. (Wald's Identity) Let N denote a random variable on $(S_w, \mathcal{B}_w, P)$ whose value $N(t)$, for every $t \in E_0^c$, is equal to the number of steps that t takes to reach either a positive value or $-L$ for the first time. Let further T_N denote the random variable on $(S_w, \mathcal{B}_w, P)$ whose value, for every $t \in E_0^c$, is equal to the value that t visits after $N(t)$ steps. Then for all $\theta \in \mathbb{R}$ we have

$$\mathcal{E}\left(\mathbb{M}(\theta)^{-N} \exp(\theta T_N) \right) = 1. \tag{7.14}$$

Note that in particular Theorem 7.5 states that $\mathbb{M}(\theta)^{-N} \exp(\theta T_N)$ is integrable on S_w with respect to P for every $\theta \in \mathbb{R}$.

Wald's identity (7.14) for $\theta = \theta^*$ becomes

$$\mathcal{E}\left(\exp(\theta^* T_N) \right) = 1,$$

that is,

$$\sum_{k=-L-c+1}^{-L} Q_k(L)\exp(\theta^* k) + \sum_{k=1}^{d} Q_k(L)\exp(\theta^* k) = 1.$$

Letting $L \to \infty$ in this identity gives identity (7.13), as required.

Identity (7.13) means that $\sum_{j=0}^{\infty} b_j = 1$. It also shows that the sequence $\{c_j\}$ is bounded (see Exercise 7.7). Hence the Renewal Theorem can be applied to the three sequences $\{a_j\}, \{b_j\}, \{c_j\}$ introduced above. We then obtain that the limit $\lim_{\alpha \to \infty} V(\alpha)$ exists and is equal to $V = A/\mu$, where $A = \sum_{j=0}^{\infty} a_j$ and $\mu = \sum_{j=0}^{\infty} jb_j$. We will now find A. We have

$$A = \sum_{j=0}^{\infty} a_j = \sum_{j=0}^{d} \exp(\theta^* j) \sum_{k=j+1}^{d} Q_k.$$

Multiplying both parts of this identity by $\exp(\theta^*) - 1$ we obtain

$$A(\exp(\theta^*) - 1) = \sum_{k=1}^{d} Q_k \exp(\theta^* k) - (Q_1 + \ldots + Q_d) = 1 - (Q_1 + \ldots + Q_d) = \overline{Q},$$

where we again used identity (7.13). Hence

$$A = \frac{\overline{Q}}{\exp(\theta^*) - 1},$$

and therefore

$$V = \frac{\overline{Q}}{(\exp(\theta^*) - 1) \sum_{k=1}^{d} kQ_k \exp(\theta^* k)}. \tag{7.15}$$

Now from (7.9) we obtain

$$F_{\tilde{Y}}^*(\alpha) \sim V \exp(-\theta^* \alpha),$$

with V given by formula (7.15). Therefore formula (7.5) yields

$$F_Y^*(\alpha) \sim C \exp(-\theta^* (\alpha + 1)),$$

where

$$C = \frac{\overline{Q}\left(1 - \sum_{j=1}^{c} R_{-j} \exp(-\theta^* j)\right)}{(1 - \exp(-\theta^*)) \sum_{k=1}^{d} kQ_k \exp(\theta^* k)}. \tag{7.16}$$

Thus, Y is geometric-like (note that $C > 0$).

Next, let N' be the following random variable on $(S_w, \mathcal{B}_w, P)$: for a trajectory $t \in E_0^c$ define $N'(t)$ to be the number of steps that t takes to reach the value -1. We will find $\mathcal{E}(N')$. Let $T_{N'}(t)$ be the value that t visits after N' steps. One can show that Theorem 7.5 holds for N' and $T_{N'}$ in place of N and T_N as well, and therefore for all $\theta \in \mathbb{R}$ we have

$$\mathcal{E}\left(\mathbb{M}(\theta)^{-N'}\exp(\theta T_{N'})\right) = 1. \tag{7.17}$$

Since N' is a random variable that takes infinitely many values, the expression in the left-hand side of identity (7.17) is an infinite series. It is possible to prove that this series can be differentiated term by term, and hence we obtain

$$\frac{d}{d\theta}\mathcal{E}\left(\mathbb{M}(\theta)^{-N'}\exp(\theta T_{N'})\right) = \mathcal{E}\left(\frac{d}{d\theta}\left(\mathbb{M}(\theta)^{-N'}\exp(\theta T_{N'})\right)\right)$$

$$= \mathcal{E}\left(-N'\mathbb{M}(\theta)^{-N'-1}\exp(\theta T_{N'})\frac{d}{d\theta}\mathbb{M}(\theta) + \mathbb{M}(\theta)^{-N'}T_{N'}\exp(\theta T_{N'})\right).$$

Therefore (7.17) implies

$$\mathcal{E}\left(-N'\mathbb{M}(\theta)^{-N'-1}\exp(\theta T_{N'})\frac{d}{d\theta}\mathbb{M}(\theta) + \mathbb{M}(\theta)^{-N'}T_{N'}\exp(\theta T_{N'})\right) = 0,$$

which for $\theta = 0$ gives

$$-\mathcal{E}(N')\sum_{j=-c}^{d} jp_j + \mathcal{E}(T_{N'}) = 0,$$

that is,

$$\mathcal{E}(N') = \frac{\mathcal{E}(T_{N'})}{\displaystyle\sum_{j=-c}^{d} jp_j}.$$

Clearly, the probability distribution of $T_{N'}$ is $\{R_{-j}, j = 1, \ldots, c\}$. Therefore,

$$\mathcal{E}(T_{N'}) = -\sum_{j=1}^{c} jR_{-j},$$

and hence

$$\mathcal{E}(N') = -\frac{\displaystyle\sum_{j=1}^{c} jR_{-j}}{\displaystyle\sum_{j=-c}^{d} jp_j}. \tag{7.18}$$

7.3 Significance of Scores

In this section we will apply the theory of random walks from Sect. 7.2 to determine the behavior of F_s, the distribution function of the random variable s on the probability space $(S, \mathcal{B}, P)$ introduced in Sect. 7.1. Let $(x, y) \in S$. Every local ungapped alignment between x and y can be extended as far as possible in either direction and thus defines a particular *overlap* between x and y. There are a total of $N_1 + N_2 - 1$ such overlaps. Figure 7.2 shows some overlaps between two DNA sequences of lengths 8 and 10.

```
AGTCGAGC                      AGTCGAGC
        CTCGACGACG                    CTCGACGACG


AGTCGAGC                      AGTCGAGC
       CTCGACGACG                   CTCGACGACG


   AGTCGAGC                   AGTCGAGC
      CTCGACGACG              CTCGACGACG
```

Fig. 7.2.

We now fix a particular overlap $\mathcal{O}$ of some length $N \leq \min\{N_1, N_2\}$, the same for all $(x, y) \in S$, and consider a random variable $s_{\mathcal{O}}$ on $(S, \mathcal{B}, P)$ with $s_{\mathcal{O}}((x, y))$ defined as the score of an ungapped local alignment that has the highest score among all ungapped local alignments between x and y that fit in the overlap $\mathcal{O}$. We will now determine the behavior of $F_{s_{\mathcal{O}}}$. Let S_N be the collection of all pairs (x, y) of sequences of length N of letters from $\mathcal{Q}$. We will define a probability measure on S_N analogously to that on S: for $(x, y) \in S$ with $x = x_1 \ldots x_N$, $y = y_1 \ldots y_N$ set

$$P\big(\{(x, y)\}\big) = \prod_{i=1}^{N} p_{x_i} p'_{y_i}.$$

Since S_N is a finite set and since the probability measure P is defined for every elementary event, it can be extended to the collection $\mathcal{B}_N$ of all events in S_N. Thus, we obtain a probability space $(S_N, \mathcal{B}_N, P)$. The random variable $s_{\mathcal{O}}$ was originally introduced for the probability space $(S, \mathcal{B}, P)$, but can also be naturally defined on $(S_N, \mathcal{B}_N, P)$. It is easy to observe that the distributions of $s_{\mathcal{O}}$ on these two probability spaces coincide (see Exercise 7.8). Therefore, for the purposes of studying $F_{s_\mathcal{O}}$ we can consider $s_\mathcal{O}$ on $(S_N, \mathcal{B}_N, P)$, which will be a more convenient setup for us.

Consider the random walk associated with the substitution matrix $(s(a, b))$ and the frequencies $\{p_a\}$, $\{p'_b\}$, as at the beginning of Sect. 7.2. Let $(S_w, \mathcal{B}_w, P)$ be the corresponding probability space. We will now consider a "truncated" probability space defined as follows. Let $S_{w,N}$ be the collection of all possible trajectories t of length N. We define a probability measure on $S_{w,N}$ as follows. For a trajectory $t = t_1, \ldots, t_N$, with $t_j \in T$, set

$$P\big(\{t\}\big) = \prod_{j=1}^N p_{t_j}.$$

Since P is defined for all elementary events in $S_{w,N}$ and since $S_{w,N}$ is finite, P can be extended to the collection $\mathcal{B}_{w,N}$ of all events in $S_{w,N}$. In our considerations N is fixed, but if for the moment we allow $N \to \infty$, then the resulting sequence of probability spaces $\{(S_{w,N}, \mathcal{B}_{w,N}, P)\}$ can be thought of as an approximation of the probability space $(S_w, \mathcal{B}_w, P)$. Indeed, if we consider only the initial N elements $t_1, \ldots, t_N$ in every t from a cylinder event $E_{t_{i_1}^0, \ldots, t_{i_m}^0}$ defined in (7.2), the probability of the resulting event $E_{t_{i_1}^0, \ldots, t_{i_m}^0, N} \subset S_{w,N}$ tends to the probability of $E_{t_{i_1}^0, \ldots, t_{i_m}^0}$, as $N \to \infty$. In fact, $P(E_{t_{i_1}^0, \ldots, t_{i_m}^0, N}) = P(E_{t_{i_1}^0, \ldots, t_{i_m}^0})$, if $N \geq i_m$.

Let Y_N be a random variable on $(S_{w,N}, \mathcal{B}_{w,N}, P)$ analogous to the random variable Y on $(S_w, \mathcal{B}_w, P)$ considered in Sect. 7.2. Namely, if a trajectory $t \in S_{w,N}$ visits negative values, define $Y_N(t)$ to be the maximal value achieved by the trajectory t until it reaches -1, in other words, $Y_N(t)$ in this case is the height of the excursion between the first ladder point 0 and the second ladder point (the first negative value visited by t); if a trajectory t never visits negative values, define $Y_N(t)$ to be the maximal value achieved by t. One can show that $Y_N \overset{D}{\to} Y$ (see Exercise 7.9). In particular, if both N and α are large, we have

$$F_{Y_N}^*(\alpha) \approx C \exp(-\theta^*(\alpha + 1)),$$

where θ^* and C are found from formulas (7.6) and (7.16) respectively.

We will now relate the distribution of Y_N on $(S_{w,N}, \mathcal{B}_{w,N}, P)$ to that of $s_\mathcal{O}$ on $(S_N, \mathcal{B}_N, P)$. Each element $(x, y) \in S_N$ obviously generates a trajectory $t((x, y)) \in S_{w,N}$, and every $t \in S_{w,N}$ is generated by some element of S_N. Note, however, that the mapping $(x, y) \mapsto t((x, y))$ is not one-to-one; it is possible that for $(x', y') \neq (x, y)$, we have $t((x', y')) = t((x, y))$, for example, $t(y, x) = t(x, y)$ for all $(x, y) \in S_N$. Consider a random variable Y'_N on $(S_N, \mathcal{B}_N, P)$ defined as $Y'_N((x, y)) = Y_N(t((x, y)))$. It is straightforward to verify that the distribution of Y'_N on $(S_N, \mathcal{B}_N, P)$ coincides with that of Y_N on $(S_{w,N}, \mathcal{B}_{w,N}, P)$ (see Exercise 7.10). Hence for large N and α we have

$$F_{Y'_N}^*(\alpha) \approx C \exp(-\theta^*(\alpha + 1)). \tag{7.19}$$

Further, let $t \in S_{w,N}$ be a trajectory that visits negative values. We will modify it into a *path* $\bar{t}$ in the following way. Suppose that t visits its first negative value after m steps. Then we leave the portion of t up to and including the $(m-1)$th step unchanged, set the value of $\bar{t}$ at the mth step to be equal to 0, and continue along the steps taken by t starting from 0 until $\bar{t}$ reaches its next negative value, after which we will repeat the above procedure, etc. If t never visits any negative values, we set $\bar{t} = t$. The path $\bar{t}$ never visits any negative values. Note that it is possible that $\bar{t_1} = \bar{t_2}$, if $t_1 \neq t_2$. Figure 7.3 shows the path for a portion of the trajectory in Fig. 7.1.

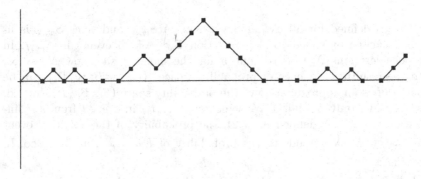

Fig. 7.3.

As for random walk trajectories, one can introduce the notion of an *excursion* of a path $\bar{t}$ as a part of $\bar{t}$ starting at 0 until the highest point visited by $\bar{t}$ before the next 0. In this terms, for $(x,y) \in S_N$, the value $s_\mathcal{O}((x,y))$ is the maximum of the heights of the excursions of the path $\overline{t((x,y))}$. The maximum of the heights of excursions of $\overline{t((x,y))}$ between the first 0 and the second ladder point of $t((x,y))$ is $Y'_N((x,y))$. If for every element $(x,y) \in S_N$ the trajectory $t((x,y))$ had the same number of ladder points n and the last step of $t((x,y))$ was a ladder point as well, then we would have

$$s_\mathcal{O}((x,y)) = \max\{Y_1((x,y)), \ldots, Y_n((x,y))\} = Y_{\max}((x,y)), \qquad (7.20)$$

where Y_j are iid random variables on $(S_N, \mathcal{B}_N, P)$, each having the distribution of Y'_N (in fact, Y'_N itself would have been one of Y_j). In reality, of course, the number of ladder points of $t((x,y))$ depends on (x,y), and the last step of $t((x,y))$ may not be a ladder point (the latter effect is sometimes called the *edge effect*). In our considerations we will ignore these complications, and as an approximation to the distribution of $s_\mathcal{O}$ we will study the distribution of the random variable Y_{max} in the right-hand side of identity (7.20) with $n = N/\mathcal{E}(N')$, where $\mathcal{E}(N')$ is found from formula (7.18). In fact, one can show that, as $N \to \infty$, the distribution functions of $s_{\mathcal{O}}$ and Y_{max} become arbitrarily close.

We will calculate an approximation to $F^*_{Y_{max}}(\alpha)$, where α is a large integer. We will also assume that N is large and therefore use the approximations

$$F^*_{Y_j}(\alpha) \approx C \exp(-\theta^*(\alpha + 1)), \quad j = 1, \ldots, n, \tag{7.21}$$

that follow from (7.19). From formulas (7.21) we have

$$F^*_{Y_{max}}(\alpha) = 1 - F_{Y_{max}}(\alpha) = 1 - \prod_{j=1}^{n} F_{Y_j}(\alpha) = 1 - \prod_{j=1}^{n}(1 - F^*_{Y_j}(\alpha))$$

$$\approx 1 - \left(1 - C \exp(-\theta^*(\alpha + 1))\right)^n.$$

Let $\beta = C \exp(-\theta^*(\alpha+1))$. If α is large, we can approximate $1 - \beta$ by $\exp(-\beta)$ and hence we have

$$F^*_{Y_{max}}(\alpha) \approx 1 - \exp\left(-nC \exp(-\theta^*(\alpha + 1))\right).$$

Therefore for $F^*_{s_{\mathcal{O}}}(\alpha)$ with large N and α we obtain

$$F^*_{s_{\mathcal{O}}}(\alpha) \approx 1 - \exp\left(-nC \exp(-\theta^*(\alpha + 1))\right),$$

or

$$F_{s_{\mathcal{O}}}(\alpha) \approx \exp\left(-nC \exp(-\theta^*(\alpha + 1))\right). \tag{7.22}$$

Recall that in Sect. 7.1 we formulated the problem of the significance of sequence alignment scores as the problem to calculate $F^*_s(s_0 - 1)$, where s is the random variable on the probability space $(S, \mathcal{B}, P)$ introduced in Sect. 7.1, and s_0 is a large integer equal to the score of an optimal local alignment between two fixed sequences x^0 and y^0. Clearly, for all $(x, y) \in S$ we have

$$s((x, y)) = \max_{\mathcal{O}} s_{\mathcal{O}}((x, y)),$$

where the maximum is taken over all possible overlaps $\mathcal{O}$. Suppose for the moment that the random variables $s_{\mathcal{O}}$ are independent and that for each $s_{\mathcal{O}}$ approximation (7.22) held for large α with $n = N_{\mathcal{O}}/\mathcal{E}(N')$, where $N_{\mathcal{O}}$ is the length of the overlap $\mathcal{O}$. Then we have

$$F_s(\alpha) = \prod_{\mathcal{O}} F_{s_{\mathcal{O}}}(\alpha) \approx \exp\left(-\frac{C}{\mathcal{E}(N')} \exp(-\theta^*(\alpha+1)) \sum_{\mathcal{O}} N_{\mathcal{O}}\right).$$

Suppose for convenience that $N_1 \geq N_2$. Then we obtain

$$\sum_{\mathcal{O}} N_{\mathcal{O}} = 2 \sum_{j=1}^{N_2-1} N_2 + N_2(N_1-N_2+1) = N_2(N_2-1) + N_2(N_1-N_2+1) = N_1 N_2,$$

which gives

$$F_s(\alpha) \approx \exp\left(-\frac{C}{\mathcal{E}(N')} N_2 N_2 \exp(-\theta^*(\alpha+1))\right),$$

or

$$F_s^*(\alpha) \approx 1 - \exp\left(-\frac{C}{\mathcal{E}(N')} N_1 N_2 \exp(-\theta^*(\alpha+1))\right). \tag{7.23}$$

To derive formula (7.23) we assumed that the random variables $s_{\mathcal{O}}$ were independent and that for each $s_{\mathcal{O}}$ approximation (7.22) holds for large α. However, $s_{\mathcal{O}}$ are in fact not independent. Further, approximation (7.22) depends on the assumption that the length of the overlap is large, whereas, of course, there are very short overlaps. There exists a general theory that deals with these and other complications [DKZ1], [DKZ2]. This theory is beyond the scope of this book; we only state here that under some assumptions it can be shown that the right-hand side of formula (7.23) indeed serves as a reasonable approximation to $F_s^*(\alpha)$ (see a discussion in [EG]). We also note in passing that a better approximation can be achieved by taking into account the edge effect.

Since s_0 is a large integer, we will use the approximation for $F_s^*(s_0 - 1)$ found from (7.23) with $\alpha = s_0 - 1$ to test the significance of the score of any optimal local alignment between x^0 and y^0. Note that for very large s_0 (such that $\left(C/\mathcal{E}(N')\right) N_1 N_2 \exp(-\theta^* s_0)$ is close to 0) we can rewrite the approximation to $F_s^*(s_0 - 1)$ in the form

$$F_s^*(s_0 - 1) \approx \frac{C}{\mathcal{E}(N')} N_1 N_2 \exp(-\theta^* s_0). \tag{7.24}$$

We will now introduce some notation standard for sequence similarity searches. First of all, θ^* is usually denoted by λ. Further, let

$$K = \frac{C}{\mathcal{E}(N')} \exp(-\lambda). \tag{7.25}$$

In this notation formulas (7.23) and (7.24) take the forms

$$F_s^*(s_0 - 1) \approx 1 - \exp\left(-K N_1 N_2 \exp(-\lambda(s_0 - 1))\right), \tag{7.26}$$

and

$$F_s^*(s_0 - 1) \approx KN_1N_2 \exp(-\lambda(s_0 - 1)), \qquad (7.27)$$

respectively, where in formula (7.27) we assume that $KN_1N_2 \exp(-\lambda(s_0 - 1))$ close to 0. Note that the right-hand side in formula (7.1) is analogous to that in formula (7.26) with $M = N_1N_2$, which is the total number of the *starts* of all possible local alignments between two sequences of lengths N_1 and N_2 respectively, but not the total *number* of local alignments, as M was taken to be in (7.1). This inconsistency arises because of our assumption in Sect. 7.1 that the random variables $S\left(A_{i,j,N}(x,y)\right)$ are independent.

The value $F_s^*(s_0 - 1)$ is called the *P-value* and is sometimes denoted by $P(s_0 - 1)$. The expression $KN_1N_2 \exp(-\lambda(s_0 - 1))$ is called the *E-value* and is sometimes denoted by $E(s_0 - 1)$. One can show that the E-value is approximately the mean number of local alignments between $(x, y) \in S$ with score $\geq s_0$. Using this notation we can rewrite formulas (7.26) and (7.27) as

$$P(s_0 - 1) \approx 1 - \exp(-E(s_0 - 1)), \qquad (7.28)$$

and, if $E(s_0 - 1)$ is close to 0,

$$P(s_0 - 1) \approx E(s_0 - 1).$$

To use formula (7.26) or formula (7.27) we must compute the constants λ and K. Recall that λ is found from equation (7.6) which for the case of the random walk associated with the matrix $(s(a,b))$ and frequencies $\{p_a\}$, $\{p_b'\}$ becomes

$$\sum_{a,b \in \mathcal{Q}} p_a p_b' \exp(\lambda s(a,b)) = 1. \qquad (7.29)$$

The above equation is usually dealt with numerically, and an approximate value of λ is generated. Once λ has been determined, K is found from formula (7.25) using (7.16) and (7.18). The latter two formulas contain the probabilities Q_k and R_{-j} that are in fact not easy to compute analytically. However, there exist rapidly converging methods that allow to calculate K with high degree of accuracy. Methods for computing approximate values of λ and K are beyond the scope of this book, but below we give an example where we find these values analytically for a simple case.

Example 7.6. Let $\mathcal{Q}$ be the DNA alphabet, $p_a = 1/4$, $p_b' = 1/4$ for all $a, b \in \mathcal{Q}$, and the substitution matrix be as in Example 2.1, that is,

$$(s(a,b)) = \begin{pmatrix} 1 & -1 & -1 & -1 \\ -1 & 1 & -1 & -1 \\ -1 & -1 & 1 & -1 \\ -1 & -1 & -1 & 1 \end{pmatrix}. \qquad (7.30)$$

Clearly, the above substitution matrix and frequencies $\{p_a\}$, $\{p_b'\}$ satisfy conditions (iv) and (vi) from Sect. 7.2. Condition (v) is satisfied as well, since

$$\sum_{a,b\in Q} s(a,b)p_a p'_b = -\frac{1}{2} < 0. \tag{7.31}$$

The constant λ is easy to find from equation (7.29). Indeed, for this example the equation becomes

$$\exp(\lambda)\frac{1}{4} + \exp(-\lambda)\frac{3}{4} = 1,$$

and its only positive solution is $\lambda = \ln 3$.

To find K, we need to determine Q_1 and R_{-1} (see formulas (7.16), (7.18) and (7.25)). Clearly, $R_{-1} = 1$. In order to find Q_1 consider the associated random walk. We have $T = \{-1,0,1\}$, $p_{-1} = 3/4$, $p_0 = 0$, $p_1 = 1/4$. For this random walk $c = d = 1$, that is, it is an example of a *simple random walk*. It is not hard to show (see Exercise 7.11) that

$$Q_1 = \sum_{n=1}^{\infty} B_{2n-1} \left(\frac{1}{4}\right)^n \left(\frac{3}{4}\right)^{n-1}, \tag{7.32}$$

where B_{2n-1} is the number of trajectories $t = t_1,\dots,t_{2n-1}$ in the sample space $S_{w,2n-1}$ for which $\sum_{j=1}^{2n-1} t_j = 1$ and $\sum_{j=1}^{m} t_j \le 0$ for $m = 1,\dots, 2n-2$. Similarly, we have (see Exercise 7.11)

$$R_{-1} = \sum_{n=1}^{\infty} B_{2n-1} \left(\frac{1}{4}\right)^{n-1} \left(\frac{3}{4}\right)^n, \tag{7.33}$$

and therefore

$$Q_1 = \frac{1}{3}R_{-1} = \frac{1}{3}.$$

Now formulas (7.16), (7.18), (7.31), (7.25) yield that $K = 1/9$. For more information on simple random walks see, for example, [EG].

We will now give an example of using the above values of λ and K for assessing the significance of the score of a local alignment. Suppose we are given two sequences $x^0 = ACATGCTG$, $y^0 = CATTGCGA$. It is easy to see that the frequency of each letter from the DNA alphabet in either sequence is equal to $1/4$, and therefore the above calculations for λ and K are valid for these sequences, if we align them by means of substitution matrix (7.30). It is easy to find all optimal local ungapped alignments between them by applying the Smith-Waterman algorithm described in Sect. 2.3. Looking for ungapped alignments means that we utilize the linear gap model with $d = \infty$. There are two optimal alignments

$$x^0 : C\ A\ T$$
$$y^0 : C\ A\ T,$$

$$x^0 : T\ G\ C$$
$$y^0 : T\ G\ C,$$

each having score equal to 3.

Thus, we have $s_0 = 3$, $N_1 = N_2 = 8$, $\lambda = \ln 3$, $K = 1/9$. From (7.28) we then obtain that the P-value is approximately equal to 0.54 which is larger than 0.05. Therefore, on the basis of this analysis neither of the above optimal alignments is ragarded as significant. Of course, one should bear in mind that we gave this example for illustration purposes only, and that in order for approximation (7.28) to work, the values of s_0, N_1, N_2 must be much larger.

In practice a query sequence x^0 is compared not just to a single sequence y^0, but to all sequences z in a particular database D of sequences. Both BLAST and FASTA searches involve such comparisons. In this case we are looking for an optimal alignment between segments in x^0 and segments in all possible sequences $z \in D$. Suppose that a highest-scoring alignment with a large score s_0 occurs between a segment in x^0 and a segment contained in some sequence $y^0 \in D$. How do we decide in this case whether the similarity found is significant or not? One approach is to just assess the local alignment between x^0 and y^0 in the way it was done above. Such assessment, however, would ignore the fact that the alignment was found by comparing x^0 to *all* sequences in D, not just the single sequence y^0. In reality assessment is done as follows. Let $|D|$ be the sum of the lengths of all sequences in D. We then treat the optimal local alignment as an optimal local alignment between x^0 and the sequence of length $|D|$ obtained by joining together all the sequences contained in D (this treatment ignores some edge effects that can be assumed to be small if all sequences in D are sufficiently long). Then the corresponding E-value (that we denote by $E_D(s_0 - 1)$) is

$$E_D(s_0 - 1) = KN_1|D| \exp(-\lambda s_0),$$

and the corresponding P-value (that we denote by $P_D(s_0 - 1)$) found from (7.28) is

$$P_D(s_0 - 1) \approx 1 - \exp\left(-KN_1|D| \exp(-\lambda(s_0 - 1))\right),$$

that is $P_D(s_0-1)$ is bigger than the P-value $P(s_0-1)$ found by from comparing x^0 to the sequence y^0 alone. If $P_D(s_0-1)$ is bigger than say 0.05, we conclude that the match that we have found is not significant. However, it may happen that in this case the P-value $P(s_0 - 1)$ is less than 0.05. Therefore, using $P(s_0-1)$ instead of $P_D(s_0-1)$ leads to overestimating the significance of the scores of optimal local alignments.

In BLAST computations, significance is assessed as follows. First, $P(s_0-1)$ is computed, next, $E_D(s_0 - 1)$ is approximated as follows

$$E_D(s_0 - 1) \approx E_D'(s_0 - 1) = \frac{P(s_0 - 1)|D|}{N_2},$$

and, finally, an approximation to $P_D(s_0 - 1)$ is found as

$$P_D(s_0 - 1) \approx 1 - \exp(-E'_D(s_0 - 1)).$$

We note that BLAST attempts to assess not only the highest-scoring local alignments, but other high scoring local alignments as well, for example, second best, third best, etc. Assessing the significance of the scores of all such alignments requires certain modifications of the procedure discussed above which are beyond the scope of this book. The interested reader is referred to [KA] and a discussion in [EG].

Exercises

7.1. Prove that the collection of events $\mathcal{F}_w$ defined in Sect. 7.2 is a semi-algebra.

7.2. Show that formula (7.3) defines a probability measure on the semi-algebra of events $\mathcal{F}_w$ from Exercise 7.1.

7.3. Consider the probability spaces $(S_w, \mathcal{B}_w, P)$ and $(T, \mathcal{B}_T, P)$ defined in Sect. 7.2. Show that $(S_w, \mathcal{B}_w, P)$ is the infinite Cartesian power of $(T, \mathcal{B}_T, P)$.

7.4. Show that the functions Y and $\tilde{Y}$ defined in Sect. 7.2 are random variables on the probability space $(S_w, \mathcal{B}_w, P)$.

7.5. Show that the events E_{-j} and E_k, $j = 1, \ldots, c$, $k \in \mathbb{N}$, defined in (7.4) and (7.7) respectively are measurable, that is, belong to $\mathcal{B}_w$.

7.6. For a discrete random variable f that takes infinitely many values $r_1, r_2, \ldots$, with $r_i \neq r_j$ for $i \neq j$, and has a probability distribution $\{p_f(r_j), j = 1, 2 \ldots\}$ define the *moment-generating function* as follows

$$\mathbb{M}_f(\theta) = \mathcal{E}(\exp(\theta f)) = \sum_{j=1}^{\infty} \exp(\theta r_j) p_f(r_j), \quad \theta \in \mathbb{R}.$$

This function is only defined on a subset of $\mathbb{R}$ for which the series in the right-hand side of the above formula converges. Let $(S, \mathcal{B}(\mathcal{F}), P)$ be the probability space from Part 3 of Example 6.14. Define the following random variable on it

$$f(e) = \begin{cases} 0, & \text{if } e = 0, \\ n, & \text{if } \dfrac{1}{n+1} < e \leq \dfrac{1}{n}. \end{cases}$$

Find the domain of definition of $\mathbb{M}_f$.

7.7. Show that the sequence $\{c_j\}$ introduced in Sect. 7.2 is bounded.

7.8. Show that the distribution of the random variable $s_\mathcal{O}$ on the probability space $(S_N, \mathcal{B}_N, P)$ coincides with its distribution on the probability space $(S, \mathcal{B}, P)$ (see Sect. 7.3 for details).

7.9. Show that $Y_N \overset{D}{\to} Y$ (see Sect. 7.3 for details).

7.10. Prove that the distribution of Y_N' on $(S_N, \mathcal{B}_N, P)$ coincides with that of Y_N on $(S_{w,N}, \mathcal{B}_{w,N}, P)$ (see Sect. 7.3 for details).

7.11. Prove formulas (7.32) and (7.33).

7.12. Find all optimal ungapped local alignments between the sequences $x^0 = ACATGCTG$, $y^0 = GCATGCTA$ using substitution matrix (7.30) and assess the significance of the score of each alignment, assuming that approximation (7.28) holds for the P-value.

7.13. Let $x^0 = AGCTGC$, $y^0 = GATTGACTA$. For these sequences and substitution matrix (7.30) verify that conditions (iv)-(vi) from Sect. 7.2 hold. Further, find all optimal ungapped local alignments between x^0, y^0 and assess the significance of the score of each optimal alignment assuming that approximation (7.28) holds for the P-value.

8

Elements of Statistics

In this chapter we attempt to give a mathematically rigorous exposition of some aspects of statistics. As in Chap. 6, we do not concentrate on proofs here. Instead, the emphasis is on the main constructions that we illustrate by many examples. All proofs can be found, for instance, in [W].

8.1 Statistical Modeling

We start with the following definition.

Definition 8.1. *A* statistical model *is a family of probability spaces* $\{(S_\theta, \mathcal{B}_\theta, P_\theta)\}$ *and a family of random variables* $\{f_\theta\}$ *with common range* $\mathcal{W} \subset \mathbb{R}$, *each defined on the respective space for* $\theta \in \mathcal{P}$, *where* $\mathcal{P}$ *is an index set. The variable* θ *denotes the* parameters *of the model, the set* $\mathcal{W}$ *is called the* range *of the model, and the index set* $\mathcal{P}$ *the* parameter space *of the model. Hence a statistical model can be thought of as a family* $\{(S_\theta, \mathcal{B}_\theta, P_\theta, f_\theta, \mathcal{W}, \mathcal{P})\}$.

When studying and applying statistical models, one is primarily interested in the distributions of f_θ. Therefore, often statistical models are not specified in full as in Definition 8.1, but only the family $\{F_\theta = F_{f_\theta}\}$, $\theta \in \mathcal{P}$, of the distribution functions of f_θ is given. Once the family $\{F_\theta\}$, $\theta \in \mathcal{P}$, is specified, one can construct a statistical model in the sense of Definition 8.1, for example, by setting $S_\theta = \mathbb{R}$, $\mathcal{B}_\theta = \mathcal{B}(\mathcal{F}_0)$ (the σ-algebra of Borel sets in $\mathbb{R}$), $P_\theta = P_{F_\theta}$, $f_\theta(x) = x$, $\mathcal{W} = \mathbb{R}$, for $\theta \in \mathcal{P}$ (see Sect. 6.7). This model, however, is not always useful.

We also remark that one can consider more general statistical models by allowing f_θ to be vector-valued. For example, one can define a natural *product* of two or more models. We will not consider such generalizations in this chapter, but most of what follows can be easily adjusted to accommodate them.

We will now introduce an important special class of statistical models. Suppose that S_θ and $\mathcal{B}_\theta$ do not depend on θ, that is, for all $\theta \in \mathcal{P}$, we have $S_\theta = S$ and $\mathcal{B}_\theta = \mathcal{B}$, with some S and $\mathcal{B}$. Suppose, in addition, that S is a countable set (finite or infinite) and $\mathcal{B}$ is the σ-algebra of all events in S. Let further P_θ, for $\theta \in \mathcal{P}$, be a family of probability measures on $\mathcal{B}$. In this case one is often interested in P_θ, not in the distribution of any particular random variable on $(S, \mathcal{B}, P_\theta)$. To formally satisfy the definition of statistical model above, we will construct a discrete random variable f_θ on $(S, \mathcal{B}, P_\theta)$ for each value of $\theta \in \mathcal{P}$, such that the probability distribution of f_θ is $\{P_\theta(e), e \in S\}$. This can be done as follows. Enumerate all elementary events in S, that is, find a (non-unique) one-to-one correspondence $\varphi : S \to \mathbb{N}$ between points in S and points in a set $\mathcal{D} \subset \mathbb{N}$, where $\mathcal{D}$ is either a finite set $\{1, \ldots, N\}$ for some N, or all of $\mathbb{N}$. We then set

$$f_\theta = \varphi, \quad \text{for all } \theta \in \mathcal{P}. \tag{8.1}$$

Clearly, f_θ is a discrete random variable on $(S, \mathcal{B}, P_\theta)$ with probability distribution $\{P_\theta(e), e \in S\}$, as required. Many statistical models we will be interested in are of this type. For convenience we will call such models *reduced statistical models*. The range of a reduced statistical model is the corresponding set $\mathcal{D}$.

We will now give some examples of statistical models.

Example 8.2.

1. Suppose we are given a family of Markov chains with an end state arising from a fixed *a priori* connectivity. The transition probabilities are allowed to vary subject to the *a priori* connectivity and to the condition of non-trivial connectedness. Consider the associated probability spaces as constructed in Part 5 of Example 6.11. Since sequences of infinite length do not contribute to the probabilities of events, we consider the smaller sample space that consists only of finite sequences. This sample space is countable, and we consider the probability spaces arising from it. Note that the probability measures defined in formula (6.4) depend on the transition probabilities of the models. We arrange all the transition probabilities in a vector of real numbers and call this vector θ. Clearly, θ varies over a subset $\mathcal{P}$ of $\mathbb{R}_+^K = \{(x_1, \ldots, x_K) \in \mathbb{R}^K : x_j \geq 0, j = 1, \ldots K\}$, for some K. The parameter space $\mathcal{P}$ is given by the conditions that the transition probabilities from each state to any possible state sum up to 1, the *a priori* connectivity assumptions and by the condition of non-trivial connectedness (see, for example, the model from Exercise 8.4). If we introduce f_θ as in (8.1), we obtain a reduced statistical model with range $\mathbb{N}$.

2. Suppose we are given a family of HMMs, arising from a fixed *a priori* Markov chain connectivity. Consider the associated probability spaces as constructed in Part 6 of Example 6.11, and, as in Part 1 above, concentrate only on sequences of finite length. We will denote the subset of $S^\mathbf{a}$ that consists of pairs of finite sequences by $\hat{S}^\mathbf{a}$ and the subset of $S^\mathbf{b}$ that consists of finite sequences by $\hat{S}^{\mathbf{b}}$. Then we obtain two reduced statistical model. The two different models are suitable for different problems (see Part 2 of Example 8.5). Note that the parameters of the models are the transition and emission probabilities. Hence the parameter space $\mathcal{P}^{\mathbf{a}} = \mathcal{P}^{\mathbf{b}} = \mathcal{P}$ in this case is the product of the parameter space from Part 1 above and a subset of $\mathbb{R}_+^{|\mathcal{Q}|N}$, where N is the number of non-zero states of the model and $|\mathcal{Q}|$ is the number of elements in the alphabet $\mathcal{Q}$, given by the condition that the emission probabilities at each state sum up to 1 (see, for example, the model from Exercise 8.5). In particular, in this case the parameter space is again a subset of a Euclidean space. The range of each of the models is $\mathbb{N}$.

3. Fix an evolutionary model and consider all rooted phylogenetic trees with N labeled leaves. The associated probability spaces as constructed in Part 7 of Example 6.11 define a reduced statistical model. Note that in this case the sample space is finite (it contains 4^N elementary events) and the parameters of the model are the tree and the parameters arising from the evolutionary model (for instance, in the case of the Kimura and HKY models such parameters arise from β – see Sect. 5.4). In particular, the parameter space in this case is not a subset of any Euclidean space. The range of the model is the finite set $\{1, \ldots, 4^N\}$.

In statistical modeling one is interested in a large number of "independent runs of a model". This concept is formalized in the following definition.

Definition 8.3. *Let $\{(S_\theta, \mathcal{B}_\theta, P_\theta, f_\theta, \mathcal{W}, \mathcal{P})\}$ be a statistical model. The random sample of size n from the model is the family $\{f_\theta^{\times \infty, n}\}$ of the samples of size n from the infinite Cartesian powers of the random variables f_θ, with $\theta \in \mathcal{P}$.*

Recall that for every $\theta \in \mathcal{P}$, $f_\theta^{\times \infty, n}$ is defined on the infinite Cartesian power of $(S_\theta, \mathcal{B}_\theta, P_\theta)$, its range is $\mathcal{W}^n$, and its components are iid random variables whose distributions coincide with that of f_θ.

We are now ready to define the concept of statistical modeling.

Definition 8.4. *Suppose that we are given a vector $w = (w_1, \ldots, w_n) \in \mathcal{W}^n$ (usually in some way derived from real data). We say that $\{(S_\theta, \mathcal{B}_\theta, P_\theta, f_\theta, \mathcal{W}, \mathcal{P})\}$ is a statistical model for w or w is statistically modeled by $\{(S_\theta, \mathcal{B}_\theta, P_\theta, f_\theta, \mathcal{W}, \mathcal{P})\}$, if one tries to "fit" the model to w by choosing a value $\theta_0 \in \mathcal{P}$ that is in some sense "optimal" for w.*

The vagueness of Definition 8.4 is due to the fact that the optimality of model fit is not well-defined and is often understood intuitively. As a result, there are numerous fitting procedures used for different models. Some of them are model-specific, but some of them are quite general. In the next section we will discuss a universal way of model fitting by means of *parameter estimation*. The vector w in Definition 8.4 usually arises from real data, and hence statistical modeling can be thought of as a process of fitting a model to a dataset.

We will now give some examples of statistical modeling arising from the models in Example 8.2.

Example 8.5.

1. Suppose we have DNA data for n prokaryotic genes. Consider the *a priori* connectivity shown in Fig. 3.2 and associate with it a statistical model as explained in Part 1 of Example 8.2. Each of the given sequences $x^1, \ldots, x^n$ can be regarded as an element of the sample space S (we can add 0 at the beginning and end of each sequence). Consider the mapping $\varphi : S \to \mathbb{N}$ (see the discussion preceding formula (8.1)) and set $w_j = \varphi(x^j)$, $j = 1, \ldots n$. Let $w = (w_1, \ldots, w_n)$. This DNA statistical model can be fitted to the vector w, as it was done in Sect. 3.1, namely, by using formula (3.2). Not only does this intuitively seem to be the right way to do fitting, but it in fact agrees with the general theory of parameter estimation discussed in the next section. Of course, this procedure can be used for arbitrary Markov chains.

2. Suppose we are given a family of HMMs as specified in Part 2 of Example 8.2. Consider the two statistical models that arise from it.

    a. Let $(x^1, \pi^1), \ldots, (x^n, \pi^n)$ be pairs of sequences, where for each $j = 1, \ldots, n$, x^j is a sequence of elements from $\mathcal{Q}$, π^j is a path, and the lengths of x^j and π^j are equal. Each pair can be regarded as an element of the sample space $\hat{S}^{\mathbf{a}}$. Consider the corresponding mapping $\varphi : \hat{S}^{\mathbf{a}} \to \mathbb{N}$ and set $w_j = \varphi((x^j, \pi^j))$, $j = 1, \ldots n$. Let $w = (w_1, \ldots, w_n)$. This statistical model can be fitted to the vector w as it was done in Sect. 3.6 in the case when paths are known. This way of model fitting also agrees with the general theory of parameter estimation discussed in the next section.

    b. Let $x^1, \ldots, x^n$ be sequences with elements from $\mathcal{Q}$. Each of the sequences can be regarded as an element of the sample space $\hat{S}^{\mathbf{b}}$. Consider the corresponding mapping $\varphi : \hat{S}^{\mathbf{b}} \to \mathbb{N}$ and set $w_j = \varphi(x^j)$, $j = 1, \ldots n$. Let $w = (w_1, \ldots, w_n)$. This statistical model can be fitted to the vector w as it was done in Sect. 3.6, by either the Baum-Welch or Viterbi training. Note that the Baum-Welch training attempts to maximize the product of probabilities $P(x^1) \times \ldots \times P(x^n)$ (we called it the "likelihood of the training data" in Sect. 3.6). For every fixed $\theta \in \mathcal{P}$, in terms of the probability distribution of the sample of size n from the model, this product can be expressed as

$$P(x^1) \times \ldots \times P(x^n) = \prod_{j=1}^{n} p_{f_\theta}(w_j) = p_{f_\theta^{\times \infty, n}}(w).$$

Hence, the Baum-Welch training attempts to determine a value of θ for which the probability $p_{f_\theta^{\times \infty, n}}(w)$ is maximal. Note, however, that the Baum-Welch training may lead only to a local maximum of $p_{f_\theta^{\times \infty, n}}(w)$.

3. Suppose that we are given a reduced alignment of N DNA sequences of length n (see Sect. 5.1). Let $c^1, \ldots, c^n$ be the columns of the alignment. As before, each of the columns can be regarded as an element of the sample space S. Consider the corresponding mapping $\varphi : S \to \{1, \ldots, 4^N\}$ and set $w_j = \varphi(c^j)$, $j = 1, \ldots n$. Let $w = (w_1, \ldots, w_n)$. In this case the statistical model can be fitted to w as it was done in Sect. 5.5, that is, by maximizing the likelihood of the dataset given the tree $L(D|\mathcal{T})$ defined there. Note that we have

$$L(D|\mathcal{T}) = \prod_{j=1}^{n} p_{f_\theta}(w_j) = p_{f_\theta^{\times\infty,n}}(w),$$

that is, by maximizing $L(D|\mathcal{T})$ we maximized $p_{f_\theta^{\times\infty,n}}(w)$.

In the next section we will see that maximizing $p_{f_\theta^{\times\infty,n}}(w)$ as in Parts 2.b and 3 of Example 8.5, is a special case of a standard model fitting procedure called the *maximum likelihood parameter estimation*. We will see that the model fitting procedures discussed in Parts 1 and 2.a are examples of maximum likelihood parameter estimation as well.

8.2 Parameter Estimation

Let $\{(S_\theta, \mathcal{B}_\theta, P_\theta, f_\theta, \mathcal{W}, \mathcal{P})\}$ be a statistical model. In this section we will discuss a universal approach to statistical modeling called *parameter estimation*. There are two important types of parameter estimation: *point estimation* and *set estimation*. In this book we will be only concerned with point estimation. For simplicity we will assume first that $\mathcal{P} \subset \mathbb{R}^q$, that is, θ is a vector in $\mathbb{R}^q$: $\theta = (\theta_1, \ldots, \theta_q)$. We will use the term *parameter* when referring to a particular component of θ.

Point estimation is based on producing a collection of vector-valued functions $\{g_k(x_1, \ldots, x_k)\}$, with $g_k : \mathcal{W}^k \to \mathbb{R}^q$, $k = 1, 2, \ldots$, such that, for every $\theta \in \mathcal{P}$ and $k = 1, 2, \ldots$, the vector-valued function $g_k^\theta = g_k(f_\theta^{\times\infty,k})$ is a vector-valued random variable on $(S_\theta, \mathcal{B}_\theta, P_\theta)^\infty$ and its distribution is in some sense concentrated around θ for large k (this last requirement will be made precise in Definition 8.6). Such a collection $\{g_k(x_1, \ldots, x_k)\}$ is called a *point estimator for θ*. Of course, to construct a point estimator for θ, it is sufficient to construct, for each $j = 1, \ldots, q$, a *point estimator for the parameter θ_j*, that is, a collection of functions $\{g_k^j(x_1, \ldots, x_k)\}$, with $g_k^j : \mathcal{W}^k \to \mathbb{R}$, $k = 1, 2, \ldots$, such that, for every $\theta \in \mathcal{P}$ and $k = 1, 2, \ldots$, the function $g_k^{j,\theta} = g_k^j(f_\theta^{\times\infty,k})$ is a random variable on $(S_\theta, \mathcal{B}_\theta, P_\theta)^\infty$ and its distribution is in some sense concentrated around θ_j for large k.

We will mention briefly that for set estimation with $q = 1$ (in which case the procedure is called *interval estimation*), two collections of functions $\{\underline{g}_k(x_1, \ldots, x_k)\}$ and $\{\overline{g}_k(x_1, \ldots, x_k)\}$ with $\underline{g}_k(x_1, \ldots, x_k) < \overline{g}_k(x_1, \ldots, x_k)$

for each k are produced, such that for all $\theta \in \mathcal{P}$, we have with specified probability

$$\theta \in \left(\underline{g}_k^\theta, \overline{g}_k^\theta \right),$$

and the above interval is in some sense as short as possible, if k is large. The collections $\{\underline{g}_k(x_1, \ldots, x_k)\}$ and $\{\overline{g}_k(x_1, \ldots, x_k)\}$ are called *interval estimators*.

From now on we will only consider point estimators and will be interested in estimators of the following type.

Definition 8.6. For a fixed $j = 1, \ldots, q$, a collection of functions $\{g_k^j(x_1, \ldots, x_k)\}$, with $g_k^j : \mathcal{W}^k \to \mathbb{R}$, $k = 1, 2, \ldots$, is called a *consistent estimator for θ_j* if, for all $\theta \in \mathcal{P}$ and $k = 1, 2, \ldots$, the function $g_k^{j,\theta} = g_k^j(f_\theta^{\times \infty, k})$ is a random variable on $(S_\theta, \mathcal{B}_\theta, P_\theta)^\infty$ and $g_k^{j,\theta} \xrightarrow{P} \theta_j$ for all $\theta \in \mathcal{P}$. If the components of vector-valued functions $g_k(x_1, \ldots, x_k)$, with $g_k : \mathcal{W}^k \to \mathbb{R}^q$, $k = 1, 2, \ldots$ form consistent estimators for $\theta_1, \ldots, \theta_q$, the collection $\{g_k(x_1, \ldots, x_k)\}$ is called a *consistent estimator for θ*.

Everywhere below we will be attempting to construct consistent estimators, but sometimes certain "almost consistent" estimators will be also acceptable (see, e.g., Exercises 8.4, 8.5).

If, for some $n \in \mathbb{N}$, we are given a vector $w = (w_1, \ldots, w_n) \in \mathcal{W}^n$ as in Definition 8.4, then the numbers $g_n^j(w_1, \ldots, w_n)$ are called *estimates for the parameters θ_j*, $j = 1, \ldots q$, respectively. In this case the vector $g_n(w_1, \ldots, w_n)$ is called an *estimate for θ*. Thus, one way to model the vector w by $\{(S_\theta, \mathcal{B}_\theta, P_\theta, f_\theta, \mathcal{W}, \mathcal{P})\}$, is to choose θ_0 from Definition 8.4 to be equal to $g_n(w_1, \ldots, w_n)$. This is a very popular way to perform model fitting. Of course, it depends on choosing a "good" estimator in the first place. Further in this section we will describe one fairly general way of producing "good" estimators.

What would one consider to be a "good" estimator? Above we have already described the consistency property. Now we will introduce another desirable property of estimators.

Definition 8.7. Let $\{g_k^j(x_1, \ldots, x_k)\}$ be an estimator for θ_j. Assume that for every $\theta \in \mathcal{P}$ the corresponding random variables $g_k^{j,\theta}$ are integrable on $(S_\theta, \mathcal{B}_\theta, P_\theta)^\infty$. Then, for every $\theta \in \mathcal{P}$, the *bias* of the estimator is the sequence of numbers $\{\mathcal{E}(g_k^{j,\theta}) - \theta_j\}$. If, for all $\theta \in \mathcal{P}$ and all $k = 1, 2, \ldots$, we have $\mathcal{E}(g_k^{j,\theta}) = \theta_j$, the estimator is called *unbiased*; otherwise it is called *biased*. Let the components of vector-valued functions $g_k(x_1, \ldots, x_k)$, with $g_k : \mathcal{W}^k \to \mathbb{R}^q$, $k = 1, 2, \ldots$ be estimators for $\theta_1, \ldots, \theta_q$. Then the collection $\{g_k(x_1, \ldots, x_k)\}$ is called an *unbiased estimator for θ*, if the components of $g_k(x_1, \ldots, x_k)$ are unbiased estimators for $\theta_1, \ldots, \theta_q$ respectively; otherwise it is called a *biased estimator for θ*, and its bias for every fixed $\theta \in \mathcal{P}$ is the sequence of vectors $\{\mathcal{E}(g_k^\theta) - \theta\}$.

It is generally desirable to have a consistent unbiased estimator. A further good property of an unbiased estimator is that it has a low variance. It is not always possible to find unbiased estimators, and even if it is possible, such estimators are not always preferred. For a biased estimator it is natural to require that it has a low *mean square error*

$$MSE\left(g_k^\theta\right) = \mathcal{E}\left(\left(g_k^\theta - \theta\right)^2\right) = Var\left(g_k^\theta\right) + \left(\mathcal{E}(g_k^\theta) - \theta\right)^2, \qquad (8.2)$$

for all $\theta \in \mathcal{P}$, as $k \to \infty$, where the above formulas are understood component-wise. In some cases the MSE of a biased estimator is less than the MSE, that is, the variance, of an unbiased estimator. In such cases the biased estimator might be preferred to the unbiased one.

Finally, it is also desirable that an estimator has, at least approximately, a well-known distribution, for example, a normal distribution. In this case an existing theory for the known distribution can be applied when using such an estimator.

We will now give examples of estimators.

Example 8.8. Let $\mathcal{P} = \mathbb{R} \times \mathbb{R}_+ \times \mathcal{P}'$, where $\mathbb{R}_+ = \{x \in \mathbb{R} : x \geq 0\}$. We will write $\theta \in \mathcal{P}$ as $\theta = (\theta_1, \theta_2, \theta') = (\mu, \sigma^2, \theta')$ with $\theta' \in \mathcal{P}'$. Assume that for every $\theta \in \mathcal{P}$, f_θ has a mean equal to μ and a variance equal to σ^2, and also assume that f_θ^3 and f_θ^4 are integrable on S_θ with respect to P_θ (in fact, one can show that if f_θ^4 is integrable on S_θ, then so are f_θ, f_θ^2 and f_θ^3). We will now construct consistent estimators for $\theta_1 = \mu$ and $\theta_2 = \sigma^2$.

Set

$$g_k^1(x_1, \ldots, x_k) = \frac{1}{k} \sum_{j=1}^k x_k.$$

Then $g_k^{1,\theta} = g_k^1(f_\theta^{\times\infty,k}) = 1/k \sum_{j=1}^k \tilde{f}_j$, where $\tilde{f}_1, \ldots, \tilde{f}_k$ are the first k elements in the sequence $f^{\times\infty}$. Clearly, $g_k^{1,\theta}$ are random variables on $(S_\theta, \mathcal{B}_\theta, P_\theta)^\infty$ for all $\theta \in \mathcal{P}$ and $k = 1, 2, \ldots$. Since $\tilde{f}_j$ are iid and their distributions coincide with that of f_θ, from the Weak Law of Large Numbers (see Theorem 6.56 and Remark 6.58) we obtain for every $\theta \in \mathcal{P}$ that $g_k^{1,\theta} \xrightarrow{P} \mu$, that is, $\{g_k^1(x_1, \ldots, x_k)\}$ is a consistent estimator for μ. Next, $\mathcal{E}(g_k^{1,\theta}) = \mu$, hence $\{g_k^1(x_1, \ldots, x_k)\}$ is an unbiased estimator for μ.

We will now find $MSE(g_k^{1,\theta})$ from formula (8.2). We have

$$MSE\left(g_k^{1,\theta}\right) = Var\left(g_k^{1,\theta}\right) = \frac{1}{k^2}\mathcal{E}\left(\left(\sum_{j=1}^k \tilde{f}_j\right)^2\right) - \mu^2 = \frac{1}{k^2}\left(\sum_{j=1}^k \mathcal{E}(\tilde{f}_j^2)\right)$$

$$+ 2\sum_{i<j} \mathcal{E}(\tilde{f}_i\tilde{f}_j)\right) - \mu^2 = \frac{1}{k^2}\left(k(\sigma^2 + \mu^2) + k(k-1)\mu^2\right) - \mu^2 = \frac{\sigma^2}{k},$$

where we used identity (6.35). Clearly, $MSE(g_k^{1,\theta}) \to 0$ as $k \to \infty$, for every $\theta \in \mathcal{P}$.

Further, set

$$g_k^2(x_1, \ldots, x_k) = \frac{1}{k} \sum_{j=1}^{k} \left(x_j - g_k^1(x_1, \ldots, x_k) \right)^2.$$

Then we have

$$g_k^{2,\theta} = \frac{1}{k} \sum_{j=1}^{k} \left(\tilde{f}_j - g_k^{1,\theta} \right)^2.$$

Clearly, $g_k^{2,\theta}$ are random variables on $(S_\theta, \mathcal{B}_\theta, P_\theta)^\infty$ for all $\theta \in \mathcal{P}$ and $k = 1, 2, \ldots$. We have

$$\mathcal{E}\left(g_k^{2,\theta}\right) = \frac{1}{k} \sum_{j=1}^{k} \mathcal{E}\left((\tilde{f}_j - g_k^{1,\theta})^2 \right) = \frac{1}{k} \sum_{j=1}^{k} \mathcal{E}\left(\left((\tilde{f}_j - \mu) \right.\right.$$

$$\left.\left. -(g_k^{1,\theta} - \mu) \right)^2 \right) = \frac{1}{k} \sum_{j=1}^{k} \mathcal{E}\left(\left((\tilde{f}_j - \mu) \right.\right.$$

$$\left.\left. -\frac{1}{k} \sum_{m=1}^{k} (\tilde{f}_m - \mu) \right)^2 \right) = \frac{1}{k} \sum_{j=1}^{k} \left(\mathcal{E}\left((\tilde{f}_j - \mu)^2 \right) \right. \tag{8.3}$$

$$\left. -\frac{2}{k} \mathcal{E}\left((\tilde{f}_j - \mu)^2 \right) + \frac{1}{k^2} \sum_{m=1}^{k} \mathcal{E}\left((\tilde{f}_m - \mu)^2 \right) \right)$$

$$= \sigma^2 - \frac{2}{k}\sigma^2 + \frac{1}{k}\sigma^2 = \frac{k-1}{k}\sigma^2.$$

Further, by Chebyshev's inequality (6.38) we have

$$P\left(\left\{ e \in S_\theta^\infty : \left| g_k^{2,\theta}(e) - \frac{k-1}{k}\sigma^2 \right| \ge \varepsilon \right\} \right) \le \frac{Var\left(g_k^{2,\theta}\right)}{\varepsilon^2}. \tag{8.4}$$

For $Var\left(g_k^{2,\theta}\right)$ we obtain using (8.3)

$$Var\left(g_k^{2,\theta}\right) = \mathcal{E}\left((g_k^{2,\theta})^2 \right) - \left(\mathcal{E}(g_k^{2,\theta}) \right)^2 = \mathcal{E}\left((g_k^{2,\theta})^2 \right) - \frac{(k-1)^2}{k^2}\sigma^4. \tag{8.5}$$

To find the first term in (8.5) we calculate

$$\mathcal{E}\left(\left(g_k^{2,\theta}\right)^2\right) = \frac{1}{k^2}\left(\sum_{j=1}^{k}\mathcal{E}\left(\left(\tilde{f}_j - g_k^{1,\theta}\right)^4\right)\right.$$

$$\left. +2\sum_{j<m}\mathcal{E}\left(\left(\tilde{f}_j - g_k^{1,\theta}\right)^2\left(\tilde{f}_m - g_k^{1,\theta}\right)^2\right)\right). \tag{8.6}$$

We have

$$\mathcal{E}\left(\left(\tilde{f}_j - g_k^{1,\theta}\right)^4\right) = \mathcal{E}\left(\left(\left(\tilde{f}_j - \mu\right) - \left(g_k^{1,\theta} - \mu\right)\right)^4\right)$$

$$= \mathcal{E}\left(\left(\tilde{f}_j - \mu\right)^4 - 4\left(\tilde{f}_j - \mu\right)^3\left(g_k^{1,\theta} - \mu\right) + 6\left(\tilde{f}_j - \mu\right)^2\left(g_k^{1,\theta} - \mu\right)^2\right.$$

$$\left. -4\left(\tilde{f}_j - \mu\right)\left(g_k^{1,\theta} - \mu\right)^3 + \left(g_k^{1,\theta} - \mu\right)^4\right).$$

Let $\mathcal{E}\left(\left(f_\theta - \mu\right)^4\right) = \nu$. Then for the components in the right-hand side of the above formula we have

$$\mathcal{E}\left(\left(\tilde{f}_j - \mu\right)^4\right) = \nu,$$

$$\mathcal{E}\left(\left(\tilde{f}_j - \mu\right)^3\left(g_k^{1,\theta} - \mu\right)\right) = \frac{1}{k}\mathcal{E}\left(\left(\tilde{f}_j - \mu\right)^3\sum_{m=1}^{k}\left(\tilde{f}_m - \mu\right)\right)$$

$$= \frac{1}{k}\mathcal{E}\left(\left(\tilde{f}_j - \mu\right)^4\right) = \frac{\nu}{k},$$

$$\mathcal{E}\left(\left(\tilde{f}_j - \mu\right)^2\left(g_k^{1,\theta} - \mu\right)^2\right) = \frac{1}{k^2}\mathcal{E}\left(\left(\tilde{f}_j - \mu\right)^2\left(\sum_{m=1}^{k}\left(\tilde{f}_m - \mu\right)\right)^2\right)$$

$$= \frac{1}{k^2}\left(\mathcal{E}\left(\left(\tilde{f}_j - \mu\right)^4\right) + \sum_{m\neq j}\mathcal{E}\left(\left(\tilde{f}_j - \mu\right)^2\left(\tilde{f}_m - \mu\right)^2\right)\right) = \frac{1}{k^2}\left(\nu + (k-1)\sigma^4\right),$$

$$\mathcal{E}\left(\left(\tilde{f}_j - \mu\right)\left(g_k^{1,\theta} - \mu\right)^3\right) = \frac{1}{k^3}\mathcal{E}\left(\left(\tilde{f}_j - \mu\right)\left(\sum_{m=1}^{k}\left(\tilde{f}_m - \mu\right)\right)^3\right)$$

$$= \frac{1}{k^3}\left(\mathcal{E}\left((\tilde{f}_j - \mu)^4\right) + 3\sum_{m \neq j}\mathcal{E}\left((\tilde{f}_j - \mu)^2(\tilde{f}_m - \mu)^2\right)\right)$$

$$= \frac{1}{k^3}\left(\nu + 3(k-1)\sigma^4\right),$$

$$\mathcal{E}\left((g_k^{1,\theta} - \mu)^4\right) = \frac{1}{k^4}\mathcal{E}\left(\left(\sum_{m=1}^{k}(\tilde{f}_m - \mu)\right)^4\right)$$

$$= \frac{1}{k^4}\left(\sum_{m=1}^{k}\mathcal{E}\left((\tilde{f}_m - \mu)^4\right) + 6\sum_{l<m}\mathcal{E}\left((\tilde{f}_m - \mu)^2(\tilde{f}_l - \mu)^2\right)\right)$$

$$= \frac{1}{k^4}\left(k\nu + 3k(k-1)\sigma^4\right).$$

Using the above calculations we also have for $j \neq m$

$$\mathcal{E}\left((\tilde{f}_j - g_k^{1,\theta})^2(\tilde{f}_m - g_k^{1,\theta})^2\right) = \mathcal{E}\left(\left((\tilde{f}_j - \mu) - (g_k^{1,\theta} - \mu)\right)^2\left((\tilde{f}_m - \mu)\right.\right.$$

$$\left.\left. - (g_k^{1,\theta} - \mu)\right)^2\right) = \mathcal{E}\left((\tilde{f}_j - \mu)^2(\tilde{f}_m - \mu)^2\right) - 2\mathcal{E}\left((\tilde{f}_j - \mu)(\tilde{f}_m - \mu)^2\right.$$

$$\times (g_k^{1,\theta} - \mu)\right) - 2\mathcal{E}\left((\tilde{f}_j - \mu)^2(\tilde{f}_m - \mu)(g_k^{1,\theta} - \mu)\right) + 4\mathcal{E}\left((\tilde{f}_j - \mu)(\tilde{f}_m - \mu)\right.$$

$$\times (g_k^{1,\theta} - \mu)^2\right) + \mathcal{E}\left((\tilde{f}_j - \mu)^2(g_k^{1,\theta} - \mu)^2\right) + \mathcal{E}\left((\tilde{f}_m - \mu)^2(g_k^{1,\theta} - \mu)^2\right)$$

$$- 2\mathcal{E}\left((\tilde{f}_j - \mu)(g_k^{1,\theta} - \mu)^3\right) - 2\mathcal{E}\left((\tilde{f}_m - \mu)(g_k^{1,\theta} - \mu)^3\right) + \mathcal{E}\left((g_k^{1,\theta} - \mu)^4\right)$$

$$= \sigma^4 + \frac{2}{k^2}\left(\nu + (k-1)\sigma^4\right) - \frac{4}{k^3}\left(\nu + 3(k-1)\sigma^4\right) + \frac{1}{k^4}\left(k\nu + 3k(k-1)\sigma^4\right)$$

$$- 2\mathcal{E}\left((\tilde{f}_j - \mu)(\tilde{f}_m - \mu)^2(g_k^{1,\theta} - \mu)\right) - 2\mathcal{E}\left((\tilde{f}_j - \mu)^2(\tilde{f}_m - \mu)(g_k^{1,\theta} - \mu)\right)$$

$$+ 4\mathcal{E}\left((\tilde{f}_j - \mu)(\tilde{f}_m - \mu)(g_k^{1,\theta} - \mu)^2\right).$$

To complete the above calculation for $\mathcal{E}\left((\tilde{f}_j - g_k^{1,\theta})^2(\tilde{f}_m - g_k^{1,\theta})^2\right)$ we compute for $j \neq m$

$$\mathcal{E}\left((\tilde{f}_j - \mu)(\tilde{f}_m - \mu)^2(g_k^{1,\theta} - \mu)\right) = \frac{1}{k}\mathcal{E}\left((\tilde{f}_j - \mu)(\tilde{f}_m - \mu)^2\sum_{l=1}^{k}(\tilde{f}_l - \mu)\right)$$

$$= \frac{1}{k}\mathcal{E}\Big((\tilde{f}_j - \mu)^2(\tilde{f}_m - \mu)^2\Big) = \frac{\sigma^4}{k},$$

$$\mathcal{E}\Big((\tilde{f}_j - \mu)^2(\tilde{f}_m - \mu)(g_k^{1,\theta} - \mu)\Big) = \frac{1}{k}\mathcal{E}\left((\tilde{f}_j - \mu)^2(\tilde{f}_m - \mu)\sum_{l=1}^{k}(\tilde{f}_l - \mu)\right)$$

$$= \frac{1}{k}\mathcal{E}\Big((\tilde{f}_j - \mu)^2(\tilde{f}_m - \mu)^2\Big) = \frac{\sigma^4}{k},$$

$$\mathcal{E}\Big((\tilde{f}_j - \mu)(\tilde{f}_m - \mu)(g_k^{1,\theta} - \mu)^2\Big) = \frac{1}{k^2}\mathcal{E}\left((\tilde{f}_j - \mu)(\tilde{f}_m - \mu)\left(\sum_{l=1}^{k}(\tilde{f}_l - \mu)\right)^2\right)$$

$$= \frac{2}{k^2}\mathcal{E}\Big((\tilde{f}_j - \mu)^2(\tilde{f}_m - \mu)^2\Big) = \frac{2\sigma^4}{k^2}.$$

Hence we have

$$\mathcal{E}\Big((\tilde{f}_j - g_k^{1,\theta})^2(\tilde{f}_m - g_k^{1,\theta})^2\Big) = \sigma^4 + \frac{2}{k^2}\left(\nu + (k-1)\sigma^4\right)$$
$$- \frac{4}{k^3}\left(\nu + 3(k-1)\sigma^4\right) + \frac{1}{k^4}\left(k\nu + 3k(k-1)\sigma^4\right) - \frac{4\sigma^4}{k} + \frac{8\sigma^4}{k^2}.$$

Thus, we have calculated $\mathcal{E}\Big((\tilde{f}_j - g_k^{1,\theta})^4\Big)$ and $\mathcal{E}\Big((\tilde{f}_j - g_k^{1,\theta})^2(\tilde{f}_m - g_k^{1,\theta})^2\Big)$. Plugging the resulting expressions into formula (8.6) we obtain from (8.5)

$$Var\Big(g_k^{2,\theta}\Big) = \frac{(k-1)^2\nu}{k^3} - \frac{(k^2 - 4k + 3)\sigma^4}{k^3} \to 0, \quad \text{as } k \to \infty, \qquad (8.7)$$

for every $\theta \in \mathcal{P}$.

From (8.4) and (8.7) for every $\theta \in \mathcal{P}$ and sufficiently large k we have

$$P\left(\Big\{e \in S_\theta^\infty : \Big|g_k^{2,\theta}(e) - \sigma^2\Big| \geq \varepsilon\Big\}\right) \leq P\left(\Big\{e \in S_\theta^\infty : \Big|g_k^{2,\theta}(e) - \frac{k-1}{k}\sigma^2\Big|\right.$$

$$\left. \geq \varepsilon - \frac{\sigma^2}{k}\Big\}\right) \leq P\left(\Big\{e \in S_\theta^\infty : \Big|g_k^{2,\theta}(e) - \frac{k-1}{k}\sigma^2\Big| \geq \frac{\varepsilon}{2}\Big\}\right) \to 0, \quad \text{as } k \to \infty.$$

which tends to 0 as $k \to \infty$. Hence $g_k^{2,\theta} \xrightarrow{P} \sigma^2$ for every $\theta \in \mathcal{P}$, which means that $\{g_k^2(x_1,\ldots,x_k)\}$ is a consistent estimator for σ^2. Formula (8.3) shows that the bias of $\{g_k^2(x_1,\ldots,x_k)\}$ is $\{-\sigma^2/k\}$. In particular, the bias of $\{g_k^2(x_1,\ldots,x_k)\}$ tends to 0, as $k \to \infty$, for every $\theta \in \mathcal{P}$.

We will now find $MSE\Big(g_k^{2,\theta}\Big)$ from formula (8.2). Using formulas (8.3) and (8.7) we obtain

$$MSE\left(g_k^{2,\theta}\right) = Var\left(g_k^{2,\theta}\right) + \left(\mathcal{E}(g_k^{2,\theta}) - \theta\right)^2 = \frac{(k-1)^2\nu}{k^3} - \frac{(k^2 - 5k + 3)\sigma^4}{k^3}.$$

Clearly, $MSE\left(g_k^{2,\theta}\right) \to 0$ as $k \to \infty$, for every $\theta \in \mathcal{P}$.

One can also produce an unbiased estimator for $\theta_2 = \sigma^2$ as follows

$$\hat{g}_k^2(x_1, \ldots, x_k) = \frac{1}{k-1}\sum_{j=1}^{k}\left(x_j - g_k^1(x_1, \ldots, x_k)\right)^2.$$

It can be proved analogously that $\{\hat{g}_k^2(x_1, \ldots, x_k)\}$ is a consistent estimator for σ^2, and a calculation similar to that in (8.3) shows that it is unbiased. For this estimator we have

$$MSE\left(\hat{g}_k^{2,\theta}\right) = Var\left(\hat{g}_k^{2,\theta}\right) = \frac{\nu}{k} - \frac{(k-3)\sigma^4}{k(k-1)}.$$

Thus, $MSE\left(\hat{g}_k^{2,\theta}\right) \to 0$ as $k \to \infty$, for every $\theta \in \mathcal{P}$.

Interestingly, there is an estimation procedure that in many cases satisfies most of, and in some cases all of the criteria for "good" estimators as stated in the discussion following Definition 8.7. This is the procedure of *maximum likelihood estimation*. First, we will define the *likelihood* given a statistical model. It will be defined in the situation when, for every $\theta \in \mathcal{P}$, either f_θ is continuous, or F_{f_θ} is the distribution function of a discrete random variable (note that in the latter case f_θ itself may not be discrete). We will call such models *regular statistical models*.

For a regular model, denote by $\mathcal{P}_c$ the set of all $\theta \in \mathcal{P}$ for which f_θ is continuous and by $\mathcal{P}_d$ the complement to $\mathcal{P}_c$ in $\mathcal{P}$ (that is, the set of all $\theta \in \mathcal{P}$ for which F_{f_θ} is the distribution function of a discrete random variable).

Definition 8.9. *Let* $\{(S_\theta, \mathcal{B}_\theta, P_\theta, f_\theta, \mathcal{W}, \mathcal{P})\}$ *be a regular statistical model and* $\theta \in \mathcal{P}_d$. *The likelihood of order k at the point θ given the model is a function* $L_{k,d}^\theta : \mathbb{R}^k \to \mathbb{R}$ *defined in terms of the random sample of size k from the model as follows*

$$L_{k,d}^\theta(x) = P\left(\left\{e \in S_\theta^\infty : f_\theta^{\times \infty, k}(e) = x\right\}\right), \tag{8.8}$$

with $x \in \mathbb{R}^k$.

Since for each $\theta \in \mathcal{P}_d$ the components of the random sample of size k are iid and their distributions are identical to that of f_θ, formula (8.8) for $x = (x_1, \ldots, x_k) \in \mathbb{R}^k$ can be rewritten as

$$\begin{aligned}L_{k,d}^\theta(x) &= P\left(\{e \in S_\theta^\infty : f_\theta(e) = x_1\}\right) \times \cdots \\ &\times P\left(\{e \in S_\theta^\infty : f_\theta(e) = x_k\}\right) = (F_{f_\theta}(x_1) - F_{f_\theta}(x_1 - 0)) \times \cdots \\ &\times (F_{f_\theta}(x_k) - F_{f_\theta}(x_k - 0)).\end{aligned} \tag{8.9}$$

We also remark here that if $\mathcal{P}_d = \mathcal{P}$ and the random variables f_θ are themselves discrete and have a common range $\mathcal{W}$, then, for $r = (r_1, \ldots, r_k) \in \mathcal{W}^k$ formulas (8.8) and (8.9) can be rewritten in terms of the probability distributions of $f_\theta^{\times \infty, k}$ and f_θ as follows

$$L_{k,d}^\theta(r) = p_{f_\theta^{\times \infty, k}}(r) = p_{f_\theta}(r_1) \times \ldots \times p_{f_\theta}(r_k). \tag{8.10}$$

Next we will consider the continuous case.

Definition 8.10. *Let* $\{(S_\theta, \mathcal{B}_\theta, P_\theta, f_\theta, \mathcal{W}, \mathcal{P})\}$ *be a regular statistical model and* $\theta \in \mathcal{P}_c$. *The* likelihood *of order* k *at the point* θ *given the model is a function* $L_{k,c}^\theta : \mathbb{R}^k \to \mathbb{R}$ *defined in terms of the density function of the random sample of size* k *from the model as follows*

$$L_{k,c}^\theta(x) = \varrho_{f_\theta^{\times \infty, k}}(x), \tag{8.11}$$

with $x \in \mathbb{R}^k$.

Using Definition 6.52, we can rewrite formula (8.11) as

$$L_{k,c}^\theta(x) = \varrho_{f_\theta}(x_1) \times \ldots \times \varrho_{f_\theta}(x_k), \tag{8.12}$$

where $x = (x_1, \ldots, x_k)$. Recall also that ϱ_{f_θ} is not uniquely defined as a function on $\mathbb{R}$ and that for every $\theta \in \mathcal{P}_c$ there is in fact a family of density functions of f_θ such that any two functions from the family coincide a.e. on $\mathbb{R}$ (see the discussion following Definition 6.45). Therefore, for the purposes of Definition 8.10 we assume that a particular "natural" choice of density function ϱ_{f_θ} has been made for each $\theta \in \mathcal{P}_c$. Clearly, the likelihoods $L_{k,c}^\theta$ constructed from two different density functions for fixed $k \in \mathbb{N}$ and $\theta \in \mathcal{P}_c$, coincide a.e. on $\mathbb{R}^k$.

In the remainder of this section we will only consider regular models. For such models *maximum likelihood estimators* are constructed as follows.

Definition 8.11. *For any* $k \in \mathbb{N}$ *and* $x = (x_1, \ldots, x_k) \in \mathcal{W}^k$ *set* $g_k(x_1, \ldots, x_k)$ *to be a value of* θ *selected as follows: if* $L_{k,d}^\theta(x) \neq 0$ *for some* $\theta \in \mathcal{P}_d$, *then choose* $\theta \in \mathcal{P}_d$ *for which* $L_{k,d}^\theta(x)$ *is maximal; if* $L_{k,d}^\theta(x) = 0$ *for all* $\theta \in \mathcal{P}_d$, *then choose* $\theta \in \mathcal{P}_c$ *for which* $L_{k,c}^\theta(x)$ *is maximal. Any collection of functions* $\{g_k(x_1, \ldots, x_k)\}$ *so constructed is called a* maximum likelihood estimator *for* θ.

Definition 8.11 requires some clarification. First of all, the likelihoods $L_{k,d}^\theta(x)$ and $L_{k,c}^\theta(x)$ must attain their maxima over $\mathcal{P}_d$ and $\mathcal{P}_c$ respectively for every $k \in \mathbb{N}$ and $x \in \mathcal{W}^k$. This condition is satisfied if, for example, $\mathcal{P}_d$ and $\mathcal{P}_c$ are compact subsets of $\mathbb{R}^q$ and $L_{k,d}^\theta(x)$ and $L_{k,c}^\theta(x)$ are continuous functions of θ for every k and x. Secondly, it is possible that for some k and x, a value of θ for which the corresponding likelihood is maximal is not unique. In Definition

8.11 the value of the function g_k at the point x was taken to be *any* such value of θ. Hence *a priori* there may exist more than one maximum likelihood estimators. Thirdly, in order for $\{g_k(x_1, \ldots, x_k)\}$ to be an estimator in the usual sense, it is necessary for the vector-valued functions $g_k^\theta = g_k(f_\theta^{\times \infty, k})$ defined on $(S_\theta, \mathcal{B}_\theta, P_\theta)^\infty$ to be random variables for all $k \in \mathbb{N}$ and $\theta \in \mathcal{P}$. This is known to be true under broad assumptions. Fourthly, maximum likelihood estimators are known to be consistent under certain conditions. However, as we will see below, such estimators may be biased. Fifthly, under certain conditions, maximum likelihood estimators are known to be asymptotically normally distributed. More precisely, the sequence of $\mathbb{R}^q$-valued random variables $\{\sqrt{k}(g_k^\theta - \theta)\}$ converges in distribution to an $\mathbb{R}^q$-valued continuous random variable with density function of the form

$$\varrho(y) = \frac{\sqrt{\det A}}{(2\pi)^{q/2}} \exp\left(-\frac{1}{2} \sum_{i,j=1}^{q} a_{ij} y_i y_j\right),$$

where $A = (a_{ij})$ is a $q \times q$ invertible positive-definite symmetric matrix, and $y = (y_1, \ldots, y_q) \in \mathbb{R}^q$. Finally, it must be stressed that the maximum likelihood estimation procedure produces estimators for *all* of the parameters θ_j, $j = 1, \ldots, q$ at the same time. For a precise formulation of conditions under which maximum likelihood estimators are consistent and asymptotically normally distributed, see Remark 8.17.

Maximum likelihood estimators are often used in the situation when either $\mathcal{P}_d = \emptyset$, or $\mathcal{P}_c = \emptyset$, and the parameter space $\mathcal{P}$ is either a domain or the closure of a domain in $\mathbb{R}^q$, that is, we have either $\mathcal{P} = \Omega$, or $\mathcal{P} = \overline{\Omega}$, where $\Omega \subset \mathbb{R}^q$ is an open connected set. Let $L_k^\theta(x)$ denote the corresponding likelihoods. If in this case, for every k and $x \in \mathcal{W}^k$, $L_k^\theta(x)$ is differentiable in Ω as a function of θ and its maximum over $\mathcal{P}$ occurs at a point $\theta_0 \in \Omega$ (note that θ_0 depends on k and x), then θ_0 is a critical point of $L_k^\theta(x)$ and we have

$$\frac{\partial L_k^\theta(x)}{\partial \theta_j}(\theta_0) = 0, \qquad j = 1, \ldots, q. \tag{8.13}$$

Equations (8.13) are called the *maximum likelihood equations* and are an important tool for constructing maximum likelihood estimators. Of course, these equations may have solutions other than θ_0. For example, further analysis is required to distinguish θ_0 from those points in Ω, at which $L_k^\theta(x)$ has either local extrema or saddle points. For instance, to distinguish local maxima from local minima one can use the Hessian matrix of $L_k^\theta(x)$ (if $L_k^\theta(x) \in C^2(\Omega)$).

If, as above, $\mathcal{P} = \Omega$, or $\mathcal{P} = \overline{\Omega}$ and, for all k and x, $L_k^\theta(x)$ is differentiable in Ω as a function of θ, it is customary to start looking for maximum likelihood estimators by solving the maximum likelihood equations (8.13). One must bear in mind, however, that, for some k and x, the function $L_k^\theta(x)$ may not attain its maximum in $\mathcal{P}$ at all (for example, if Ω is unbounded, $L_k^\theta(x)$ may grow as x approaches ∞) and, even if it does, it may attain its maximum at a boundary point of Ω (in the case $\mathcal{P} = \overline{\Omega}$). Therefore, for some k and x, none of the solutions to the maximum likelihood equations may be related to maximum likelihood estimators. We also remark that the maximum likelihood equations often make sense for regular models where neither of $\mathcal{P}_d$ and $\mathcal{P}_c$ is empty and where $\mathcal{P}$ is obtained from a domain Ω by adding some (but not all) boundary points to it.

Suppose $\{g_k(x_1, \ldots, x_k)\}$ is a maximum likelihood estimator. If, for some $n \in \mathbb{N}$, we are given a vector $w \in \mathcal{W}^n$ as in Definition 8.4, then the vector $g_n(w_1, \ldots, w_n) \in \mathbb{R}^q$ is the corresponding *maximum likelihood estimate for θ*.

We will now give examples of maximum likelihood estimators and estimates.

Example 8.12.

1. Suppose that $q = 1$, and denote the parameter $\theta = \theta_1$ by λ. Let $\mathcal{P} = \overline{\Omega} = \{\lambda \in \mathbb{R} : \lambda \geq 0\}$, with $\Omega = \{\lambda \in \mathbb{R} : \lambda > 0\}$, and assume that f_λ for each λ is discrete, takes values $0, 1, 2, \ldots$ (hence $\mathcal{W} = \mathbb{Z}_+ = \{0, 1, 2, \ldots\}$) and has the Poisson distribution with mean λ

$$\left\{ p_{f_\lambda}(j) = \frac{\exp(-\lambda)\lambda^j}{j!}, \quad j = 0, 1, 2, \ldots \right\},$$

(see Sect. 6.9). Clearly, in this example we are dealing with a regular model, where $\mathcal{P}_c = \emptyset$. Fix $k \in \mathbb{N}$ and choose $x = (x_1, \ldots, x_k) \in \mathbb{Z}_+^k$. Then from formula (8.10) we have

$$L_k^\lambda(x) = \frac{\exp(-\lambda)\lambda^{x_1}}{x_1!} \times \ldots \times \frac{\exp(-\lambda)\lambda^{x_k}}{x_k!} = \frac{\exp(-k\lambda)\lambda^{\sum_{j=1}^k x_j}}{\prod_{j=1}^k x_j!}.$$

We will now consider two cases. Suppose first that $x_1 = \ldots = x_k = 0$. Then $L_k^\lambda(x) = \exp(-k\lambda)$. In this case L_k^λ attains it maximum at the boundary point $\lambda = 0$ of Ω. Suppose now that $\sum_{j=1}^k x_j > 0$. Clearly, $L_k^\lambda(x)$ is infinitely many times differentiable in Ω for all k and x, and we can attempt to find the solutions to the maximum likelihood equations (8.13), that become the following single equation in this case

$$\frac{\exp(-k\lambda)\lambda^{\sum_{j=1}^k x_j - 1}}{\prod_{j=1}^k x_j!} \left(-k\lambda + \sum_{j=1}^k x_j \right) = 0.$$

Clearly, the only solution to this equation for fixed k and x is

$$\lambda = \lambda_0 = \frac{1}{k} \sum_{j=1}^k x_j. \tag{8.14}$$

Calculating the second derivative of $L_k^\lambda(x)$ with respect to λ at the point λ_0 we obtain the negative number

$$-\frac{k \exp(-k\lambda_0)\lambda_0^{k\lambda_0 - 1}}{\prod\limits_{j=1}^{k} x_j!}.$$

Therefore, λ_0 is a point of local maximum for $L_k^\lambda(x)$. Further, since λ_0 is the only critical point of $L_k^\lambda(x)$ in Ω and since $L_k^\lambda(x)$ is continuous on $\overline{\Omega}$, λ_0 is in fact a point of global maximum of $L_k^\lambda(x)$ on $\overline{\Omega}$.

We now observe that formula (8.14) gives a value that maximizes $L_k^\lambda(x)$ for all $x \in \mathbb{Z}_+^k$, including the case $x_1 = \ldots = x_k = 0$. This formula leads to precisely the collection of functions $\{g_k^1(x_1, \ldots, x_k)\}$ that we considered in Example 8.8 in the general case. Hence the maximum likelihood procedure in this case gives a consistent unbiased estimator for λ.

2. Suppose again that $q = 1$, and denote the parameter $\theta = \theta_1$ by p. Let $\mathcal{P} = \overline{\Omega} = \{p \in \mathbb{R} : 0 \le p \le 1\}$, with $\Omega = \{p \in \mathbb{R} : 0 < p < 1\}$, and assume that f_p for each p is discrete, takes values $0, 1$ (hence $\mathcal{W} = \{0, 1\}$) and its distribution is that of the Bernoulli trial with mean p

$$\{p_{f_p}(0) = 1 - p, \, p_{f_p}(1) = p\}.$$

(see Sect. 6.9). As above, here we are dealing with a regular model with $\mathcal{P}_c = \emptyset$. Fix $k \in \mathbb{N}$ and choose $x = (x_1, \ldots, x_k) \in \{0, 1\}^k$. Then from formula (8.10) we have

$$L_k^p(x) = p^{x_1}(1-p)^{1-x_1} \times \ldots \times p^{x_k}(1-p)^{1-x_k} = p^{\sum_{j=1}^k x_j}(1-p)^{k - \sum_{j=1}^k x_j}.$$

We will now consider three cases. Assume first that $x_1 = \ldots = x_k = 0$. Then $L_k^p(x) = (1-p)^k$, and hence $L_k^p(x)$ takes its maximum at the boundary point $p = 0$ of Ω. Suppose now that $x_1 = \ldots = x_k = 1$. In this case $L_k^p(x) = p^k$, and hence $L_k^p(x)$ takes its maximum at the boundary point $p = 1$ of Ω. Assume finally that $0 < \sum_{j=1}^k x_j < k$. Clearly, $L_k^p(x)$ is infinitely many times differentiable in Ω for all k and x, and we can attempt to find the solutions to the maximum likelihood equations (8.13) that become the following single equation in this case

$$p^{\sum_{j=1}^k x_j - 1}(1-p)^{k-1-\sum_{j=1}^k x_j}\left(\sum_{j=1}^k x_j - kp\right) = 0.$$

Clearly, the only solution to this equation for fixed k and x is

$$p = p_0 = \frac{1}{k}\sum_{j=1}^k x_j. \tag{8.15}$$

Calculating the second derivative of $L_k^p(x)$ with respect to p at the point p_0 we obtain the negative number

$$-kp_0^{kp_0-1}(1-p_0)^{k-kp_0-1}.$$

Therefore, p_0 is a point of local maximum for $L_k^p(x)$. Further, since p_0 is the only critical point of $L_k^p(x)$ in Ω and since $L_k^p(x)$ is continuous on $\overline{\Omega}$, p_0 is in fact a point of global maximum of $L_k^p(x)$ on $\overline{\Omega}$.

We now observe that formula (8.15) gives a value that maximizes $L_k^p(x)$ for all $x \in \{0,1\}^k$, including the cases $x_1 = \ldots = x_k = 0$ and $x_1 = \ldots = x_k = 1$. As in Part 1 above, this formula leads to precisely the collection of functions $\{g_k^1(x_1, \ldots, x_k)\}$ that we considered in Example 8.8 in the general case. Hence the maximum likelihood procedure in this case gives a consistent unbiased estimator for p.

3. Suppose again that $q = 1$, and denote the parameter $\theta = \theta_1$ by V. Let $\mathcal{P} = \Omega = \{V \in \mathbb{R} : V > 0\}$, and assume that f_V for each V is continuous, has range $(0, \infty)$ (that is, $\mathcal{W} = (0, \infty)$) and is uniformly distributed with the density function

$$\varrho_{f_V}(z) = \begin{cases} 0, & \text{if } z \leq 0 \text{ or } z > V, \\ \dfrac{1}{V}, & \text{if } 0 < z \leq V \end{cases}$$

(see Sect. 6.9). In this example we are dealing with a regular model with $\mathcal{P}_d = \emptyset$. Fix $k \in \mathbb{N}$ and choose $x = (x_1, \ldots, x_k) \in (0, \infty)^k$. Then from formula (8.12) we have

$$L_k^V(x) = \varrho_{f_V}(x_1) \times \ldots \times \varrho_{f_V}(x_k).$$

Defining $x_{\max} = \max\{x_1, \ldots, x_k\}$ we have

$$L_k^V(x) = \begin{cases} 0, & \text{if } V < x_{\max}, \\ \dfrac{1}{V^k}, & \text{if } V \geq x_{\max}. \end{cases}$$

Note that $L_k^V(x)$ is not differentiable with respect to V at $x_{\max}$ (but it is infinitely many times differentiable at any other point in Ω). It is easy to see that $L_k^V(x)$ takes its maximal value at $V = x_{\max}$.

Set $g_k(x_1, \ldots, x_k) = x_{\max}$ and show that $\{g_k(x_1, \ldots, x_k)\}$ is a consistent estimator for V. Consider $g_k^V = g_k(f_V^{\times \infty, k})$. Since the maximum of a finite number of independent random variables is a random variable, g_k^V is a random variable for all $k \in \mathbb{N}$ and $V \in \Omega$. We will now find its distribution.

From the independence of the components of $f_V^{\times \infty, k} = (\tilde{f}_1, \ldots, \tilde{f}_k)$, for $y \in \mathbb{R}$ we have

$$F_{g_k^V}(z) = P\left(\{e \in S_V^\infty : g_k^V(e) \leq z\}\right) = P\left(\{e \in S_V^\infty : \tilde{f}_1(e) \leq z, \dots \tilde{f}_k(e) \leq z\}\right)$$

$$= \prod_{j=1}^{k} P\left(\{e \in S_V^\infty : \tilde{f}_j(e) \leq z\}\right) = (F_{f_V}(z))^k.$$

Differentiating the above identity with respect to z at $z \neq 0, V$, we find for the density function of g_k^V

$$\varrho_{g_k^V}(z) = k\varrho_{f_V}(z)(F_{f_V}(z))^{k-1},$$

which gives

$$\varrho_{g_k^V}(z) = \begin{cases} 0, & \text{if } z \leq 0 \text{ or } z > V, \\[2mm] \dfrac{kz^{k-1}}{V^k}, & \text{if } 0 < z \leq V. \end{cases}$$

It is therefore easy to find the mean and the variance of g_k^V from formulas (6.26) and (6.27) respectively. Indeed, we have

$$\mathcal{E}\left(g_k^V\right) = \frac{k}{V^k} \int_0^V z^k \, dz = \frac{kV}{k+1},$$

$$Var\left(g_k^V\right) = \frac{k}{V^k} \int_0^V z^{k+1} \, dz - \frac{k^2 V^2}{(k+1)^2} = \frac{kV^2}{(k+1)^2(k+2)}.$$

As in Example 8.8 we obtain from Chebyshev's inequality (6.38) for $\varepsilon > 0$

$$P\left(\left\{e \in S_V^\infty : \left|g_k^V(e) - \frac{kV}{k+1}\right| \geq \varepsilon\right\}\right) \leq \frac{Var\left(g_k^V\right)}{\varepsilon^2} \to 0, \quad \text{as } k \to \infty,$$

for every V. Therefore, for every V and sufficiently large k we have

$$P\left(\{e \in S_V^\infty : |g_k^V(e) - V| \geq \varepsilon\}\right) \leq P\left(\left\{e \in S_V^\infty : \left|g_k^V(e) - \frac{kV}{k+1}\right|\right.\right.$$

$$\left.\left. \geq \varepsilon - \frac{V}{k+1}\right\}\right) \leq P\left(\left\{e \in S_V^\infty : \left|g_k^V(e) - \frac{kV}{k+1}\right| \geq \frac{\varepsilon}{2}\right\}\right),$$

which tends to 0 as $k \to \infty$. Thus, we have proved that $g_k^V \xrightarrow{P} V$ for every V, which means that $\{g_k(x_1, \dots, x_k)\}$ is a consistent estimator for V. The bias of this estimator is $\{-V/(k+1)\}$; it tends to zero as $k \to \infty$, for every V. Further, for $MSE(g_k^V)$ from formula (8.2) we have

$$MSE\left(g_k^V\right) = \frac{2V^2}{(k+1)(k+2)} \to 0, \quad \text{as } k \to \infty,$$

for every V.

4. Suppose now that $q = 2$, and denote the parameters θ_1 and θ_2 by μ and σ^2 respectively, as we did in Example 8.8. Let $\mathcal{P} = \overline{\Omega} = \{(\mu, \sigma^2) \in \mathbb{R}^2 : \sigma^2 \geq 0\}$, with $\Omega = \{(\mu, \sigma^2) \in \mathbb{R}^2 : \sigma^2 > 0\}$. Suppose that $\mathcal{W} = \mathbb{R}$ and assume that $f_{(\mu,\sigma^2)}$ with $\sigma^2 > 0$ is continuous and has the normal distribution with mean μ and variance σ^2

$$\varrho_{f_{(\mu,\sigma^2)}}(z) = \frac{1}{\sqrt{2\pi\sigma^2}} \exp\left(-\frac{(z-\mu)^2}{2\sigma^2}\right).$$

Assume further that $f_{(\mu,0)}$ is a random variable with range $\mathbb{R}$ and the following distribution function

$$F_{f_{(\mu,0)}}(z) = \begin{cases} 0, & \text{if } z < \mu, \\ \\ 1, & \text{if } z \geq \mu. \end{cases}$$

Note that $F_{f_{(\mu,0)}}$ is the distribution function of a discrete random variable, but $f_{(\mu,0)}$ itself is not discrete. Thus, in this example we are dealing with a regular model where $\mathcal{P}_c = \Omega$ and $\mathcal{P}_d = \partial\Omega$.

Fix $k \in \mathbb{N}$ and choose $x = (x_1, \ldots, x_k) \in \mathbb{R}^k$. Let first $\sigma^2 > 0$. Then from formula (8.12) we have

$$L_{k,c}^{(\mu,\sigma^2)}(x) = \prod_{j=1}^{k} \frac{1}{\sqrt{2\pi\sigma^2}} \exp\left(-\frac{(x_j - \mu)^2}{2\sigma^2}\right)$$

$$= \frac{1}{(\sqrt{2\pi\sigma^2})^k} \exp\left(-\frac{1}{2\sigma^2}\sum_{j=1}^{k}(x_j - \mu)^2\right). \tag{8.16}$$

Let now $\sigma^2 = 0$. In this case formula (8.9) gives

$$L_{k,d}^{(\mu,0)}(x) = \begin{cases} 1, & \text{if } x_1 = \ldots = x_k = \mu, \\ \\ 0, & \text{if } x_i \neq \mu \text{ for some } 1 \leq i \leq k. \end{cases}$$

Suppose first that $x_1 = \ldots = x_k = a$. In this case the only value of μ for which $L_{k,d}^{(\mu,0)}(x) \neq 0$ is $\mu = a$, that is, we have

$$g_k(x_1, \ldots, x_k) = (a, 0). \tag{8.17}$$

Let now $x_i \neq x_j$ for some $i, j = 1, \ldots, k$. Then $L_{k,d}^{(\mu,0)}(x) = 0$ for all μ, and we will concentrate on formula (8.16). Since $L_{k,c}^{(\mu,\sigma^2)}(x)$ is infinitely many times differentiable in Ω, we can find its critical points there. Differentiating the right-hand side of (8.16) with respect to μ and σ^2 we obtain

$$\frac{\partial L_{k,c}^{(\mu,\sigma^2)}(x)}{\partial \mu} = \frac{1}{(\sqrt{2\pi\sigma^2})^k \sigma^2} \exp\left(-\frac{1}{2\sigma^2} \sum_{j=1}^{k} (x_j - \mu)^2\right) \sum_{j=1}^{k} (x_j - \mu),$$

$$\frac{\partial L_{k,c}^{(\mu,\sigma^2)}(x)}{\partial \sigma^2} = \frac{1}{2(\sqrt{2\pi\sigma^2})^k (\sigma^2)^2} \exp\left(-\frac{1}{2\sigma^2} \sum_{j=1}^{k} (x_j - \mu)^2\right)$$

$$\times \left(-k\sigma^2 + \sum_{j=1}^{k} (x_j - \mu)^2\right).$$

From the maximum likelihood equations

$$\frac{\partial L_{k,c}^{(\mu,\sigma^2)}(x)}{\partial \mu} = 0,$$

$$\frac{\partial L_{k,c}^{(\mu,\sigma^2)}(x)}{\partial \sigma^2} = 0$$

we obtain

$$\mu = \mu_0 = \frac{1}{k} \sum_{j=1}^{k} x_j,$$

$$\sigma^2 = \sigma_0^2 = \frac{1}{k} \sum_{j=1}^{k} \left(x_j - \frac{1}{k} \sum_{m=1}^{k} x_m\right)^2. \tag{8.18}$$

We will now show that (μ_0, σ_0^2) is a point of maximum for $L_{k,c}^{(\mu,\sigma^2)}(x)$. Calculating its Hessian matrix at (μ_0, σ_0^2) we obtain

$$\frac{\partial^2 L_{k,c}^{(\mu,\sigma^2)}(x)}{\partial \mu^2}(\mu_0, \sigma_0^2) = -\frac{k}{(\sqrt{2\pi\sigma_0^2})^k \sigma_0^2} \exp\left(-\frac{1}{2\sigma_0^2} \sum_{j=1}^{k} (x_j - \mu_0)^2\right)$$

$$= -\frac{k}{(\sqrt{2\pi\sigma_0^2})^k \sigma_0^2} \exp\left(-\frac{n}{2}\right) < 0,$$

$$\frac{\partial^2 L_{k,c}^{(\mu,\sigma^2)}(x)}{\partial \mu \partial \sigma^2}(\mu_0, \sigma_0^2) = 0,$$

$$\frac{\partial^2 L_{k,c}^{(\mu,\sigma^2)}(x)}{\partial(\sigma^2)^2}(\mu_0,\sigma_0^2) = -\frac{k}{2(\sqrt{2\pi\sigma_0^2})^k(\sigma_0^2)^2}\exp\left(-\frac{1}{2\sigma_0^2}\sum_{j=1}^{k}(x_j-\mu_0)^2\right)$$

$$= -\frac{k}{2(\sqrt{2\pi\sigma_0^2})^k(\sigma_0^2)^2}\exp\left(-\frac{n}{2}\right) < 0.$$

Hence the Hessian matrix of $L_{k,c}^{(\mu,\sigma^2)}(x)$ is negative-definite at (μ_0,σ_0^2) and therefore (μ_0,σ_0^2) is a point of local maximum of $L_{k,c}^{(\mu,\sigma^2)}(x)$. Since this is the only critical point in Ω, it is in fact a point of global maximum of $L_{k,c}^{(\mu,\sigma^2)}(x)$ on Ω.

Now we note that (8.17) is a special case of (8.18) and hence formulas (8.18) give a maximum likelihood estimator $\{g_k(x_1,\ldots,x_k)\}$. Note that $g_k(x_1,\ldots,x_k) = \left(g_k^1(x_1,\ldots,x_k),g_k^2(x_1,\ldots,x_k)\right)$, where $\{g_k^1(x_1,\ldots,x_k)\}$ and $\{g_k^2(x_1,\ldots,x_k)\}$ are the estimators for μ and σ^2 found in Example 8.8. Hence $\{g_k(x_1,\ldots,x_k)\}$ is a consistent estimator for $\theta = (\mu,\sigma^2)$ with bias $\{(0,-\sigma^2/k)\}$ and MSE $\{(\sigma^2/k,(k-1)^2\nu/k^3 - (k^2 - 5k + 3)\sigma^4/k^3)\}$.

5. Consider the statistical model from Part 1 of Example 8.2 and fit it to a dataset represented by a vector w in the way described in Part 1 of Example 8.5. It is possible to show that the resulting parameter values are certain maximum likelihood estimates for the model (note that there are many maximum likelihood estimators in this case – see, e.g., Exercise 8.4). Below we will prove this statement for the simplest possible case when X (the set of all non-zero states of the models) consists of a single state. For the case when X consists of two states, see Exercise 8.4.

Consider the *a priori* Markov chain connectivity shown in Fig. 8.1, where the arrows are labeled with the corresponding transition probabilities. For the

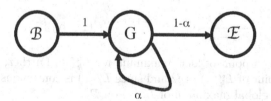

Fig. 8.1.

corresponding statistical model we have

$$S = \{0 \overbrace{G \ldots G}^{m \text{ times}} 0, \, m \in \mathbb{N}\},$$

the range is $\mathbb{N}$, $\theta = \theta_1 = \alpha$, and the parameter space is $\mathcal{P} = [0,1) = \{0\} \cup \Omega$, with $\Omega = (0,1)$ (the point 1 is not included to ensure that the Markov chains are non-trivially connected).

Fix $x = (x_1, \ldots, x_k) \in \mathbb{N}^k$ and let $x^j = \varphi^{-1}(x_j)$ and $x^j = 0 \overbrace{G \ldots G}^{m_j \text{ times}} 0$, $j = 1, \ldots, k$. It then follows from formulas (6.4) and (8.10) that

$$L_k^\alpha(x) = \alpha^{\sum_{j=1}^{k} m_j - k} (1 - \alpha)^k.$$

We will now consider two cases. Suppose first that $m_j = 1$ for $j = 1, \ldots, k$. Clearly, in this case $L_k^\alpha(x) = (1 - \alpha)^k$, and thus $L_k^\alpha(x)$ attains its maximum at the boundary point $\alpha = 0$ of Ω. Suppose now that $\sum_{j=1}^{k} m_j > k$. The function $L_k^\alpha(x)$ is infinitely many times differentiable in Ω for all k and x, and we can attempt to find the solution to the maximum likelihood equations (8.13) that become the following single equation in this case

$$\alpha^{\sum_{j=1}^{k} m_j - k - 1} (1 - \alpha)^{k-1} \left(\alpha \sum_{j=1}^{k} m_j - \sum_{j=1}^{k} m_j + k \right) = 0.$$

The solution to this equation for fixed k and x is

$$\alpha = \alpha_0 = \frac{\sum\limits_{j=1}^{k} m_j - k}{\sum\limits_{j=1}^{k} m_j}. \tag{8.19}$$

Calculating the second derivative of $L_k^\alpha(x)$ with respect to α at the point α_0 we obtain the negative number

$$-\frac{k}{1 - \alpha_0} \alpha_0^{((k+1)\alpha_0 - 1)/(1 - \alpha_0)} (1 - \alpha_0)^{k-1}.$$

Therefore, α_0 is a point of local maximum for $L_k^\alpha(x)$. Further, since α_0 is the only critical point of $L_k^\alpha(x)$ in Ω and since $L_k^\alpha(x)$ is continuous on $\mathcal{P}$, α_0 is in fact a point of global maximum of $L_k^\alpha(x)$ on $\mathcal{P}$.

Observe that formula (8.19) gives a value that maximizes $L_k^\alpha(x)$ for all $x \in \mathbb{N}^k$, including the case $m_j = 1$ for $j = 1, \ldots, k$. Thus the maximum likelihood estimator for α is $\{g_k(x_1, \ldots, x_k)\}$, where $g_k(x_1, \ldots, x_k)$ is given by the right-hand side of formula (8.19). If we fit the model to a dataset represented by a vector $w \in \mathbb{N}^n$ by setting $\alpha = g_n(w_1, \ldots, w_n)$, we obtain exactly the value given by formula (3.2).

6. Consider the two statistical models from Part 2 of Example 8.2.

a. Suppose that model fitting is performed as in Part 2 of Example 8.5. It is possible to prove that the resulting parameter values are certain maximum likelihood estimates for the model (note that there are many maximum likelihood estimators in this case – see, e.g., Exercise 8.5). Below we will prove this statement for the simplest possible case when the underlying Markov chains of the HMMs have only one non-zero state (that is, when X consists of a single point) and when Q is the two-letter alphabet $\{A, B\}$. For the case when X consists of two states, see Exercise 8.5.

The *a priori* connectivity of the underlying Markov chains of the HMMs is shown in Fig. 8.1, where the arrows are labeled with the corresponding transition probabilities. Let state G emit A with probability ν and B with probability $1 - \nu$.

For the corresponding statistical model we have

$$\hat{S}^{\mathbf{a}} = \{(x, \overbrace{G \ldots G}^{m \text{ times}}) : x \text{ is a sequence of length } m \text{ of letters } A \text{ and } B, \, m \in \mathbb{N}\},$$

the range is $\mathbb{N}$, $\theta = (\theta_1, \theta_2) = (\alpha, \nu)$, and the parameter space is $\mathcal{P} = [0, 1) \times [0, 1]$ (the points in $\{1\} \times [0, 1]$ are not included to ensure that the underlying Markov chains are non-trivially connected).

Fix $x = (x_1, \ldots, x_k) \in \mathbb{N}^k$ and let $(x^j, \pi^j) = \varphi^{-1}(x_j)$; we denote by m_j the length of x^j and by l_j the number of times x^j contains the letter A, $j = 1, \ldots, k$. It then follows from formulas (6.6) and (8.10) that

$$L_k^{(\alpha, \nu)}(x) = \alpha^{\sum_{j=1}^{k} m_j - k}(1 - \alpha)^k \nu^{\sum_{j=1}^{k} l_j}(1 - \nu)^{\sum_{j=1}^{k} m_j - \sum_{j=1}^{k} l_j}.$$

Clearly, to determine the points where $L_k^{(\alpha, \nu)}(x)$ attains its maximum, it is sufficient to determine $\alpha \in [0, 1)$ at which the value of

$$\varphi_k^\alpha(x) = \alpha^{\sum_{j=1}^{k} m_j - k}(1 - \alpha)^k$$

is maximal and $\nu \in [0, 1]$ at which the value of

$$\psi_k^\nu(x) = \nu^{\sum_{j=1}^{k} l_j}(1 - \nu)^{\sum_{j=1}^{k} m_j - \sum_{j=1}^{k} l_j}$$

is maximal.

The function $\varphi_k^\alpha(x)$ can be dealt with as in Part 5 above. It attains its maximum at the point α_0 given by formula (8.19). Hence we only need to find the point of maximum for the function $\psi_k^\nu(x)$, which is done as in Part 2. We will consider three cases.

Suppose first that $l_j = 0$ for $j = 1, \ldots, k$. Then $\psi_k^\nu(x) = (1 - \nu)^{\sum_{j=1}^k m_j}$, and the maximum is attained at the boundary point $\nu = 0$. Next, if $l_j = m_j$ for $j = 1, \ldots, k$, we have $\psi_k^\nu(x) = \nu^{\sum_{j=1}^k m_j}$, and the maximum is attained at the boundary point $\nu = 1$. Assume finally that $0 < \sum_{j=1}^k l_j < \sum_{j=1}^k m_j$. The function $\psi_k^\nu(x)$ is infinitely many times differentiable in $(0, 1)$ for all k and x, and we can attempt to find the points where it attains its maximum on $(0, 1)$ by determining its critical points in $(0, 1)$. Differentiating with respect to ν we obtain

$$\frac{d\psi_k^\nu(x)}{d\nu} = \nu^{\sum_{j=1}^k l_j - 1}(1 - \nu)^{\sum_{j=1}^k m_j - \sum_{j=1}^k l_j - 1}\left(\sum_{j=1}^k l_j - \nu\sum_{j=1}^k m_j\right).$$

Clearly, the only critical point in $(0, 1)$ is

$$\nu = \nu_0 = \frac{\displaystyle\sum_{j=1}^k l_j}{\displaystyle\sum_{j=1}^k m_j}. \tag{8.20}$$

Calculating the second derivative of $\psi_k^\nu(x)$ with respect to ν at the point ν_0 we obtain the negative number

$$-\sum_{j=1}^k m_j \nu_0^{\nu_0 \sum_{j=1}^k m_j - 1}(1 - \nu_0)^{\sum_{j=1}^k m_j - \nu_0 \sum_{j=1}^k m_j - 1}.$$

Therefore, ν_0 is a point of local maximum for $\psi_k^\nu(x)$. Since ν_0 is the only critical point of $\psi_k^\nu(x)$ in $(0, 1)$ and since $\psi_k^\nu(x)$ is continuous on $[0, 1]$, ν_0 is in fact a point of global maximum of $\psi_k^\nu(x)$ on $[0, 1]$. We also observe that formula (8.20) gives a value that maximizes $\psi_k^\nu(x)$ for all $x \in \mathbb{N}^k$, including the cases $\sum_{j=1}^k l_j = 0$ and $\sum_{j=1}^k l_j = \sum_{j=1}^k m_j$.

Thus the maximum likelihood estimator for (α, ν) is $\{(g_k^1(x_1, \ldots, x_k), g_k^2(x_1, \ldots, x_k))\}$, where $g_k^1(x_1, \ldots, x_k)$ is given by the right-hand side of formula (8.19) and $g_k^2(x_1, \ldots, x_k)$ by the right-hand side of formula (8.20). If we fit the model to a dataset represented by a vector $w \in \mathbb{N}^n$ by setting $\alpha = g_n^1(w_1, \ldots, w_n)$, $\nu = g_n^2(w_1, \ldots, w_n)$, we obtain exactly the value given by formula (3.17).

b. One way to fit this model to a dataset represented by a vector w is by means of the Baum-Welch training. As we pointed out in Part 2 of Example 8.5, the Baum-Welch training attempts to maximize $L_{n,d}^\theta(w) = p_{f_\theta^{\times\infty,n}}(w)$ over $\theta \in \mathcal{P}$, but may not necessarily achieve this goal and produce only a point of local maximum for $L_{n,d}^\theta(w)$. Nevertheless, the idea behind the Baum-Welch training is based on maximum likelihood estimators.

Another way to do fitting is by means of the Viterbi training (see Sect. 3.6). Unlike the Baum-Welch training, it only attempts to maximize $P^{\mathbf{a}}(x^1, \pi^*(x^1)) \times \ldots \times P^{\mathbf{a}}(x^n, \pi^*(x^n))$, where $\pi^*(x^j)$ is a Viterbi path for x^j, $j = 1, \ldots, n$. Probably for this reason, the Viterbi training does not perform as well as the Baum-Welch training.

At the beginning of this section we assumed that the parameter space $\mathcal{P}$ is a subset of a Euclidean space. In fact, parameter estimation (maximum likelihood estimation in particular) is used in more general situations as well. For example, we may want to estimate the parameters of the model described in Part 3 of Example 8.2. Of course, there are additional complications in such cases. Suppose that $\mathcal{P}$ is a subset of a topological space $\mathbf{T}$. Then for every $k \in \mathbb{N}$, the function $g_k(x_1, \ldots, x_k)$ is defined on $\mathcal{W}^k$ and takes values in $\mathbf{T}$. The very first condition that the collection of functions $\{g_k(x_1, \ldots, x_k)\}$ must satisfy in order to be a candidate for estimation is that, for every $k \in \mathbb{N}$, the mapping $g_k^\theta = g_k(f_\theta^{\times \infty, k})$ is a random variable on the infinite power of the probability space $(S_\theta, \mathcal{B}_\theta, P_\theta)$. Thus, we need a suitable definition of random variable with values in a topological space. It must be of the following form: a mapping f defined on a probability space $(S, \mathcal{B}, P)$ with values in a topological space $\mathbf{T}$ is called a random variable if, for certain subsets A of $\mathbf{T}$ (analogues of Borel sets in $\mathbb{R}^q$), we have $f^{-1}(A) \in \mathcal{B}$. For example, if $\mathcal{B}$ is the σ-algebra of all events in S, then any mapping $f : S \to \mathbf{T}$ satisfies any reasonable definition of this kind and hence can be called a random variable. For instance, for the model in Part 3 of Example 8.2, S_θ is finite and $\mathcal{B}_\theta$ is the collection of all events in S_θ for every $\theta \in \mathcal{P}$. It is then clear from the definition of infinite power of a probability space at the end of Sect. 6.2 that every event in S_θ^∞ is measurable.

Next, in order to define a consistent estimator, we need to be able to generalize Definition 6.54 to the case when the corresponding random variables take values in a topological space. Definition 6.54 requires a distance function defined on $\mathbf{T}$, so we assume that $\mathbf{T}$ is a metric space (see Definition 5.2). For metric spaces Definition 6.54 can be generalized in the obvious way: the absolute value in formula (6.37) must be replaced by the distance between $f_k(e), f(e) \in \mathbf{T}$. For example, for the case of the model from Part 3 of Example 8.2, the parameter space is a metric space obtained by gluing together pieces of the Euclidean space of dimension $2N - 3$ (each piece is responsible for a particular topology and accommodates the corresponding branch length parameters) multiplied by an appropriate portion of another Euclidean space to accommodate the parameters arising from the evolutionary model. For this statistical model, fitting is done as described in Part 3 of Example 8.5, that is, by the maximum likelihood procedure. It can be proved (the proof is nontrivial) that in this case any maximum likelihood estimator is consistent, if a reversible evolutionary model is used [R]. In particular, it is consistent for each of the four evolutionary models described in Sect. 5.4.

If $\mathbf{T}$ is a general metric space, it is not easy to define the bias and the mean square error of an estimator, since such definitions would require to introduce the integral of a random variable with values in $\mathbf{T}$. Generalities of this kind are beyond the scope of the book.

8.3 Hypothesis Testing

In the preceding section we described a way to fit a model to data. In our considerations we assumed that a model was fixed from the start. It is an interesting and non-trivial question, however, how one can construct a model that is "suitable" for a particular dataset or how one can choose such a model from a range of available ones. This question is very hard to answer in general, and in this section we will deal with a somewhat easier one: suppose that we are given a dataset D represented by a vector w of length n and a pair of statistical models M_1 and M_2 with ranges $\mathcal{W}_1$ and $\mathcal{W}_2$ respectively, such that $w \in \mathcal{W}_j^n$, $j = 1, 2$; which of the two models is then "most suitable" for modeling the dataset D? Recall that we have already addressed this question to some extent in Sect. 5.6, where we discussed hypothesis testing for the case of phylogenetic reconstruction. Here we will give a general framework for such testing.

We will assume that M_1 is a special case of M_2. In this case we have $\mathcal{W}_1 = \mathcal{W}_2 = \mathcal{W}$, and we will write M_2 as $M_2 = \{(S_\theta, \mathcal{B}_\theta, P_\theta, f_\theta, \mathcal{W}, \mathcal{P})\}$ with M_1 corresponding to a subset $\mathcal{P}' \subset \mathcal{P}$. Clearly, M_2 models D at least "as well as M_1 does", and we wish to know whether or not M_2 models D "significantly better" than M_1. This setup is usually formalized as follows. We consider the *null hypothesis*

$$H_0 : \text{the dataset } D \text{ obeys model } M_1,$$

and the *alternative hypothesis*

$$H_A : \text{the dataset } D \text{ obeys model } M_2,$$

where "D obeys M_j" means that we have chosen to model w by means of M_j. The null hypothesis is a special case of the alternative one, that is, the hypotheses are *nested*. We will now define what it means to *statistically test the null hypothesis against the alternative one*.

Definition 8.13. *A statistical test of the null hypothesis H_0 against the alternative hypothesis H_A or a statistical test of the significance of the alternative hypothesis H_A is a choice of a Borel set $W_n \subset \mathcal{W}^n \subset \mathbb{R}^n$ such that, if $w \in W_n$, H_0 is rejected in favor of H_A, and if $w \in \mathcal{W}^n \setminus W_n$, H_0 is accepted.*

If H_0 is rejected in favor of H_A, we say that M_2 models D *significantly better* than M_1. If H_0 is accepted, we say that M_2 *does not model D significantly* *better than* M_1 or that M_1 *models* D *as well as* M_2 *does*. Note that a statistical test of this type does not provide information on "how well" D is modeled by either M_1 or M_2, it only concerns their relative modeling performance.

One usually associates two types of error with hypothesis testing: a *type I error* is committed if H_0 is rejected when it is true and a *type II error* is committed if H_0 is accepted when it is false and H_A is true. More precisely, the following probabilities

$$E_I(\theta) = P\left(\{e \in S_\theta^\infty : f_\theta^{\times\infty,n}(e) \in W_n\}\right) = P_{F_{f_\theta^{\times\infty,n}}}(W_n),$$

for $\theta \in \mathcal{P}'$, and

$$E_{II}(\theta) = P\left(\{e \in S_\theta^\infty : f_\theta^{\times\infty,n}(e) \in \mathcal{W}^n \setminus W_n\}\right) = 1 - P_{F_{f_\theta^{\times\infty,n}}}(W_n),$$

for $\theta \in \mathcal{P} \setminus \mathcal{P}'$, are called the *probabilities of committing a type I error and a type II error* respectively (in the above formulas we used (6.34)). Further, the function

$$Q_n(\theta) = 1 - E_{II}(\theta) = P_{F_{f_\theta^{\times\infty,n}}}(W_n)$$

defined on $\mathcal{P} \setminus \mathcal{P}'$ is called the *power of the test*. One can think of $Q_n(\theta)$ as the probability of rejecting H_0 when it is false and H_A is true.

Fix $0 < \alpha < 1$ and assume that W_n is chosen to satisfy

$$E_I(\theta) \leq \alpha, \tag{8.21}$$

for all $\theta \in \mathcal{P}'$. The number α is called the *significance level of the test*. One always wants to minimize the probability of a type I error and hence α is chosen to be small (usually 0.01 or 0.05). At the same time, one wants to maximize the power of the test $Q_n(\theta)$ for all $\theta \in \mathcal{P} \setminus \mathcal{P}'$. In particular, a statistical test is called *consistent*, if $Q_n(\theta) \to 1$, as $n \to \infty$, for every $\theta \in \mathcal{P} \setminus \mathcal{P}'$.

We will now consider a particular testing procedure called the *likelihood ratio test*. Suppose that M_1 and M_2 are regular statistical models for which respective maximum likelihood estimators $\{g_{1,k}(x_1, \ldots, x_k)\}$ and $\{g_{2,k}(x_1, \ldots, x_k)\}$ exist, and for $k \in \mathbb{N}$ and $x = (x_1, \ldots, x_k) \in \mathcal{W}^k$ denote by $L_{0,k}^{\max}(x)$ and $L_{A,k}^{\max}(x)$ the values of the likelihoods of order k given models M_1, M_2, calculated at x for parameter values $g_{1,k}(x_1, \ldots, x_k)$ and $g_{2,k}(x_1, \ldots, x_k)$ respectively. For every $y \in \mathcal{W}^n$ we now consider $L_{0,n}^{\max}(y)$ and $L_{A,n}^{\max}(y)$ and define the *likelihood ratio of order* n

$$\Delta_n(y) = \frac{L_{0,n}^{\max}(y)}{L_{A,n}^{\max}(y)}.$$

In the above formula we assume that $L_{A,n}^{\max}(y) \neq 0$ for every $y \in \mathcal{W}^n$. However, everything that follows applies (after making minor adjustments) in the situation when

$$L_{A,n}^{\max}(w) \neq 0,$$

and

$$P_{F_{f_\theta^{\times\infty,n}}}(E_n^\infty) = 0 \quad \text{for all } \theta \in \mathcal{P}', \tag{8.22}$$

where $E_n^\infty = \{y \in \mathcal{W}^n : L_{A,n}^{\max}(y) = 0\}$ (in particular, E_n^∞ is assumed to be measurable).

Clearly, $\Delta_n(y) \geq 0$ for all y (note that if $\mathcal{P}_d = \emptyset$ or $\mathcal{P}_c = \emptyset$, then, in addition, $\Delta_n(y) \leq 1$ for all y). For $K \geq 0$ define

$$W_n(K) = \{y \in \mathcal{W}^n : \Delta_n(y) \leq K\}. \tag{8.23}$$

Suppose that $W_n(K)$ is a Borel set in $\mathbb{R}^n$ for all $K \geq 0$. Then $\Delta_n(f_\theta^{\times\infty,n})$ is a random variable on $(S_\theta, \mathcal{B}_\theta, P_\theta)^\infty$ for every $\theta \in \mathcal{P}'$. Then by property (iv) of distribution functions in Sect. 6.7 and using formula (6.34) we have for every $\theta \in \mathcal{P}'$

$$P_{F_{f_\theta^{\times\infty,n}}}(W_n(K)) = F_{\Delta_n(f_\theta^{\times\infty,n})}(K) \to F_{\Delta_n(f_\theta^{\times\infty,n})}(0) = P_{F_{f_\theta^{\times\infty,n}}}(E_0),$$

as $K \to 0$, where

$$E_n^0 = \{y \in \mathcal{W}^n : \Delta_n(y) = 0\}. \tag{8.24}$$

Suppose now that, for all $\theta \in \mathcal{P}'$, $F_{\Delta_n(f_\theta^{\times\infty,n})}$ is continuous on $[0,\infty)$ and $P_{F_{f_\theta^{\times\infty,n}}}(E_n^0) = 0$. Then we can choose the largest number $K_\alpha(\theta)$ such that

$$F_{\Delta_n(f_\theta^{\times\infty,n})}(K_\alpha(\theta)) = P_{F_{f_\theta^{\times\infty,n}}}(W_n(K_\alpha(\theta))) = \alpha. \tag{8.25}$$

Assume further that $\inf_\theta K_\alpha(\theta) = K_\alpha > 0$ and set

$$W_n = W_n(K_\alpha). \tag{8.26}$$

This choice guarantees that condition (8.21) holds. Thus, under the likelihood ratio test, the null hypothesis is rejected, if $\Delta_n(w) \leq K_\alpha$. The number K_α is called the *significance point of the test*. To determine the significance point K_α, one needs to know the distribution of $\Delta_n(f_\theta^{\times\infty,n})$ for every $\theta \in \mathcal{P}'$. This family of distributions is called the *null hypothesis distribution of the likelihood ratio* Δ_n.

In the above considerations we assumed, in particular, that $F_{\Delta_n(f_\theta^{\times\infty,n})}$ is continuous on $[0,\infty)$ for all $\theta \in \mathcal{P}'$. In general, $F_{\Delta_n(f_\theta^{\times\infty,n})}$ is only upper semi-continuous on $\mathbb{R}$, and therefore a number satisfying identity (8.25) may not exist. If this is the case for some $\theta_0 \in \mathcal{P}'$, then we construct $K_\alpha(\theta_0)$ as follows. First, we find the (unique) number $K_\alpha'(\theta_0)$ such that $F_{\Delta_n(f_{\theta_0}^{\times\infty,n})}(K_\alpha'(\theta_0)) > \alpha$ and $F_{\Delta_n(f_{\theta_0}^{\times\infty,n})}(K_\alpha'(\theta_0) - 0) < \alpha$. We now set $K_\alpha(\theta_0)$ to be a number less than $K_\alpha'(\theta_0)$ and close to it. Of course, there is a great deal of arbitrariness in this choice of $K_\alpha(\theta_0)$. Note that instead of satisfying identity (8.25) with $\theta = \theta_0$, it satisfies the inequality

$$F_{\Delta_n(f_{\theta_0}^{\times \infty, n})}(K_\alpha(\theta_0)) = P_{F_{f_{\theta_0}^{\times \infty, n}}}(W_n(K_\alpha(\theta_0))) < \alpha.$$

This way we can choose $K_\alpha(\theta)$ for all $\theta \in \mathcal{P}'$. If $\inf_\theta K_\alpha(\theta) = K_\alpha > 0$, we define the set W_n as in (8.26). As before, this choice of W_n guarantees that condition (8.21) holds.

Instead of using the likelihood ratio Δ_n for hypothesis testing, we can use some other function $\lambda_n : \mathcal{W}^n \to \mathbb{R}$ that satisfies all the conditions that we needed above to produce a significance point. Such a function λ_n used for hypothesis testing is called a *test statistic*. A variety of test statistics are utilized, but the likelihood ratio is perhaps the most frequently used one. One of its advantages is that it is known in some cases to produce a test for which the power $Q_n(\theta)$ is maximal possible (this holds, for example, when $\mathcal{P}$ is a two-point set and $\mathcal{P}'$ is a one-point subset of $\mathcal{P}$). Secondly, under certain conditions, the likelihood ratio test is consistent (for a precise statement see Remark 8.17). Thirdly, in many cases one can determine the asymptotic null hypothesis distribution of Δ_n, that is, the distribution of $\Delta_n(f_\theta^{\times \infty, n})$, for $\theta \in \mathcal{P}'$, as $n \to \infty$. Specifically, one can show that under certain assumptions $-2\ln(\Delta_n(f_\theta^{\times \infty, n}))$ converges in distribution to a random variable that has the chi-square distribution with ν degrees of freedom (see Sect. 6.9), where $\nu = \dim \mathcal{P} - \dim \mathcal{P}'$. Because of the importance of this last fact, we will give a precise formulation below. We will only consider the case of regular models with either $\mathcal{P}_c = \emptyset$ and f_θ discrete for all $\theta \in \mathcal{P}$, or $\mathcal{P}_d = \emptyset$. More general cases can be also dealt with, but we do not go in such details here.

First, we need the following definitions.

Definition 8.14. Let $\{(S_\theta, \mathcal{B}_\theta, P_\theta, f_\theta, \mathcal{W}, \mathcal{P})\}$ be a regular model with $\mathcal{P}_c = \emptyset$. Assume that $\mathcal{P} = \mathcal{P}_d = \Omega$, where $\Omega \subset \mathbb{R}^q$ is a domain, and we will write $\theta = (\theta_1, \ldots, \theta_q)$. Let f_θ be discrete for all $\theta \in \Omega$, and consider the corresponding probability distributions $\left\{p_{f_\theta}(r) = P\Big(\{e \in S_\theta : f_\theta(e) = r\}\Big), r \in \mathcal{W}\right\}$. Then the model is called smooth, if the following conditions hold

(i) $p_{f_\theta}(r)$ is twice differentiable with respect to θ in Ω for every $r \in \mathcal{W}$,

(ii) the series $\sum_{r \in \mathcal{W}} \partial p_{f_\theta}(r)/\partial \theta_i$ and $\sum_{r \in \mathcal{W}} \partial^2 p_{f_\theta}(r)/\partial \theta_i \partial \theta_j$ absolutely converge to 0, for every $i, j = 1, \ldots, q$ and $\theta \in \Omega$,

(iii) a maximum likelihood estimator $\{g_k(x_1, \ldots, x_k)\}$ exists for θ,

(iv) there is a unique solution to the maximum likelihood equations (8.13) for every $k \in \mathbb{N}$ and $x = (x_1, \ldots, x_k) \in \mathcal{W}^k$ (this solution is necessarily equal to $g_k(x_1, \ldots, x_k)$).

Definition 8.15. Let $\{(S_\theta, \mathcal{B}_\theta, P_\theta, f_\theta, \mathcal{W}, \mathcal{P})\}$ be a regular model with $\mathcal{P}_d = \emptyset$. Assume that $\mathcal{P} = \mathcal{P}_c = \Omega$, where $\Omega \subset \mathbb{R}^q$ is a domain, and we will write

$\theta = (\theta_1, \ldots, \theta_q)$. Then the model is called smooth, if the following conditions hold

(i) $\varrho_{f_\theta}(z)$ is twice differentiable with respect to θ in Ω for every $z \in \mathbb{R}$,

(ii) $\partial \varrho_{f_\theta}(z)/\partial \theta_i$ and $\partial^2 \varrho_{f_\theta}(z)/\partial \theta_i \partial \theta_j$ are Lebesgue integrable on $\mathbb{R}$, for every $i, j = 1, \ldots, q$ and $\theta \in \Omega$,

(iii) $\int_{-\infty}^{\infty} \partial \varrho_{f_\theta}(z)/\partial \theta_i \, d\mu = 0$, for every $i = 1, \ldots, q$ and $\theta \in \Omega$,

(iv) $\int_{-\infty}^{\infty} \partial^2 \varrho_{f_\theta}(z)/\partial \theta_i \partial \theta_j \, d\mu = 0$, for every $i, j = 1, \ldots, q$ and $\theta \in \Omega$,

(v) a maximum likelihood estimator $\{g_k(x_1, \ldots, x_k)\}$ exists for θ,

(vi) there is a unique solution to the maximum likelihood equations (8.13) for every $k \in \mathbb{N}$ and $x = (x_1, \ldots, x_k) \in \mathcal{W}^k$ (this solution is necessarily equal to $g_k(x_1, \ldots, x_k)$).

We are now ready to formulate the following theorem.

Theorem 8.16. Let M_2 be a smooth statistical model and let model M_1 be given by the conditions $\theta_1 = \theta_1^0, \ldots, \theta_r = \theta_r^0$ for some $\theta_1^0, \ldots, \theta_r^0 \in \mathbb{R}$ with $0 < r \le q$. Suppose that M_1 is smooth as well. Let $L_{A,n}^{\max}(y) \neq 0$ for all $y \in \mathcal{W}^n$ and $n \in \mathbb{N}$. Assume further that for all $K \ge 0$ and $n \in \mathbb{N}$ the set $W_n(K)$ defined in (8.23) is Borel and that for the set E_n^0 defined in (8.24) we have $P_{F_{f_\theta}^{\times \infty, n}}(E_n^0) = 0$ for every $\theta \in \mathcal{P}'$ and $n \in \mathbb{N}$.

Then the sequence of random variables $\{-2 \ln (\Lambda_n (f_\theta^{\times \infty, n}))\}$ converges in distribution to a random variable that has the chi-square distribution with r degrees of freedom, for every $\theta \in \mathcal{P}'$.

Strictly speaking, $-2 \ln (\Lambda_n (f_\theta^{\times \infty, n}))$ is not defined at the points of S_θ^∞ where the value of $f_\theta^{\times \infty, n}$ belongs to E_n^0. However, the probability of this event is assumed to be zero, and this complication does not affect the assertion of the theorem. We can set the values of $-2 \ln (\Lambda_n (f_\theta^{\times \infty, n}))$ at the points of E_n^0 arbitrarily, say equal to 0. Also note that, for a similar reason, instead of requiring that $L_{A,n}^{\max}(y) \neq 0$ for all $y \in \mathcal{W}^n$ and $n \in \mathbb{N}$, it is sufficient to require that condition (8.22) holds for all $n \in \mathbb{N}$.

Remark 8.17. It can be shown that for a smooth statistical model the corresponding maximum likelihood estimator is consistent and asymptotically normally distributed. Further, under the assumptions of Theorem 8.16, the likelihood ratio test is consistent. Thus, the assumption of smoothness is sufficient for maximum likelihood estimators and the corresponding likelihood ratios to have all the good properties. They also can be shown to hold for more general models.

We will now give examples of applying the likelihood ratio test.

Example 8.18.
1. Let M_2 be the statistical model from Part 3 of Example 8.12, and let M_1 be given by setting $V = V^*$ for some $V^* > 0$, that is, by setting $\mathcal{P}'$ to be a single point $\mathcal{P}' = \{V^*\} \subset \mathcal{P}$. Fix $n \in \mathbb{N}$ and for every $y \in \mathcal{W}^n$ we will now calculate $L_{A,n}^{\max}(y)$ and $L_{0,n}^{\max}(y)$.

The value of $L_{A,n}^{\max}(y)$ can be found from our calculations in Part 3 of Example 8.12. We have

$$L_{A,n}^{\max}(y) = \frac{1}{y_{\max}^n},$$

where $y_{\max} = \max\{y_1, \ldots, y_n\}$. For $L_{0,n}^{\max}(y)$ we clearly have

$$L_{0,n}^{\max}(y) = \begin{cases} 0, & \text{if } y_{\max} > V^*, \\ \dfrac{1}{V^{*n}}, & \text{if } y_{\max} \leq V^*. \end{cases}$$

Therefore we obtain

$$\Delta_n(y) = \begin{cases} 0, & \text{if } y_{\max} > V^*, \\ \left(\dfrac{y_{\max}}{V^*}\right)^n, & \text{if } y_{\max} \leq V^*. \end{cases}$$

Clearly, for every $K \geq 0$ the set $W_n(K)$ defined in (8.23) is a Borel set in $\mathbb{R}^n$. Hence $\Delta_n\left(f_{V^*}^{\times\infty,n}\right)$ is a random variable on $(S_{V^*}, \mathcal{B}_{V^*}, P_{V^*})^\infty$ for every n. Further, we have $E_n^0 = \{y \in \mathcal{W}^n : \Delta_n(y) = 0\} = \{y \in \mathbb{R}^n : y_1 > 0, \ldots, y_n > 0, \text{ and } y_{\max} > V^*\}$ for every n (see (8.24)). It is not hard to observe that $P_{F_{f_{V^*}^{\times\infty,n}}}(E_n^0) = 0$.

Next, for $n \in \mathbb{N}$ we need to find the null hypothesis distribution of the likelihood ratio, that is, the distribution of the random variable $\Delta_n\left(f_{V^*}^{\times\infty,n}\right)$. For $z \geq 0$ we have

$$F_{\Delta_n\left(f_{V^*}^{\times\infty,n}\right)}(z) = P\left(\left\{e \in S_{V^*}^\infty : \Delta_n\left(f_{V^*}^{\times\infty,n}(e)\right) \leq z\right\}\right)$$

$$= P\left(\left\{e \in S_{V^*}^\infty : g_n^{V^*}(e) > V^*\right\}\right) + P\left(\left\{e \in S_{V^*}^\infty : g_n^{V^*}(e)\right.\right.$$

$$\left.\left. \leq \min\{V^*, z^{1/n}V^*\}\right\}\right) = F_{g_n^{V^*}}^*(V^*) + F_{g_n^{V^*}}\left(\min\{V^*, z^{1/n}V^*\}\right) \tag{8.27}$$

$$= 1 - F_{g_n^{V^*}}(V^*) + F_{g_n^{V^*}}\left(\min\{V^*, z^{1/n}V^*\}\right).$$

The random variables g_k^V were determined and studied in Part 3 of Example 8.12. In particular, for the distribution function of $g_n^{V^*}$ we have

$$F_{g_n^{V*}}(z) = \begin{cases} 0, & \text{if } z \leq 0, \\ \dfrac{z^n}{V^{*n}}, & \text{if } 0 < z \leq V^*, \\ 1, & \text{if } z > V^*. \end{cases}$$

Therefore formula (8.27) gives for $z \geq 0$

$$F_{\Delta_n\left(f_{V*}^{\times\infty,n}\right)}(z) = \begin{cases} z, & \text{if } 0 \leq z \leq 1, \\ 1, & \text{if } z > 1. \end{cases} \tag{8.28}$$

Let $0 < \alpha < 1$ be the significance level of the test. Since $F_{\Delta_n\left(f_{V*}^{\times\infty,n}\right)}$ is continuous on $[0, \infty)$ and strictly increasing on $[0, 1]$, the significance point $K_\alpha(V^*) = K_\alpha$ is found from the identity

$$F_{\Delta_n\left(f_{V*}^{\times\infty,n}\right)}(K_\alpha) = \alpha,$$

(see (8.25)), which gives $K_\alpha = \alpha$. If a vector $w \in W^n$ represents a real dataset to which we wish to apply the likelihood ratio test, then the null hypothesis is rejected, if $\Delta_n(w) \leq \alpha$.

The distribution of $-2\ln\left(\Delta_n\left(f_{V*}^{\times\infty,n}\right)\right)$ is easily obtained from formula (8.28). Namely, we have

$$F_{-2\ln\left(\Delta_n\left(f_{V*}^{\times\infty,n}\right)\right)}(z) = \begin{cases} 1 - \exp\left(-\dfrac{z}{2}\right), & \text{if } z \geq 0, \\ 0, & \text{if } z < 0. \end{cases}$$

Differentiating $F_{-2\ln\left(\Delta_n\left(f_{V*}^{\times\infty,n}\right)\right)}(z)$ for $z \neq 0$ we obtain the density function of $-2\ln\left(\Delta_n\left(f_{V*}^{\times\infty,n}\right)\right)$

$$\varrho_{-2\ln\left(\Delta_n\left(f_{V*}^{\times\infty,n}\right)\right)}(z) = \begin{cases} \dfrac{1}{2}\exp\left(-\dfrac{z}{2}\right), & \text{if } z > 0, \\ 0, & \text{if } z \leq 0, \end{cases}$$

where we set $\varrho_{-2\ln\left(\Delta_n\left(f_{V*}^{\times\infty,n}\right)\right)}(0) = 0$. Thus, $-2\ln\left(\Delta_n\left(f_\theta^{\times\infty,n}\right)\right)$ has the exponential distribution with $\lambda = 1/2$ (see Sect. 6.9), that is, the chi-square distribution with 2 degrees of freedom. Note that, according to Theorem 8.16, we could expect to obtain asymptotically the chi-square distribution with 1 degree of freedom, so the assertion of the theorem does not hold for this example. The reason for this is that model M_2 is *not* smooth. Indeed, consider ϱ_{fv} from Part 3 of Example 8.12. It is differentiable at every $V_0 > 0$ for $z \neq V_0$, and we have

$$\frac{\partial \varrho_{fv}(z)}{\partial V}(V_0) = \begin{cases} 0, & \text{if } z \leq 0 \text{ or } z > V_0, \\ -\dfrac{1}{V_0^2}, & \text{if } 0 < z < V_0. \end{cases}$$

The above function is not Lebesgue integrable on $\mathbb{R}$, that is, condition (ii) of Definition 8.15 does not hold.

2. Let M_2 be the statistical model from Part 4 of Example 8.12 with $\sigma^2 = \sigma^{*2} > 0$. More precisely, M_2 is described as follows. Set $q = 1$ and denote the parameter $\theta = \theta_1$ by μ. Let $\mathcal{P} = \mathcal{W} = \mathbb{R}$, and assume that f_μ is continuous and has the normal distribution with mean μ and variance σ^{*2}

$$\varrho_{f_\mu}(z) = \frac{1}{\sqrt{2\pi\sigma^{*2}}} \exp\left(-\frac{(z-\mu)^2}{2\sigma^{*2}}\right).$$

Suppose now that M_1 is given by setting $\mu = \mu^*$ for some $\mu^* \in \mathbb{R}$, that is, by setting $\mathcal{P}' = \{\mu^*\} \subset \mathcal{P}$.

Arguing in the same way as in Part 4 of Example 8.12 we obtain a maximum likelihood estimator $\{g_{2,k}(x_1,\ldots,x_k)\}$ for M_2, where

$$g_{2,k}(x_1,\ldots,x_k) = \frac{1}{k}\sum_{j=1}^{k} x_j$$

(see (8.18)). Therefore for $n \in \mathbb{N}$ and $y \in \mathbb{R}^n$ we obtain

$$L_{A,n}^{\max}(y) = \frac{1}{\left(\sqrt{2\pi\sigma^{*2}}\right)^n} \exp\left(-\frac{1}{2\sigma^{*2}}\sum_{j=1}^{n}\left(y_j - \frac{1}{n}\sum_{m=1}^{n} y_m\right)^2\right).$$

Also, it is clear that

$$L_{0,n}^{\max}(y) = \frac{1}{\left(\sqrt{2\pi\sigma^{*2}}\right)^n} \exp\left(-\frac{1}{2\sigma^{*2}}\sum_{j=1}^{n}(y_j - \mu^*)^2\right).$$

It then follows that

$$\Delta_n(y) = \exp\left(-\frac{1}{2n\sigma^{*2}}\left(\sum_{j=1}^{n} y_j - n\mu^*\right)^2\right).$$

Clearly, for every $K \geq 0$ and $n \in \mathbb{N}$ the set $W_n(K)$ defined in (8.23) is a Borel set in $\mathbb{R}^n$. Hence $\Delta_n\left(f_{\mu^*}^{\times\infty,n}\right)$ is a random variable on $(S_{\mu^*}, \mathcal{B}_{\mu^*}, P_{\mu^*})^\infty$ for every n. Further, we have $E_n^0 = \emptyset$.

We will now attempt to determine the asymptotic null hypothesis distribution of the likelihood ratio, that is, the distribution of $\Delta_n\left(f_{\mu^*}^{\times\infty,n}\right)$, as $n \to \infty$. Let $f_{\mu^*}^{\times\infty,n} = (\tilde{f}_1, \ldots, \tilde{f}_n)$. Then we have

$$-2\ln\left(\Delta_n\left(f_{\mu^*}^{\times\infty,n}\right)\right) = \left(\frac{\sum_{j=1}^{n} \tilde{f}_j - n\mu^*}{\sqrt{n\sigma^{2*}}}\right)^2.$$

By the Central Limit Theorem (Theorem 6.64) we have

$$\frac{\sum_{j=1}^{n} \tilde{f}_j - n\mu^*}{\sqrt{n\sigma^{*2}}} \xrightarrow{D} f,$$

where f is a random variable whose distribution is the standard normal distribution. Clearly,

$$\left(\frac{\sum_{j=1}^{n} \tilde{f}_j - n\mu^*}{\sqrt{n\sigma^{*2}}}\right)^2 \xrightarrow{D} f^2.$$

By Proposition 6.49, f^2 has the chi-square distribution with 1 degree of freedom. Hence, $-2\ln\left(\Delta_n\left(f_{\mu^*}^{\times\infty,n}\right)\right)$ converges in distribution to a random variable that has the chi-square distribution with 1 degree of freedom, which agrees with Theorem 8.16. Hence for every $0 < z \leq 1$ there exists $N_z \in \mathbb{N}$ such that for $n \geq N_z$ we have

$$F_{\Delta_n\left(f_{\mu^*}^{\times\infty,n}\right)}(z) \approx \Phi(z) = \frac{1}{\sqrt{2\pi}} \int_{-2\ln z}^{\infty} \frac{1}{\sqrt{t}} \exp\left(-\frac{t}{2}\right) dt. \qquad (8.29)$$

Formula (8.29) can be used to find an approximate significance point of the test. Let $0 < \alpha < 1$ be the significance level. Since $\Phi(z)$ is continuous and strictly increasing on $[0,1]$, the approximate significance point $K_\alpha^{\text{approx}}(V^*) = K_\alpha^{\text{approx}}$ is found from the identity

$$\Phi(K_\alpha^{\text{approx}}) = \alpha,$$

(see (8.25)). Let a vector $w \in \mathcal{W}^n$ represent a real dataset to which we wish to apply the likelihood ratio test; then, if we use the approximate significance point K_α^{approx}, the null hypothesis is rejected, if $\Delta_n(w) \leq K_\alpha^{\text{approx}}$.

Example 8.18 illustrates several important points that one should keep in mind when using the likelihood ratio test. First of all, it may not be possible to determine the null hypothesis distribution of the likelihood ratio for a given n, in which case one can attempt to use an asymptotic distribution. The chi-square distribution with an appropriate number of degrees of freedom (as stated in Theorem 8.16) is often used, but one has to be aware that if the conditions of Theorem 8.16 are not satisfied, an asymptotic distribution may not exist at all, or, if it does, may not be of the chi-square type, or, if it is, may have a wrong number of degrees of freedom, as in Part 1 of Example 8.18. Even if an asymptotic distribution exists and is determined correctly, using it for hypothesis testing can lead to poor results, since the asymptotic distribution only produces an approximate significance point. Under very general assumptions one can show that the approximate significance point approaches the true significance point as $n \to \infty$, hence the more data we have (that is, the longer the vector w is), the more accurate results the asymptotic distribution produces. From this point of view it is desirable to have some information on the speed of convergence in Theorem 8.16. This is, however, a non-trivial issue and we do not discuss it here. We also note that in both parts of Example 8.18 the null hypothesis is *simple*, that is, $\mathcal{P}'$ consists of a single point. In general, one has to determine an asymptotic distribution of $\Delta_n \left(f_\theta^{\times \infty, n} \right)$ for *every* $\theta \in \mathcal{P}'$, which is an additional complication.

With the increase of computing power it has become possible to obtain approximations to the null hypothesis distribution of the likelihood ratio by *data simulation*, rather than by using asymptotic distributions supplied by Theorem 8.16. This is done as follows. Suppose, as before, that we are given a vector $w \in \mathcal{W}^n$. Then for every $\theta \in \mathcal{P}'$ we randomly generate a large number of values of $f_\theta^{\times \infty, n}$, say vectors $w^1(\theta), \ldots, w^N(\theta) \in \mathcal{W}^n$, with N independent of θ. We then calculate $\Delta_n(w^j(\theta))$ for $j = 1, \ldots, N$ and choose $K_\alpha(\theta)$ as the largest value for which the ratio of the number of elements in the set $\{j : \Delta_n(w^j(\theta)) \le K_\alpha(\theta)\}$ and N does not exceed α. This procedure leads to an approximate significance point of the test K_α^{approx}, if the set $\mathcal{P}'$ is finite. Then the null hypothesis is rejected if $\Delta(w) \le K_\alpha^{\text{approx}}$. If, however, $\mathcal{P}'$ is infinite, the following simplified testing procedure is used.

Let $\{g_{1,k}(x_1, \ldots, x_k)\}$ be a maximum likelihood estimator for θ under model M_1, and let $\theta_0 = g_{1,n}(w_1, \ldots, w_n)$ be the corresponding maximum likelihood estimate. Then it seems reasonable to base a decision to either accept or reject the null hypothesis on the number $K_\alpha(\theta_0)$ alone, rather than on $K_\alpha = \inf_\theta K_\alpha(\theta)$, as we did earlier. With this approach we set $W_n = W_n(K_\alpha(\theta_0))$ and hence the null hypothesis is rejected, if $\Delta(w) \le K_\alpha(\theta_0)$. Of course, with this approach we can only guarantee that inequality (8.21) holds for $\theta = \theta_0$. To determine $K_\alpha(\theta_0)$ we need to know the distribution of the random variable $\Delta_n \left(f_{\theta_0}^{\times \infty, n} \right)$. An approximation to this distribution can be obtained either from Theorem 8.16, or from data simulation by generating a large number of values of $f_{\theta_0}^{\times \infty, n}$. This method of hypothesis testing is called the *parametric bootstrap method* and was described in the special case of the comparison of evolutionary models in Sect. 5.6. The parametric bootstrap method has been investigated and found satisfactory by many authors. It is widely used in statistics.

8.4 Significance of Scores for Global Alignments

One can also apply the likelihood ratio test in the case when the null and alternative hypotheses are not nested. Theorem 8.16 does not generally hold in this case, so one has to determine the null hypothesis distribution of the likelihood ratio statistic, or an approximation to this distribution, in a different way. Certainly, one can proceed by simulation, as was described in the previous section. However, in some cases it is still possible to obtain a theoretical asymptotic distribution. In this section we give an example of such a situation. We consider the problem of testing the significance of the score of a global ungapped pairwise alignment of sequences of letters from an alphabet $\mathcal{Q}$.

We define models M_1 and M_2 as follows. Set $S = \{(a, b) : a, b \in \mathcal{Q}\}$ and let $\mathcal{B}$ be the σ-algebra of all events in S. For elementary events in S define

$$P\Big(\{(a, b)\}\Big) = p_a p'_b,$$

where $p_a > 0$ and $p'_b > 0$, $a, b \in \mathcal{Q}$, are such that $\sum_{a \in \mathcal{Q}} p_a = \sum_{b \in \mathcal{Q}} p'_b = 1$. Let M_1 be the reduced statistical model associated with the probability space $(S, \mathcal{B}, P)$. This model is simple, that is, the parameter space of M_1 is a single point. Model M_2 is the reduced model associated with the probability space $(S, \mathcal{B}, \hat{P})$, where

$$\hat{P}\Big(\{(a, b)\}\Big) = p_{ab}.$$

Here $p_{ab} > 0$, for $a, b \in \mathcal{Q}$, are such that

$$\sum_{a, b \in \mathcal{Q}} p_{ab} = 1. \tag{8.30}$$

Clearly, M_2 is also a simple model. The range of the models is the set $\mathcal{W} = \{1, \ldots, |\mathcal{Q}|^2\}$, where $|\mathcal{Q}|$ is the number of elements in the alphabet $\mathcal{Q}$. The corresponding random variables for the models are $f = \hat{f} = \varphi$ (see formula (8.1)).

Let $w = (w_1, \ldots, w_n) \in \mathcal{W}^n$. Then, unless $p_{ab} = p_a p'_b$ for all $a, b \in \mathcal{Q}$, the corresponding null and alternative hypotheses are not nested. Set $(a_j, b_j) = \varphi^{-1}(w_j)$ for $j = 1, \ldots, n$ and $x^0 = a_1 \ldots a_n$ and $y^0 = b_1 \ldots b_n$. One can think of the null hypothesis as the statement that x^0 and y^0 are unrelated randomly generated sequences and of the alternative hypothesis as the statement that the residues occupying the same positions in x^0 and y^0 are in some way related. In these terms the rejection of the null hypothesis means that we should treat x^0 and y^0 as related sequences, and the acceptance of the null hypothesis means that we may treat them as unrelated. For this reason model M_1 is sometimes called a *random model*, and model M_2 with $p_{ab} \neq p_a p'_b$ for some $a, b \in \mathcal{Q}$, a *match model*.

We have

$$\Delta_n(w) = \prod_{j=1}^{n} \frac{p_{a_j} p'_{b_j}}{p_{a_j b_j}}.$$

We will now make a specific choice of $\{p_{ab}\}$, provided $\{p_a\}$ and $\{p'_b\}$ have been chosen. Suppose there exists a substitution matrix whose elements $s(a, b)$, $a, b \in \mathcal{Q}$, are integers satisfying conditions (iv)-(vi) from Sect. 7.2 for $\{p_a\}$ and $\{p'_b\}$ and set

$$p_{ab} = p_a p'_b \exp(\lambda s(a, b)), \tag{8.31}$$

where $\lambda > 0$ is found from equation (7.29). Equation (7.29) guarantees that condition (8.30) is satisfied. Note that a solution to (7.29) exists due to our assumptions on the substitution matrix.

It follows from (8.31) that

$$\Delta_n(w) = \exp\left(-\lambda S(x^0, y^0)\right),$$

where $S(x^0, y^0)$ denotes the score of the (only) global alignment between the sequences x^0 and y^0. Since the likelihood ratio statistic is expressed in terms of the score of the corresponding global alignment, the likelihood ratio test in this case is the test of the significance of the score $S(x^0, y^0)$.

Clearly, in this case the set $W_n(K)$ defined in (8.23) is a Borel set in $\mathbb{R}^n$ and $E_n^0 = \emptyset$ (see (8.24)). We now fix a significance level $0 < \alpha < 1$ and attempt to determine the significance point K_α from the equation

$$F_{\Delta_n(f \times \infty, n)}(K_\alpha) = \alpha.$$

For $z \geq 0$ we have

$$F_{\Delta_n(f \times \infty, n)}(z) = P\left(\left\{(x, y) \in S_n : S(x, y) \geq -\frac{\ln z}{\lambda}\right\}\right),$$

where the probability in the right-hand side is calculated in the sense of the probability space $(S_n, \mathcal{B}_n, P)$ introduced in Sect. 7.3. Thus we need to determine the distribution of $S(x, y)$ considered as a random variable on $(S_n, \mathcal{B}_n, P)$.

We have

$$S(x, y) = \sum_{i=1}^{n} s(x_i, y_i),$$

for all $(x, y) \in S_n$, that is, S is the sum of iid random variables, each of which is distributed as the step size corresponding to the random walk associated with the substitution matrix and random model (see Sect. 7.2). The mean value μ and variance σ^2 of the step size are

$$\mu = \sum_{a,b \in \mathcal{Q}} s(a, b) p_a p'_b,$$

$$\sigma^2 = \sum_{a,b \in \mathcal{Q}} (s(a, b) - \mu)^2 p_a p'_b,$$

where $\mu < 0$ and $\sigma^2 > 0$ due to our assumptions on the substitution matrix. Hence, by the Central Limit Theorem (Theorem 6.64), the distribution function of S becomes arbitrarily close to that of a random variable which is normally distributed with mean $n\mu$ and variance $n\sigma^2$, uniformly on $\mathbb{R}$, as $n \to \infty$. Therefore, if n is large, one can find an approximate significance point of the test K_α^{approx} from the equation

$$\frac{1}{\sqrt{2\pi\sigma^2}} \int_{-\ln K_\alpha^{\text{approx}}/\lambda}^{\infty} \exp\left(-\frac{(z-n\mu)^2}{2n\sigma^2}\right) dz = \alpha.$$

We then set $W_n = W_n(K_\alpha^{\text{approx}})$. Therefore, with this test, H_0 is rejected in favor of H_A, if $\mathcal{S}(x^0, y^0) \geq -\ln K_\alpha^{\text{approx}}/\lambda$. Note that since $\mu < 0$, we have $K_\alpha^{\text{approx}} \to \infty$ as $n \to \infty$.

Above we assumed that model M_1 is simple. One could, however, allow $\{p_a\}$ and $\{p'_b\}$ to be parameters subject to the conditions $p_a \geq 0$, $p'_b \geq 0$ for all $a, b \in \mathcal{Q}$, and $\sum_{a \in \mathcal{Q}} p_a = \sum_{b \in \mathcal{Q}} p'_b = 1$. These parameters can be estimated from the data w (equivalently, the sequences x^0 and y^0). A natural way to estimate the parameters for M_1 is, for every $a \in \mathcal{Q}$, to set p_a to be the frequency of the occurrence of a in the sequence x^0 and, for every $b \in \mathcal{Q}$, to set p'_b to be the frequency of the occurrence of b in the sequence y^0, as we did in Sect. 7.1 for the case of local alignments (here we assume that the composition of the sequences x^0 and y^0 shows enough variation to ensure that $p_a > 0$ and $p'_b > 0$ for all $a, b \in \mathcal{Q}$). One can prove that the estimates obtained in this way are in fact maximum likelihood estimates. Of course, $\{p_a\}$ and $\{p'_b\}$ are the parameters of model M_2 as well (we assume that the substitution matrix used to define M_2 in (8.31) satisfies all the necessary requirements for the values of $\{p_a\}$ and $\{p'_b\}$ estimated for M_1), so any choice of parameter values for M_1 is automatically a choice of parameter values for M_2.

If the parameters of M_1 and M_2 are estimated from the data as described above and a significance point is produced as explained, then we in fact follow the procedure of the parametric bootstrap method introduced at the end of Sect. 8.3, that is, the distribution of the likelihood ratio is studied not for all possible values of the parameters of M_1, but for the single value derived from the data. Note that a similar approach was taken in Chap. 7 for the case of local alignments. However, the problem of the significance of scores for local alignments is much harder, as it does not fit into the general framework of hypothesis testing described in this chapter.

The above discussion applies only to testing the significance of the scores of ungapped alignments. If gaps are allowed in alignments, then this approach does not work. In this case an approximation to the null hypothesis distribution of the likelihood ratio can be produced by data simulation. Similarly, for the problem of testing the significance of the scores of local gapped alignments, an approximation to the distribution of the random variable s introduced in Sect. 7.1 can be found from data simulation. It turns out that for local gapped alignments the general form of formulas (7.26), (7.27) remains true, but the constants K and λ are no longer found from formulas (7.25) and (7.29).

Exercises

8.1. Consider a statistical model with $q = 1$, and denote the parameter $\theta = \theta_1$ by p. Let $\mathcal{P} = \{p \in \mathbb{R} : 0 \leq p < 1\}$, and assume that f_p for each p is discrete, takes values $0, 1, 2, \ldots$ (hence $\mathcal{W} = \mathbb{Z}_+$) and has the geometric distribution with mean $p/(1-p)$ (see Sect. 6.9). Find a maximum likelihood estimator for p.

8.2. Consider a statistical model with $q = 1$, and denote the parameter $\theta = \theta_1$ by p. Let $\mathcal{P} = \{p \in \mathbb{R} : 0 \leq p \leq 1\}$, and assume that f_p for each p is discrete, takes values $0, \ldots, n$ (hence $\mathcal{W} = \{0, \ldots, n\}$) and has the binomial distribution with mean np (see Sect. 6.9). Find a maximum likelihood estimator for p.

8.3. Consider a statistical model with $q = 1$, and denote the parameter $\theta = \theta_1$ by λ. Let $\mathcal{P} = \{\lambda \in \mathbb{R} : \lambda > 0\}$, and assume that f_λ for each λ is continuous, takes values in $(0, \infty)$ (that is, $\mathcal{W} = (0, \infty)$) and has the exponential distribution with mean $1/\lambda$ (see Sect. 6.9). Find a maximum likelihood estimator for λ.

8.4. Consider the family of Markov models arising from the *a priori* connectivity in Fig. 8.2. Denote the parameters associated with the transition probabilities as shown. For the corresponding statistical model we have

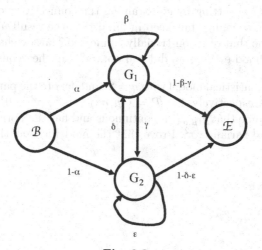

Fig. 8.2.

$\theta = (\alpha, \beta, \gamma, \delta, \varepsilon)$, and $\mathcal{P} = [0, 1]^5 \setminus \mathcal{R}$, where

$$\mathcal{R} = \{(\alpha, \beta, \gamma, \delta, \varepsilon) \in [0, 1]^5 : \beta + \gamma = 1, \, \delta + \varepsilon = 1\} \cup$$
$$\{(\alpha, \beta, \gamma, \delta, \varepsilon) \in [0, 1]^5 : \alpha \neq 0, \, \beta = 1, \, \gamma = 0\} \cup$$
$$\{(\alpha, \beta, \gamma, \delta, \varepsilon) \in [0, 1]^5 : \alpha = 0, \, \beta = 1, \, \gamma = 0, \, \delta \neq 0\} \cup$$
$$\{(\alpha, \beta, \gamma, \delta, \varepsilon) \in [0, 1]^5 : \alpha \neq 1, \, \delta = 0, \varepsilon = 1\} \cup$$
$$\{(\alpha, \beta, \gamma, \delta, \varepsilon) \in [0, 1]^5 : \alpha = 1, \, \gamma \neq 0, \, \delta = 0, \, \varepsilon = 1\}$$

(by removing the set $\mathcal{R}$ from $[0, 1]^5$ we ensure that the Markov chains are non-trivially connected).

Show that for this model formula (3.2) gives a maximum likelihood estimator for θ. [If you find it hard to prove the general statement, prove it in a more restricted setting by fixing some of the components of θ, but keep in mind that the resulting a priori connectivity must be that of a non-trivially connected Markov chain.] How many maximum likelihood estimators did you obtain? Are they consistent?

8.5. Let the family of underlying Markov chains of a family of HMMs be as in Exercise 8.4, and $\mathcal{Q}$ be the two-letter alphabet $\{A, B\}$. Let

$$q_1(A) = \nu, \, q_1(B) = 1 - \nu,$$
$$q_2(A) = \eta, \, q_2(B) = 1 - \eta.$$

For the corresponding statistical model we have $\theta = (\alpha, \beta, \gamma, \delta, \varepsilon, \nu, \eta)$, and the parameter space is the product of that from Exercise 8.1 and $[0, 1]^2$.

Show that for this model formula (3.17) gives a maximum likelihood estimator for θ. [If you find it hard to prove the general statement, prove it in a more restricted setting by constraining the connectivity of the family of underlying Markov chains, but keep in mind that the resulting a priori connectivity must be that of a non-trivially connected Markov chain.] How many maximum likelihood estimators did you obtain? Are they consistent?

8.6. Consider a statistical model with $q = 2$, and denote the parameters θ_1 and θ_2 by μ and σ^2 respectively. Let $\mathcal{P} = \{(\mu, \sigma^2) \in \mathbb{R}^2 : \sigma^2 > 0\}$. Suppose that $\mathcal{W} = \mathbb{R}$ and assume that $f_{(\mu, \sigma^2)}$ is continuous and has the normal distribution with mean μ and variance σ^2. Prove that the model is smooth.

9

Substitution Matrices

9.1 The General Form of a Substitution Matrix

Up to this moment, in problems related to sequence alignment, we have assumed that substitution matrices were given. In this chapter we will show how such matrices are actually constructed. As we have seen in Chap. 7 and Sect. 8.4, in order to assess the significance of the score of an optimal local or global ungapped alignment between sequences x^0 and y^0, one needs to have a substitution matrix that has integer entries and satisfies conditions (iv)-(vi) from Sect. 7.2 with $\{p_a\}$ and $\{p'_b\}$ derived from x^0 and y^0 respectively as the frequencies with which letters from the relevant alphabet $\mathcal{Q}$ occur in the sequences. Of course, it is impossible to construct a matrix that satisfies the above conditions for *all* choices of $\{p_a\}$ and $\{p'_b\}$. Therefore, we will attempt to construct substitution matrices that satisfy these conditions for *some* "generic" choice of $\{p_a\}$ and $\{p'_b\}$, that is, for some "generic" choice of random model. Such substitution matrices will be called *admissible*. We will always assume that $p_a = p'_a > 0$ for all $a \in \mathcal{Q}$ and call the numbers $\{p_a\}$ the *background frequencies*. In the following sections we will explain how background frequencies can be constructed, and for the moment we assume that they have already been chosen.

As we have seen in Sect. 8.4, the elements of every admissible substitution matrix can be represented in the form

$$s(a, b) = C \times \ln \left(\frac{p_{ab}}{p_a p_b} \right), \tag{9.1}$$

for some choice of $\{p_{ab}\}$ (all of which are positive), that is, for some choice of match model, and some $C > 0$. From the uniqueness of a solution to equation (7.6) (see also (7.29)), it follows that representation (9.1) is unique.

Conversely, if the elements of a substitution matrix are integers represented in the form (9.1) with $p_{ab} \neq p_a p_b$ for some $a, b \in \mathcal{Q}$, that satisfy condition (vi) from Sect. 7.2, then the matrix is admissible. Indeed, since $p_{ab} \neq p_a p_b$

for some $a, b \in Q$, there exist $\hat{a}, \hat{b} \in Q$ such that $p_{\hat{a}\hat{b}} > p_{\hat{a}}p_{\hat{b}}$ and hence the maximal element $\hat{s}$ of the substitution matrix is positive. Further, since $p_a > 0$ for all $a \in Q$, condition (iv) follows. Next, condition (v) is a consequence of the well-known Kullback-Leibler inequality, as stated in the following theorem.

Theorem 9.1. (The Kullback-Leibler Inequality) *Let $0 < c_j < 1$, $0 < d_j < 1$ for $j = 1, \ldots, n$ be such that $\sum_{j=1}^{n} c_j = \sum_{j=1}^{n} d_j = 1$. Then we have*

$$\sum_{j=1}^{n} \ln\left(\frac{c_j}{d_j}\right) d_j \leq 0, \tag{9.2}$$

and the equality holds if and only if $c_j = d_j$ for $j = 1, \ldots, n$.

Theorem 9.1 easily follows from Jensen's inequality and the concavity of $\ln x$ (see Exercise 9.1). Applying inequality (9.2) to $\{p_{ab}\}$ and $\{p_a p_b\}$ yields condition (v).

The general approach to constructing admissible substitution matrices that we will take in the forthcoming sections is as follows. First, we make particular choices of a random model $\{p_a\}$ and a match model $\{p_{ab}\}$. Then we consider the matrix given by (9.1) making a particular choice of C. To ensure that the resulting matrix is symmetric, $\{p_{ab}\}$ will be chosen to satisfy $p_{ab} = p_{ba}$ for all $a, b \in Q$. Not all entries of the matrix may be integers, and, even if they are, the matrix may not satisfy condition (vi) from Sect. 7.2. Therefore, we may need to first round up the entries to the nearest integers and then alter the integers slightly to fit condition (vi). In most cases this procedure leads to an admissible substitution matrix with the same random model and a slightly altered match model $\{\tilde{p}_{ab}\}$ and constant $\tilde{C}$.

The above approach for obtaining substitution matrices in principle can be applied to sequences of letters from any alphabet Q. For DNA and RNA sequences very simple substitution matrices are usually effective. For example, a popular substitution matrix used for DNA sequences has 5 on the main diagonal and -4 elsewhere. However, when the sequences under consideration code for protein, in most cases it turns out to be more efficient to compare the protein translations than to compare the DNA or RNA sequences directly. Sometimes one may wish to compare non-coding DNA regions, and for such purposes more sensitive substitution matrices than the one mentioned above are required. For example, in this case one can set for a random model $p_A = p_C = p_G = p_T = 1/4$, and for a match model one can choose $\{p_{ab}\}$ in such a way that the pairs representing transitions are more likely than the pairs representing transversions.

In this chapter we concentrate on protein coding sequences and for this reason discuss only amino acid substitution matrices. Thus in the next two sections we assume that Q is either the 20-letter amino acid alphabet, or the artificial 3-letter alphabet $\{A, B, C\}$ used for illustration purposes.

9.2 PAM Substitution Matrices

In this section we will describe the PAM family of substitution matrices introduced in [DSO]. A particular matrix in this family is denoted by PAMn for $n \in \mathbb{N}$. The construction of PAM matrices starts with *blocks* which are local ungapped multiple alignments of proteins whose sequences are available from existing databases. The proteins are required to be closely related, namely, each sequence in any block is required to be no more than 15% different from any other sequence in the block. The blocks are constructed on the basis of analyzing matches and mismatches in the sequences, that is, essentially by using the multidimensional identity matrix as a substitution matrix for multiple alignments. Before giving the full details of constructing PAM matrices, we sketch the procedure for choosing a match model provided a random model has been fixed.

First, a matrix $S(1) = (S(1)_{ab})$ is produced from the blocks. Its elements are non-negative and sum up to 1 in each row. Hence $S(1)$ is the matrix of transition probabilities of a Markov chain whose non-zero states are labeled by the letters of $\mathcal{Q}$. It turns out that one of the stationary probability distributions of this Markov chain is $\varphi = (p_1, \ldots, p_{20})$, and in most cases this is the only stationary probability distribution, since usually all entries of the matrix $S(n) = S(1)^n$ are non-zero if $n \geq n_0$ for some n_0 (see Sect. 5.4). We take φ as the vector of initialization probabilities. The Markov chain is meant to model the substitution process in amino acid sequences, and thus $S(1)$ describes amino acid substitutions after two steps of the chain (counting the initialization step), that is, short-term evolution. It is assumed that the short-term evolution leads to very few amino acid mutations. Specifically, it is required that the probability of the event that an amino acid produced on the first step of the chain changes on the second step is equal to 0.01. From the explicit construction of the probability space associated with a Markov chain given in Part 4 of Example 6.14, it is clear that this condition can be written as

$$\sum_{a,b \in \mathcal{Q}: a \neq b} p_a S(1)_{ab} = 0.01. \tag{9.3}$$

If we associate with the Markov chain a sequence of random variables $\{f_k\}$ as described at the beginning of Sect. 6.11, then from formula (6.36) we have for all $a, b \in \mathcal{Q}$ and $k \in \mathbb{N}$

$$P\Big(\{e \in S : f_{k+n}(e) = b, f_k(e) = a\}\Big) = (\varphi S(k-1))_a S(n)_{a,b}$$
$$= p_a S(n)_{ab}, \tag{9.4}$$

where we set $S(0)$ to be the identity matrix. Hence, the matrix $S(n)$ describes amino acid substitutions after $n + 1$ steps of the chain. If n is large, $S(n)$ models long-term evolution of amino acid sequences, and thus one is usually interested in large values of n.

We define a match model by setting

$$p_{ab} = P\Big(\{e \in S : f_{n+k}(e) = b, f_k(e) = a\}\Big), \qquad a, b \in \mathcal{Q}, \ k \in \mathbb{N},$$

which due to (9.4) is equivalent to setting

$$p_{ab} = p_a S(n)_{ab}, \qquad a, b \in \mathcal{Q}. \tag{9.5}$$

This choice of match model leads to the *PAMn matrix*. Note that the relatedness of any two amino acids imposed by the match model is by way of one of them being separated from the other by n steps of the evolutionary process modeled by the Markov chain.

As we will see later, the construction leading to the matrix $S(1)$ does not guarantee that $p_{ab} > 0$ for all $a, b \in \mathcal{Q}$. Therefore p_{ab} may need to be further adjusted to ensure that all of them are positive. In most cases of interest, however, all entries of $S(n)$ are non-zero for the relevant values of n, and thus no such adjustments are required. We will also see that $S(1)$ is constructed in such a way that $p_{ab} = p_{ba}$ for all $a, b \in \mathcal{Q}$.

Since in most cases all entries of $S(n)$ are non-zero if $n \geq n_0$ for some n_0, the sequence of matrices $\{S(n)\}$ usually converges to the matrix

$$S(\infty) = \begin{pmatrix} p_1 & \cdots & p_{20} \\ p_1 & \cdots & p_{20} \\ \vdots & \vdots & \vdots \\ p_1 & \cdots & p_{20} \end{pmatrix}$$

(see Sect. 5.4). Formula (9.1) then implies that the sequence $\{\text{PAM}n\}$ converges to the zero matrix as $n \to \infty$, if the value of C is fixed independently of n. As we mentioned above, one is usually interested in large values of n. However, n must not be "too large" for very little information is left in the PAMn matrix as $n \to \infty$. Some of the commonly used values of n are $60, 120, 200, 250$. The performance of similarity searches with PAMn matrices for different values of n has been studied (see, e.g., [EG]).

It remains to explain how a random model (that is, a set of background frequencies) is chosen and $S(1)$ is constructed. Suppose we are given a collection of blocks. For each block we determine all the most parsimonious trees, and both a random model and $S(1)$ are derived from these trees. We will illustrate the process by the following example. Suppose that $\mathcal{Q}$ is the three-letter alphabet $\{A, B, C\}$, and that we are given the following two blocks

$$\begin{array}{ll} \begin{array}{ccc} A & A & B \\ A & C & B, \\ C & C & B \end{array} & \begin{array}{ccc} A & B & C \\ A & B & B. \end{array} \end{array} \tag{9.6}$$

All the most parsimonious trees for the first block are shown in Fig. 5.4 in Sect. 5.2. The most parsimonious trees for the second block are shown in Fig. 9.1.

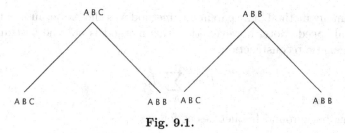

Fig. 9.1.

From these trees we will first produce a matrix $H = (H_{ab})$ of counts as follows. The matrix H is symmetric, and its columns and rows are indexed by the elements of Q. Fix $a, b \in Q$ and let first $a \neq b$. For each block we calculate the number of times a and b occur at the same position in two nodes joined by a branch of a most parsimonious tree for the block and divide by the number of trees for the block. Summing up the resulting numbers over all blocks gives the entries $H_{ab} = H_{ba}$. For the diagonal elements the procedure is exactly the same, but the resulting value is multiplied by 2. For blocks (9.6) the matrix H is shown in Fig. 9.2.

	A	B	C
A	10	0	2
B	0	13	1
C	2	1	7

Fig. 9.2.

We remark that any method of phylogenetic reconstruction that can predict sequences at all ancestral nodes could have been used above in place of the parsimony method. The parsimony method was chosen because of its computational speed. Note, however, that this method is not the best approach to phylogenetic reconstruction.

Set

$$D = \sum_{a,b \in Q} H_{ab} \tag{9.7}$$

and define background frequencies as follows

$$p_a = \frac{\sum\limits_{b \in Q} H_{ab}}{D}, \qquad a \in Q. \tag{9.8}$$

We assume that the block data used to produce the matrix H is diverse enough to ensure that $p_a > 0$ for all $a \in Q$. In fact, we assume that $p_a \geq 0.01$ for all $a \in Q$. For blocks (9.6) we obtain

$$p_A = \frac{1}{3}, \quad p_B = \frac{7}{18}, \quad p_C = \frac{5}{18}. \tag{9.9}$$

For real block databases it is often the case that the background frequencies defined in (9.8) are well approximated by the frequencies of the occurrence of the elements of Q in the original block data. Therefore these frequencies that we denote by p_a^{approx}, $a \in Q$, are often used in place of those defined in (9.8). For blocks (9.6) we have

$$p_A^{\mathrm{approx}} = \frac{1}{3}, \quad p_B^{\mathrm{approx}} = \frac{6}{15}, \quad p_C^{\mathrm{approx}} = \frac{4}{15}.$$

We will now describe the construction of $S(1)$. First, we introduce a matrix $R = (R_{ab})$

$$R_{ab} = \frac{H_{ab}}{\sum\limits_{c \in Q} H_{ac}} = \frac{H_{ab}}{D p_a}, \qquad a, b \in Q.$$

Clearly, all elements of R are non-negative and sum up to 1 in each row. Hence R is the matrix of transition probabilities of a Markov chain. The matrix $S(1)$ is obtained from R by a small modification that ensures that condition (9.3) holds. For $\lambda > 0$ we introduce a matrix $S^{[\lambda]} = \left(S_{ab}^{[\lambda]} \right)$ as follows

$$S_{ab}^{[\lambda]} = \lambda R_{ab}, \qquad\qquad a, b \in Q, \ a \neq b,$$

$$S_{aa}^{[\lambda]} = 1 - \lambda \sum_{b \in Q : b \neq a} R_{ab}, \qquad a \in Q.$$

The matrix $S^{[\lambda]}$ satisfies condition (9.3) for

$$\lambda = \lambda_0 = \frac{0.01}{\displaystyle\sum_{a,b\in Q:a\neq b} p_a R_{ab}}.$$

In the above formula we assume that the block data is diverse enough to ascertain that $\sum_{a,b\in Q:a\neq b} p_a R_{ab} \neq 0$ (note that in the critical case when all the sequences in each block are identical, the above expression is equal to 0). Since $p_a \geq 0.01$ for all $a \in Q$, the elements of the matrix $S^{[\lambda_0]}$ are nonnegative. It then follows that $S^{[\lambda_0]}$ is the matrix of transition probabilities of a Markov chain. We now define $S(1) = S^{[\lambda_0]}$.

Next, we will show that

$$p_a S(n)_{ab} = p_b S(n)_{ba}, \qquad a,b \in Q, \ a \neq b \qquad (9.10)$$

for all $n \in \mathbb{N}$ (see (9.5)). We will prove this by induction. This statement holds for $n = 1$ since for all $a \neq b$ we have $p_a S(1)_{ab} = \lambda_0 H_{ab}/D$ and the matrix H is symmetric. We now assume that (9.10) holds for all $n < N$ and prove it for $n = N$. For all $a, b \in Q$ we have

$$p_a S(N)_{ab} = \sum_{c\in Q} p_a S(N-1)_{ac} S(1)_{cb} = \sum_{c\in Q} p_c S(N-1)_{ca} S(1)_{cb}$$
$$= p_b \sum_{c\in Q} S(1)_{bc} S(N-1)_{ca} = p_b S(N)_{ba},$$

which proves (9.10).

It remains to be shown that the vector $\varphi = (p_1, \ldots, p_{20})$ is a stationary probability distribution for the Markov chain defined by $S(1)$, and hence (9.4) holds. Since H is symmetric, for every $a \in Q$ we have

$$(\varphi S(1))_a = \sum_{b\in Q} p_b S(1)_{ba} = p_a S(1)_{aa} + \sum_{b\in Q:b\neq a} p_b S(1)_{ba}$$

$$= p_a \left(1 - \frac{\lambda_0}{Dp_a} \sum_{c\in Q:c\neq a} H_{ac}\right) + \frac{\lambda_0}{D} \sum_{b\in Q:b\neq a} H_{ba} = p_a,$$

as required.

For blocks (9.6) we have

$$R = \begin{pmatrix} \dfrac{5}{6} & 0 & \dfrac{1}{6} \\[2mm] 0 & \dfrac{13}{14} & \dfrac{1}{14} \\[2mm] \dfrac{1}{5} & \dfrac{1}{10} & \dfrac{7}{10} \end{pmatrix},$$

which gives $\lambda_0 = 0.06$ and

$$
S(1) = \begin{pmatrix} \dfrac{99}{100} & 0 & \dfrac{1}{100} \\[2mm] 0 & \dfrac{697}{700} & \dfrac{3}{700} \\[2mm] \dfrac{3}{250} & \dfrac{3}{500} & \dfrac{491}{500} \end{pmatrix}.
$$

We see that the matrix $S(1)$ has zero elements. One, however, is always interested in its powers $S(n)$ for large n. It is not hard to see that for blocks (9.6) all the elements of the matrix $S(n)$ are non-zero, if $n \geq 2$. For example, for $n = 2$ we have

$$
S(2) = \begin{pmatrix} 0.98022 & 0.00006 & 0.01972 \\ 0.00005 & 0.99147 & 0.00847 \\ 0.02366 & 0.01186 & 0.96446 \end{pmatrix},
$$

where the values were rounded to the fifth decimal place. Then formula (9.5) gives for $n = 2$

$$
\begin{aligned}
&p_{AA} = 0.32674, \quad p_{AB} = p_{BA} = 0.00002, \quad p_{AC} = p_{CA} = 0.00657, \\
&p_{BB} = 0.38557, \quad p_{BC} = p_{CB} = 0.00329, \quad p_{CC} = 0.26790,
\end{aligned} \tag{9.11}
$$

where again the values were rounded to the fifth decimal place. Finally, from (9.1), (9.9) and (9.11) we obtain the following PAM2 matrix for blocks (9.6)

$$
\mathrm{PAM2} = \begin{pmatrix} 1 & -9 & -3 \\ -9 & 1 & -3 \\ -3 & -3 & 1 \end{pmatrix},
$$

where we set $C = 1$ and rounded the values to the nearest integers. Note that condition (vi) from Sect. 7.2 is satisfied for the above matrix, and therefore no further adjustments are required.

Generally one is interested in constructing the PAMn matrices for much larger values of n. Figure 9.3 shows the PAM250 matrix for the 20-letter amino acid alphabet derived from 71 blocks that were produced from databases of amino acid sequences. The name of PAM matrices originates from *accepted point mutations*. An accepted point mutation in a protein is a replacement of one amino acid by another accepted by natural selection. The amino acid differences observed in the sequences in each block from the database of blocks

	A	C	D	E	F	G	H	I	K	L	M	N	P	Q	R	S	T	V	W	Y
A	2	-2	0	0	-4	1	-1	-1	-1	-2	-1	0	1	0	-2	1	1	0	-6	-3
C	-2	12	-5	-5	-4	-3	-3	-2	-5	-6	-5	-4	-3	-5	-4	0	-2	-2	-8	0
D	0	-5	4	3	-6	1	1	-2	0	-4	-3	2	-1	2	-1	0	0	-2	-7	-4
E	0	-5	3	4	-5	0	1	-2	0	-3	-2	1	-1	2	-1	0	0	-2	-7	-4
F	-4	-4	-6	-5	9	-5	-2	1	-5	2	0	-4	-5	-5	-4	-3	-3	-1	0	7
G	1	-3	1	0	-5	5	-2	-3	-2	-4	-3	0	-1	-1	-3	1	0	-1	-7	-5
H	-1	-3	1	1	-2	-2	6	-2	0	-2	-2	2	0	3	2	-1	-1	-2	-3	0
I	-1	-2	-2	-2	1	-3	-2	5	-2	2	2	-2	-2	-2	-2	-1	0	4	-5	-1
K	-1	-5	0	0	-5	-2	0	-2	5	-3	0	1	-1	1	3	0	0	-2	-3	-4
L	-2	-6	-4	-3	2	-4	-2	2	-3	6	4	-3	-3	-2	-3	-3	-2	2	-2	-1
M	-1	-5	-3	-2	0	-3	-2	2	0	4	6	-2	-2	-1	0	-2	-1	2	-4	-2
N	0	-4	2	1	-4	0	2	-2	1	-3	-2	2	-1	1	0	1	0	-2	-4	-2
P	1	-3	-1	-1	-5	-1	0	-2	-1	-3	-2	-1	6	0	0	1	0	-1	-6	-5
Q	0	-5	2	2	-5	-1	3	-2	1	-2	-1	1	0	4	1	-1	-1	-2	-5	-4
R	-2	-4	-1	-1	-4	-3	2	-2	3	-3	0	0	0	1	6	0	-1	-2	2	-4
S	1	0	0	0	-3	1	-1	-1	0	-3	-2	1	1	-1	0	2	1	-1	-2	-3
T	1	-2	0	0	-3	0	-1	0	0	-2	-1	0	0	-1	-1	1	3	0	-5	-3
V	0	-2	-2	-2	-1	-1	-2	4	-2	2	2	-2	-1	-2	-2	-1	0	4	-6	-2
W	-6	-8	-7	-7	0	-7	-3	-5	-3	-2	-4	-4	-6	-5	2	-2	-5	-6	17	0
Y	-3	0	-4	-4	7	-5	0	-1	-4	-1	-2	-2	-5	-4	-4	-3	-3	-2	0	10

Fig. 9.3.

used to construct PAM matrices are all accepted point mutations coming from the predicted ancestral sequences on phylogenetic trees, since all the protein sequences in the blocks have been taken from currently existing species that successfully incorporated these mutations.

9.3 BLOSUM Substitution Matrices

PAM substitution matrices appeared at the time when available data consisted mainly of families of sequences of closely related proteins. From such data one could only hope to estimate the short-term evolutionary matrix $S(1)$ and then, in order to extrapolate to long-term evolutionary events, one had to raise $S(1)$ to a high power, as we described in the preceding section. When data for more distantly related proteins became available, it was shown that the matrix $S(n)$ for large values of n does not in fact reflect the true patterns of long-term amino acid substitutions [GCB]. As databases of more diverged sequences grew, it became possible to estimate such patterns directly from the sequences and avoid Markov chain extrapolation. Below we will describe a very natural procedure that leads to the BLOSUM (BLOcks SUbstitution Matrices) family of substitution matrices introduced in [HH]. A particular matrix in this family is denoted by BLOSUMr, where $0 < r < 100$.

As the name of BLOSUM matrices suggests, they are also derived from a database of blocks. As in the case of PAM matrices, the blocks are initially constructed on the basis of analyzing only matches and mismatches in the sequences. In fact, there is an automated way for constructing blocks given a substitution matrix, and initially the substitution matrix is taken to be the identity matrix. Sequences within each block are allowed to be more diverged than in the case of PAM matrices. The process of constructing a BLOSUM matrix starts with choosing a number r and then producing *clusters of sequences* in each block. We group the sequences in each block into clusters in such a way that each sequence in any cluster has $r\%$ or higher sequence identity with at least one other sequence in the cluster. We will illustrate this by the following example. Suppose that Q is the three-letter alphabet $\{A, B, C\}$, and we are given the three blocks shown below

$$
\begin{array}{ll}
x^1\ A\ B\ C\ A\ B & \qquad x^6\ \ A\ B\ C \\
x^2\ A\ B\ C\ A\ C & \qquad x^7\ \ A\ B\ C \\
x^3\ B\ B\ C\ A\ B, & \qquad x^8\ \ A\ A\ C \\
x^4\ C\ B\ C\ A\ C & \qquad x^9\ \ C\ B\ C\, ' \\
x^5\ A\ A\ A\ C\ B & \qquad x^{10}\ A\ A\ B \\
 & \qquad x^{11}\ B\ A\ B
\end{array}
\tag{9.12}
$$

$$
\begin{array}{l}
x^{12}\ A\ A\ A\ C\ B\ A\ B\ C \\
x^{13}\ B\ A\ A\ C\ B\ A\ B\ C \\
x^{14}\ A\ A\ A\ C\ B\ A\ C\ B \\
x^{15}\ A\ A\ A\ C\ B\ A\ C\ C.
\end{array}
$$

If we choose $r = 80$, we obtain two clusters in the first block: $\{x^1, x^2, x^3, x^4\}$, $\{x^5\}$, five clusters in the second block: $\{x^6, x^7\}$, $\{x^8\}$, $\{x^9\}$, $\{x^{10}\}$, $\{x^{11}\}$, and one cluster in the third block: $\{x^{12}, x^{13}, x^{14}, x^{15}\}$. Clustering is done to reduce overrepresentation of closely related sequences, and in the future each cluster will be treated essentially as a single sequence.

Next, we will produce a matrix $H = (H_{ab})$ of counts as follows. The matrix H is symmetric, and its columns and rows are indexed by the elements of Q. Fix $a, b \in Q$ and let first $a \neq b$. Next, fix a block and two particular clusters in the block. Let n and m be the numbers of sequences in the clusters. We now calculate the number of times a and b occur at the same position in the clusters, counting only those occurrences when a and b do not belong to the same cluster, and divide the result by nm. Finally, we sum up the resulting values over all pairs of clusters in each block and over all blocks afterwards. This gives the entries $H_{ab} = H_{ba}$. For the diagonal elements the procedure is exactly the same, but the resulting value is multiplied by 2. For blocks (9.12) the matrix H is shown in Fig. 9.4.

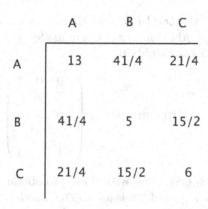

Fig. 9.4.

We now define background frequencies by formula (9.8) with D given by (9.7). As in the preceding section, we assume that the data is diverse enough to guarantee that $p_a > 0$ for all $a \in Q$. For blocks (9.12) we obtain

$$p_A = \frac{57}{140}, \quad p_B = \frac{91}{280}, \quad p_C = \frac{75}{280}. \tag{9.13}$$

For real block databases it is often the case that the background frequencies defined in (9.8) are well approximated by the frequencies of the occurrence of the elements of Q in the original block data, where each occurrence has the weight $1/n$ with n equal to the number of sequences in the cluster containing the occurrence. Therefore these frequencies that we denote by p_a^{approx}, $a \in Q$, are often used in place of those defined in (9.8). For blocks (9.12) we have

$$p_A^{\mathrm{approx}} = \frac{19}{44}, \quad p_B^{\mathrm{approx}} = \frac{13}{44}, \quad p_C^{\mathrm{approx}} = \frac{3}{11}.$$

Further, for a match model we set

$$p_{ab} = \frac{H_{ab}}{D}.$$

Here we assume that the data is diverse enough and guarantees that $p_{ab} > 0$ for all $a, b \in Q$. For blocks (9.12) this gives

$$p_{AA} = \frac{13}{70}, \quad p_{AB} = p_{BA} = \frac{41}{280}, \quad p_{AC} = p_{CA} = \frac{21}{280},$$

$$\tag{9.14}$$

$$p_{BB} = \frac{5}{70}, \quad p_{BC} = p_{CB} = \frac{15}{140}, \quad p_{CC} = \frac{6}{70}.$$

Next, we construct the *Level 1 BLOSUMr* matrix from formula (9.1) with $C = 2/\ln 2$ (sometimes other choices of C are made). For blocks (9.12), from formulas (9.13) and (9.14) we obtain

$$\text{Level 1 BLOSUM80} = \begin{pmatrix} 0 & 0 & -1 \\ 0 & -1 & 1 \\ -1 & 1 & 1 \end{pmatrix}.$$

Further, the Level 1 BLOSUMr matrix is used to obtain a new set of blocks, and the procedure is repeated. It leads to the *Level 2 BLOSUMr* matrix. Finally, the Level 2 BLOSUMr matrix is used to obtain a third database of blocks, and the final *BLOSUMr* matrix is derived from it in the same way. In fact, on the last step the clustering percentage r is allowed to be different from those on the first two steps. In the original construction r was set to be equal to 60 on the first two steps and was allowed to vary arbitrarily on the last one. The number of blocks on the first step was 2205, on the second 1961 and on the third 2106.

One often uses prior knowledge about evolutionary distances between the sequences of interest to choose which BLOSUM matrix to use. If there is no prior information, the BLOSUM62 and BLOSUM50 matrices are often used. The BLOSUM62 matrix (shown in Fig. 9.5) is standard for ungapped matching, and the BLOSUM50 matrix is often used for finding alignments containing gaps.

The relative performance of PAM and BLOSUM matrices has been studied for a variety of searches, and the broad conclusion is that using BLOSUM matrices generally leads to better results [HH].

	C	S	T	P	A	G	N	D	E	Q	H	R	K	M	I	L	V	F	Y	W
C	9	-1	-1	-3	0	-3	-3	-3	-4	-3	-3	-3	-3	-1	-1	-1	-1	-2	-2	-2
S	-1	4	1	-1	1	0	1	0	0	0	-1	-1	0	-1	-2	-2	-2	-2	-2	-3
T	-1	1	4	1	-1	1	0	1	0	0	0	-1	0	-1	-2	-2	-2	-2	-2	-3
P	-3	-1	1	7	-1	-2	-1	-1	-1	-1	-2	-2	-1	-2	-3	-3	-2	-4	-3	-4
A	0	1	-1	-1	4	0	-1	-2	-1	-1	-2	-1	-1	-1	-1	-1	-2	-2	-2	-3
G	-3	0	1	-2	0	6	-2	-1	-2	-2	-2	-2	-2	-3	-4	-4	0	-3	-3	-2
N	-3	1	0	-2	-2	0	6	1	0	0	-1	0	0	-2	-3	-3	-3	-3	-2	-4
D	-3	0	1	-1	-2	-1	1	6	2	0	-1	-2	-1	-3	-3	-4	-3	-3	-3	-4
E	-4	0	0	-1	-1	-2	0	2	5	2	0	0	1	-2	-3	-3	-3	-3	-2	-3
Q	-3	0	0	-1	-1	-2	0	0	2	5	0	1	1	0	-3	-2	-2	-3	-1	-2
H	-3	-1	0	-2	-2	-2	1	1	0	0	8	0	-1	-2	-3	-3	-2	-1	2	-2
R	-3	-1	-1	-2	-1	-2	0	-2	0	1	0	5	2	-1	-3	-2	-3	-3	-2	-3
K	-3	0	0	-1	-1	-2	0	-1	1	1	-1	2	5	-1	-3	-2	-3	-3	-2	-3
M	-1	-1	-1	-2	-1	-3	-2	-3	-2	0	-2	-1	-1	5	1	2	-2	0	-1	-1
I	-1	-2	-2	-3	-1	-4	-3	-3	-3	-3	-3	-3	-3	1	4	2	1	0	-1	-3
L	-1	-2	-2	-3	-1	-4	-3	-4	-3	-2	-3	-2	-2	2	2	4	3	0	-1	-2
V	-1	-2	-2	-2	0	-3	-3	-3	-2	-2	-3	-3	-2	1	3	1	4	-1	-1	-3
F	-2	-2	-2	-4	-2	-3	-3	-3	-3	-3	-1	-3	-3	0	0	0	-1	6	3	1
Y	-2	-2	-2	-3	-2	-3	-2	-3	-2	-1	2	-2	-2	-1	-1	-1	-1	3	7	2
W	-2	-3	-3	-4	-3	-2	-4	-4	-3	-2	-2	-3	-3	-1	-3	-2	-3	1	2	11

Fig. 9.5.

Exercises

9.1. Prove the Kullback-Leibler inequality.

9.2. To block data (9.6) add the block from Exercise 5.2 and find the corresponding PAM3 matrix from the three blocks for the three-letter alphabet $\{A, B, C\}$.

9.3. From the following block data construct the Level 1 BLOSUM60 matrix

$x^1\ A\ A\ A\ A\ B\ B\ B$
$x^2\ C\ C\ C\ A\ C\ A\ B$
$x^3\ B\ B\ C\ A\ B\ A\ C$,
$x^4\ C\ C\ C\ A\ C\ B\ C$
$x^5\ A\ A\ B\ A\ B\ A\ B$

$x^6\ \ A\ C\ C\ A$
$x^7\ \ A\ C\ B\ A$
$x^8\ \ A\ A\ A\ C$
$x^9\ \ C\ C\ B\ A$'
$x^{10}\ A\ A\ B\ B$
$x^{11}\ B\ A\ A\ C$

$x^{12}\ A\ A\ A\ C\ C\ C\ B\ B\ A\ A$
$x^{13}\ B\ A\ A\ C\ C\ A\ A\ A\ A\ A$
$x^{14}\ A\ B\ A\ B\ C\ A\ C\ C\ A\ C$
$x^{15}\ C\ A\ C\ C\ B\ A\ A\ C\ A\ A$

for the three-letter alphabet $\{A, B, C\}$.

References

[AL]   Altschul, S. F., Gish, W., Miller, W., Myers, E. W., Lipman, D. J.: Basic local alignment search tool. J. Mol. Biol., **215**, 403–410 (1990)

[B]   Baum, L. E.: An equality and associated maximization technique in statistical estimation for probabilistic functions of Markov processes. Inequalities, **3**, 1–8 (1972)

[BL]   Berger, B., Leighton, T.: Protein folding in the hydrophobic-hydrophilic (HP) model is NP-complete. J. Comput. Biol., **5**, 27-40 (1998)

[BHV]   Billera, L., Holmes, S., Vogtman, K.: Geometry of the space of phylogenetic trees. Advances in Appl. Math., **27**, 733–767 (2001)

[BLE]   Bowie, J. U. Lüthy, R., Eisenberg, D.: A method to identify protein sequences that fold into a known three-dimensional structure. Science **253**, 164–170 (1991)

[BK]   Burge, C., Karlin, S.: Prediction of complete gene structures in human genomic DNA. J. Mol. Biol., **268**, 78–94 (1997)

[CL]   Carrillo, H., Lipman, D.: The multiple sequence alignment problem in biology. SIAM J. Appl. Math., **48**, 1073–1082 (1988)

[C-SE]   Cavalli-Sforza, L., Edwards, A.: Phylogenetic analysis: models and estimation procedures. Am. J. Hum. Genetics, **19**, 233–257 (1967)

[CD]   Chan, H. S., Dill, K. A.: Compact Polymers. Macromolecules, **22**, 4559-4573 (1989)

[CY]   Crescenzi, P., Goldman, D., Papadimitriou, C., Piccolboni, A., Yannakakis, M.: On the complexity of protein folding. J. Comput. Biol., **5**, 423–465 (1998)

[DSO]   Dayhoff, M. O., Schwartz, R. M., Orcutt, B. C.: A model of evolutionary change in proteins. Atlas of Protein Sequence and Structure, **5**(Supplement 3), 345–352 (1978)

[DS]   Delcher, A. L., Harmon, D., Kasif, S., White, O., Salzberg, S. L.: Improved microbial gene identification with Glimmer. Nucleic Acids Res., **27**, 4636–4641 (1999)

[DKZ1]   Dembo, A., Karlin, S., Zeitouni, O.: Critical phenomena for sequence matching with scoring. Ann. Prob., **22**, 1993-2021 (1994)

[DKZ2]   Dembo, A., Karlin, S., Zeitouni, O.: Limit distribution of maximal non-aligned two-sequence segmental score. Ann. Prob., **22**, 2022-2039 (1994)

[Di]   Dill, K. A.: Theory for the folding and stability of globular proteins. Bio-
          chemistry, **24**, 1501–1809 (1985)

[DC]   Dill, K. A., Bromberg, S., Yue, K., Fiebig, K. M., Yee, D. P., Thomas, P.
          D., Chan, H. S.: Principles of protein folding - a perspective from simple
          exact models. Protein Sci., **4**, 561-602 (1995)

[Do]   Doob, J. L.: Stochastic Processes. John Wiley & Sons (1953)

[DEKM]   Durbin, R., Eddy, S., Krogh, A., Mitchison, G.: Biological Sequence
          Analysis. Cambridge University Press (1998)

[E]   Eddy, S. (with contributions by Birney, E.): HMMER, Version 2.3.1.
          Washington University, St. Louis (2003)

[EG]   Ewens, W. J., Grant, G. R.: Statistical Methods in Bioinformatics.
          Springer-Verlag (2001)

[F1]   Felsenstein, J.: Evolutionary trees from DNA sequences: a maximum like-
          lihood approach. J. Mol. Evol., **17**, 368–376 (1981)

[F2]   Felsenstein, J.: Confidence limits on phylogenies: an approach using the
          bootstrap. Evolution, **39**, 783–791 (1985)

[F3]   Felsenstein, J.: PHYLIP, Phylogeny Inference Package, Version 3.57. Uni-
          versity of Washington, Seattle (1995)

[FM]   Fitch, W. M., Margoliash, E.: Construction of phylogenetic trees. Science,
          155, 279–284 (1967)

[GCB]   Gonnet, G. H., Cohen, M. A., Benner, S. A.: Exhaustive matching of the
          entire protein sequence database. Science, **256**, 1443–1445 (1992)

[G]   Gotoh, O.: An improved algorithm for matching biological sequences. J.
          Mol. Biol., **162**, 705–708 (1982)

[H]   Häckel, E.: Generelle Morphologie der Organismen: Allgemeine
          Grundzuge der Organischen Formen-Wissenschaft, Mechanisch Begrun-
          det Durch die von Charles Darwin, Reformite Descendenz-Theorie. Georg
          Riemer, Berlin (1866)

[HI]   Hart, W. E., Istrail, S.: Fast protein folding in the hydrophobic-
          hydrophilic model within three-eighths of optimal. J. Comp. Biol., **3**,
          53–96 (1996)

[HKY]   Hasegawa, M., Kishino, H., Yano, T.: Dating of the human-ape splitting
          by a molecular clock of mitochondrial DNA. J. Mol. Evol., **22**, 160–174
          (1985)

[HH]   Henikoff, S., Henikoff, J. G.: Amino acid substitution matrices from pro-
          tein blocks. Proc. Nat. Acad. Sci. USA, **89**, 10915–1019 (1992)

[IJ]   Iwata, S., Lee, J. W., Okada, K., Lee, J. K., Iwata, M., Rasmussen, B.,
          Link, T. A.,Ramaswamy, S., Jap, B. K.: Complete structure of the 11-
          subunit bovine mitochondrial cytochrome bc1 complex. Science. **281**, 64–
          71 (1998)

[JC]   Jukes, T., Cantor, C.: Evolution of protein molecules. In: Munro, H. N.
          (ed) Mammalian Protein Metabolism. Academic Press, New York (1969)

[Kan]   Kanehisa, M.: Post-Genome Bioinformatics. Oxford University Press
          (2000)

[Kar]   Karlin, S.: A First Course in Stochastic Processes. Academic Press, N.Y.,
          London (1968)

[KA]   Karlin, S., Altschul, S. F.: Applications and statistics for multiple high-
          scoring segments in molecular sequences. Proc. Nat. Acad. Sci., **90**, 5873–
          5877 (1993)

[KD]   Karlin, S., Dembo, A.: Limit distributions of maximal segmental score among Markov-dependent partial sums. Adv. Appl. Prob., **24**, 113–140 (1992)

[KT]   Karlin, S., Taylor, H.: A First Course in Stochastic Processes. Academic Press, N.Y., London (1975)

[KS]   Khorasanizadeh, S., Campos-Olivas, R., Clark, C. A., Summers, M. F.: Sequence-specific 1H, 13C and 15N chemical shift assignment and secondary structure of the Htlv-I capsid protein. J. Biomol. NMR, **14**, 199–200 (1999)

[Ki]   Kimura, M.: A simple method for estimating evolutionary rate in a finite population due to mutational production of neutral and nearly neutral base substitution through comparative studies of nucleotide sequences. J. Mol. Biol., **16**, 111–120 (1980)

[KF]   Kolmogorov, A. N., Fomin, S. V.: Elements of the Theory of Functions and Functional Analysis. Graylock Press, Rochester, N.Y. (1957)

[KH]   Krogh, A. M., Brown, I. S., Mian, K., Sölander, K., Hausler, K.: Hidden Markov models in computational biology: applications to protein modelling. J. Mol. Biol., **235**, 1501–1531 (1994)

[LD]   Lau, K. F., Dill, K. A.: A lattice statistical mechanics model of the conformational and sequence spaces of proteins. Macromolecules, **22**, 3986–3997 (1989)

[LTW]   Li, H., Tang, C., Wingreen, N. S.: A protein folds atypical? Proc. Nat. Acad. Sci. USA, **95**, 4987–4990 (1998)

[LAK]   Lipman, D. J., Altschul, S. F., Kececioglu, J. D.: A tool for multiple sequence alignment. Proc. Nat. Acad. Sci. USA, **86**, 4412–4415 (1989)

[MS]   Madsen, O., Scally, M., Douady, C. J., Kao, D. J., DeBry, R. W., Adkins, R., Amrine, H. M., Stanhope, M. J., de Jong, W. W., Springer, M. S.: Parallel adaptive radiations in two major clades of placental mammals. Nature, **409**, 610–614 (2001)

[NW]   Needleman, S. B., Wunsch, C. D.: A general method applicable to the search for similarities in the amino acid sequence of two proteins. J. Mol. Biol., **48**, 443–453 (1970)

[N]   Neumaier, A.: Molecular modeling of proteins and mathematical prediction of protein structure. SIAM Rev., **39**, 407–460 (1997)

[PL]   Pearson, W. R., Lipman, D. J.: Improved tools for biological sequence comparison. Proc. Nat. Acad. Sci. USA, **4**, 2444–2448 (1988)

[RG]   Rambaut, A., Grassly, N.: Sequence-Gen: an application for the Monte Carlo simulation of DNA sequence evolution along phylogenetic trees. Comput. Appl. Biosci., **13**, 235–238 (1997)

[R]   Rogers, J. S.: Maximum likelihood estimation of phylogenetic trees is consistent when substitution rates vary according to the invariable sites plus gamma distribution. Syst. Biol., **50**, 713–722 (2001)

[SN]   Saitou, N., Nei, M.: The neighbor-joining method: a new method for reconstructing phylogenetic trees. Mol. Biol. Evol., **4**, 406–425 (1987)

[S]   Smith, R. D.: Correlations between bound N-Alkyl isocyanide orientations and pathways for ligand binding in recombinant myoglobins. Thesis, Rice, US ISSN 1047-8477 0806 (1999)

[SW]   Smith, T. F., Waterman, M. S.: Identification of common molecular subsequences. J. Mol. Biol., **147**, 195–197 (1981)

[SM]   Sokal, R. R., Michener, C. D.: A statistical method for evaluating system-
        atic relationships. Univ. of Kansas Scientific Bull., **28**, 1409–1438 (1958)

[SK]   Studier, J. A., Keppler, K. J.: A note on the neighbor-joining algorithm
        of Saitou and Nei. Mol. Biol. Evol., **5**, 729–731 (1988)

[THG]   Thompson, J. D., Higgins, D. G., Gibson, T. J.: CLUSTALW: improv-
        ing the sensitivity of progressive multiple sequence alignment through se-
        quence weighting, position specific gap penalties and weight matrix choice.
        Nucleic Acid Res., **22**, 4673–4680 (1994)

[TBB]   Thompson, J. R., Bratt, J. M., Banaszak, L. J.: Crystal structure of cel-
        lular retinoic acid binding protein I shows increased access to the binding
        cavity due to formation of an intermolecular beta-sheet. J. Mol. Biol.,
        252, 433–446 (1995)

[W]   Wilks, S. S.: Mathematical Statistics. John Wiley & Sons, N.Y., London
        (1962)

Index

HP-classification of amino acids, 85
HP-sequence of a protein, 85
SP-score of a multiple alignment, 17
α-helix, 76
β-sheet, 76
β-strand, 76
σ-additivity property of a measure, 152
σ-algebra generated by a family of
    events, 149
σ-algebra of events, 148
a priori connectivity, 26

a.c., 166
a.c. convergence, 202
a.e., 175
absolutely continuous function, 176
absorbing state of a Markov chain, 30
accepted point mutation, 278
additive distance function, 97
admissible substitution matrix, 271
affine gap model, 12
algebra generated by a family of events,
    149
algebra of events, 148
alignment of sequences, 7
alignment score, 8
almost certain, 166
almost certain convergence, 202
almost everywhere, 175
alternative hypothesis, 138, 256
amino acid, 3
amino acid alphabet, 4
amino end of a protein, 4

background frequencies, 271
backward algorithm, 41
Baum-Welch training algorithm, 48
begin state of a Markov chain, 29
Bernoulli trial, 185
bias of an estimator, 236
biased estimator for parameter(s), 236
binomial distribution, 185
biological sequence, 3
BLAST, 16
block, 273
BLOSUMr matrix, 280
bond angle energy of a protein, 83
bond energy of a protein, 83
Borel sets in $[0, 1]$, 151
Borel sets in $\mathbb{R}$, 177
bounded random variable, 166
branch and bound algorithm, 95
burial/polarity classes, 80
burial/polarity environments, 80

Cantor set, 151
carboxy end of a protein, 4
Cartesian power of a probability space,
    159
Cartesian power of a random variable,
    193
Cartesian power of a sample space, 159
Cartesian product of a sequence of
    probability spaces, 159
Cartesian product of probability spaces,
    159
Cauchy-Bunyakowski inequality, 167
Central Limit theorem, 202

Chebyshev's inequality, 197
chi-square distribution, 191
CLUSTALW, 20
cluster of sequences, 280
codon, 4
complement of an event, 147
conditional probability, 42, 160
connectivity of a Markov chain, 26
consistent estimator, 236
consistent statistical test, 257
continuous random variable, 182
continuous vector-valued random variable, 193
continuous-time finite Markov chain, 122
continuous-time finite Markov model, 122
convergence in distribution, 200
convergence in probability, 197
convergence in the mean, 199
convergence in the square mean, 199
cost function, 94
covariance, 195
cylinder event, 153, 159, 213

data simulation, 138
decoding, 36
delete state of a profile HMM, 64
deletion, 7
density function of a random variable, 182
density function of a vector-valued random variable, 193, 194
density of a random variable, 182
density of a vector-valued random variable, 193, 194
difference between events, 147
dihedral angle energy of a protein, 83
discrete random variable, 163
discrete vector-valued random variable, 193
discrete-time finite Markov chain, 25
discrete-time finite Markov model, 25
distance between two points, 96
distance function, 96
distance matrix, 96
distribution function of a random variable, 176

distribution function of a vector-valued random variable, 192
distribution of a random variable, 184
DNA alphabet, 3
DNA orientation, 3
dot matrix, 14
dynamic programming algorithm, 9

edge effect, 222
electrostatic energy of a protein, 84
elementary event, 146
emission probability of an HMM, 34
empty event, 147
end state of a Markov chain, 30
environmental descriptors, 80
environmental template method, 80
estimate for parameter(s), 236
event, 42, 146
evolutionary assumption, 131
evolutionary rate, 90, 127, 132
excursion of a path, 222
excursion of a trajectory, 213
exhausting sequence of subsets, 173
exon, 27
expected value of a random variable, 166
exponential distribution, 189
exponential of a matrix, 124
extreme value distribution, 211

FASTA, 14
Felsenstein model, 129
finite additivity property of a measure, 152
Fitch's algorithm, 95
forward algorithm, 40
four-point condition, 98
fully connected Markov chain, 26

gamma distribution, 190
gamma function, 190
gap, 8
gap extension penalty, 12
gap opening penalty, 12
gap penalty, 8
gene tree, 91
genetic code, 4
GENSCAN, 37
geometric distribution, 186
geometric-like random variable, 187

Glimmer, 32
global alignment, 9

hidden Markov model, 34
HKY model, 130
HMM, 34
HMM having no silent loops, 59
HMM with silent states, 57
HMMER, 68
homology of sequences, 7
hydrophilic amino acid, 85
hydrophobic amino acid, 85

iid random variables, 198
independent events, 161
independent identically distributed
    random variables, 198
independent random variables, 192, 199
infinite Cartesian power of a probability
    space, 159
infinite Cartesian power of a random
    variable, 199
infinite Cartesian power of a sample
    space, 159
initialization probabilities of a Markov
    chain, 26
insert column in a multiple alignment,
    64
insert state of a profile HMM, 64
insertion, 7
integrable random variable, 164
integral of a random variable, 164, 165
intersection of events, 147
interval estimation, 235
interval estimator, 236
intron, 27

Jukes-Cantor model, 127

Kimura model, 128
Kullback-Leibler inequality, 272

ladder point of a trajectory, 213
lattice HP-model, 85
leaf of a tree, 89
least squares method, 119
Lebesgue extension of a probability
    measure, 157
Lebesgue integrable function, 173, 174

Lebesgue integral, 173, 174
Lebesgue measurable event, 157
Lebesgue measurable function, 173
Lebesgue measurable set, 173
Lebesgue measure on $[0, 1]$, 172
Lebesgue measure on $\mathbb{R}$, 172
level 1 BLOSUMr matrix, 282
level 2 BLOSUM r matrix, 282
likelihood given a statistical model, 242
likelihood of a dataset given a tree, 133
likelihood of training data, 33, 45
likelihood ratio, 138, 257
likelihood ratio test, 138, 257
linear gap model, 9
linear gap model for multiple align-
    ments, 17
local alignment, 11

marginal distribution function, 192
marginal probability distribution, 193
Markov chain, 25, 122
Markov model, 25, 122
match column in a multiple alignment,
    64
match model, 266
match state of a profile HMM, 64
matrix of instantaneous change of a
    Markov chain, 125
matrix of transition probabilities of a
    Markov chain, 25
maximum likelihood equations, 244
maximum likelihood estimate, 245
maximum likelihood estimation, 242
maximum likelihood estimator, 243
mean of a random variable, 166
mean square error of an estimator, 237
mean value of a random variable, 166
measurable event, 158
measurable function, 173
messenger RNA, 5
metric space, 96
model for training data, 28, 35, 45
molecular clock condition, 113
molecular clock tree, 113
moment-generating function, 228
monotone non-decreasing function, 175
most parsimonious topology, 94
most probable path, 35
most probable state, 44

mRNA, 5
MSA, 19
multiple alignment of sequences, 16

Needleman-Wunsch algorithm, 9
neighbor-joining algorithm, 98
nested hypotheses, 138, 256
non-measurable event, 158
non-silent length of a sequence, 58
non-silent state of an HMM, 57
non-trivially connected Markov chain,
    30
normal distribution, 189
nucleic acid, 3
nucleotide, 3
null hypothesis, 138, 256
null hypothesis distribution of the
    likelihood ratio, 258

occurrence of an event, 42, 147
open reading frame, 27
operational taxonomic unit, 89
optimal alignment of sequences, 8
optimal multiple alignment of sequences,
    16
ORF, 27
orthologues, 91
OTU, 89
outer measure, 156
overlap between sequences, 220

PAMn matrix, 273
paralogues, 91
parameter estimation, 235
parameter space of a statistical model,
    231
parameter(s) of a statistical model, 231,
    235
parametric bootstrap method, 139, 265
path of a trajectory, 222
path through a Markov chain, 30, 34,
    146
Pfam, 79
PHYLIP, 122
phylogenetic reconstruction, 92
phylogenetic tree, 90
point estimation, 235
point estimator for parameter(s), 235
Poisson distribution, 187

polar aminio acid, 85
posterior decoding, 44
posterior probability, 43, 44, 49
potential energy of a protein, 82
power of a statistical test, 257
probability, 42, 152
probability distribution of a random
    variable, 181
probability distribution of a vector-
    valued random variable, 193
probability measure, 42, 152
probability of a sequence, 27, 36
probability of a sequence along a path,
    35
probability of committing a type I error,
    257
probability of committing a type II
    error, 257
probability space, 156, 158
probability with which a sequence arises
    from a model, 29, 36
probability with which a sequence arises
    from a model along a path, 36
product of events, 147
profile HMM, 64
progressive alignment of sequences, 20
protein domain, 77
protein fold, 77
protein motif, 77
protein orientation, 4
protein primary structure, 76
protein quaternary structure, 78
protein secondary structure, 76
protein super-secondary structure, 77
protein tertiary structure, 77
protein-coding gene, 3
pseudodistance function, 106
pseudodistance matrix, 106
pulley principle, 134

random model, 266
random sample from a statistical model,
    233
random variable, 162
random walk, 212
range of a statistical model, 231
reduced multiple alignment of
    sequences, 93
regular Markov chain, 124

regular statistical model, 242
Renewal theorem, 217
reversed Markov chain, 126
reversible Markov chain, 127
RNA alphabet, 5
RNA orientation, 5
RNA-coding gene, 5
root of a tree, 89
rooted tree, 89

sample from the infinite Cartesian
    power of a random variable, 199
sample space, 42, 145
scoring matrix, 9
search with a model, 28, 36
segment in a sequence, 11
semi-algebra of events, 150
Seq-Gen, 138
sequence similarity, 7
set estimation, 235
set of convergence, 203
significance level of a statistical test,
    257
significance point of a statistical test,
    258
significant sequence similarity, 210
silent state of an HMM, 57
simple function, 163
simple random walk, 226
Smith-Waterman algorithm, 11
smooth statistical model, 260
standard deviation of a random
    variable, 166
Star Alignment algorithm, 20
start codon, 4
state of a Markov chain, 25
stationary probability distribution of a
    Markov chain, 125
statistical hypothesis test, 137, 256
statistical model, 231
statistical modeling, 233
step size of a random walk, 212
stop codon, 4
Strong Law of Large Numbers, 204
subset, 147
substitution, 7
substitution matrix, 9
sum of events, 147
sum of squares, 119

symmetric difference of events, 148

taxon, 89
termination probabilities of a Markov
    chain, 30
test of significance, 137, 256
test statistic, 259
threading, 79
time-reversible Markov chain, 127
tip of a tree, 89
training data, 28, 35, 45
training data agreeing with a connectiv-
    ity, 29
trajectory of a random walk, 212
transcription, 5
transfer RNA, 5
transition, 129
transition probability of a Markov
    chain, 25
translation, 5
transversion, 129
tree generating a distance function, 97
tree relating OTUs, 90
tree topology, 90
tree-generated distance function, 96
triangle inequality, 96, 167
tRNA, 5
two-dimensional lattice HP-model, 87
type I error, 257
type II error, 257

ultrameric distance function, 110
unbiased estimator for parameter(s),
    236
underlying Markov chain of an HMM,
    34
uniform convergence, 164
uniform distribution, 185, 188
union of events, 147
unrooted tree, 89
UPGMA algorithm, 113

Van der Waals energy of a protein, 84
variance of a random variable, 166
variance-covariance matrix of a random
    variable, 195
vector of initialization probabilities, 26
vector-valued random variable, 191
Viterbi algorithm, 38

Viterbi path, 38

Viterbi training algorithm, 52

Wald's identity, 217
Weak Law of Large Numbers, 198
weighted sum of squares, 122

Universitext

Aguilar, M.; Gitler, S.; Prieto, C.: Algebraic Topology from a Homotopical Viewpoint

Aksoy, A.; Khamsi, M. A.: Methods in Fixed Point Theory

Alevras, D.; Padberg M. W.: Linear Optimization and Extensions

Andersson, M.: Topics in Complex Analysis

Aoki, M.: State Space Modeling of Time Series

Arnold, V. I.: Lectures on Partial Differential Equations

Arnold, V. I.; Cooke, R.: Ordinary Differential Equations

Audin, M.: Geometry

Aupetit, B.: A Primer on Spectral Theory

Bachem, A.; Kern, W.: Linear Programming Duality

Bachmann, G.; Narici, L.; Beckenstein, E.: Fourier and Wavelet Analysis

Badescu, L.: Algebraic Surfaces

Balakrishnan, R.; Ranganathan, K.: A Textbook of Graph Theory

Balser, W.: Formal Power Series and Linear Systems of Meromorphic Ordinary Differential Equations

Bapat, R.B.: Linear Algebra and Linear Models

Benedetti, R.; Petronio, C.: Lectures on Hyperbolic Geometry

Benth, F. E.: Option Theory with Stochastic Analysis

Berberian, S. K.: Fundamentals of Real Analysis

Berger, M.: Geometry I, and II

Bliedtner, J.; Hansen, W.: Potential Theory

Blowey, J. F.; Coleman, J. P.; Craig, A. W. (Eds.): Theory and Numerics of Differential Equations

Blyth, T. S.: Lattices and Ordered Algebraic Structures

Börger, E.; Grädel, E.; Gurevich, Y.: The Classical Decision Problem

Böttcher, A; Silbermann, B.: Introduction to Large Truncated Toeplitz Matrices

Boltyanski, V.; Martini, H.; Soltan, P. S.: Excursions into Combinatorial Geometry

Boltyanskii, V. G.; Efremovich, V. A.: Intuitive Combinatorial Topology

Bonnans, J. F.; Gilbert, J. C.; Lemarchal, C.; Sagastizbal, C. A.: Numerical Optimization

Booss, B.; Bleecker, D. D.: Topology and Analysis

Borkar, V. S.: Probability Theory

Brunt B. van: The Calculus of Variations

Carleson, L.; Gamelin, T. W.: Complex Dynamics

Cecil, T. E.: Lie Sphere Geometry: With Applications of Submanifolds

Chae, S. B.: Lebesgue Integration

Chandrasekharan, K.: Classical Fourier Transform

Charlap, L. S.: Bieberbach Groups and Flat Manifolds

Chern, S.: Complex Manifolds without Potential Theory

Chorin, A. J.; Marsden, J. E.: Mathematical Introduction to Fluid Mechanics

Cohn, H.: A Classical Invitation to Algebraic Numbers and Class Fields

Curtis, M. L.: Abstract Linear Algebra

Curtis, M. L.: Matrix Groups

Cyganowski, S.; Kloeden, P.; Ombach, J.: From Elementary Probability to Stochastic Differential Equations with MAPLE

Da Prato, G.: An Introduction to Infinite Dimensional Analysis

Dalen, D. van: Logic and Structure

Das, A.: The Special Theory of Relativity: A Mathematical Exposition

Debarre, O.: Higher-Dimensional Algebraic Geometry

Deitmar, A.: A First Course in Harmonic Analysis

Demazure, M.: Bifurcations and Catastrophes

Devlin, K. J.: Fundamentals of Contemporary Set Theory

DiBenedetto, E.: Degenerate Parabolic Equations

Diener, F.; Diener, M.(Eds.): Nonstandard Analysis in Practice

Dimca, A.: Sheaves in Topology

Dimca, A.: Singularities and Topology of Hypersurfaces

DoCarmo, M. P.: Differential Forms and Applications

Duistermaat, J. J.; Kolk, J. A. C.: Lie Groups

Dumortier.: Qualitative Theory of Planar Differential Systems

Edwards, R. E.: A Formal Background to Higher Mathematics Ia, and Ib

Edwards, R. E.: A Formal Background to Higher Mathematics IIa, and IIb

Emery, M.: Stochastic Calculus in Manifolds

Endler, O.: Valuation Theory

Engel, K.-J.; Nagel, R.: A Short Course on Operator Semigroups

Erez, B.: Galois Modules in Arithmetic

Everest, G.; Ward, T.: Heights of Polynomials and Entropy in Algebraic Dynamics

Farenick, D. R.: Algebras of Linear Transformations

Foulds, L. R.: Graph Theory Applications

Franke, J.; Hrdle, W.; Hafner, C. M.: Statistics of Financial Markets: An Introduction

Frauenthal, J. C.: Mathematical Modeling in Epidemiology

Friedman, R.: Algebraic Surfaces and Holomorphic Vector Bundles

Fuks, D. B.; Rokhlin, V. A.: Beginner's Course in Topology

Fuhrmann, P. A.: A Polynomial Approach to Linear Algebra

Gallot, S.; Hulin, D.; Lafontaine, J.: Riemannian Geometry

Gardiner, C. F.: A First Course in Group Theory

Gårding, L.; Tambour, T.: Algebra for Computer Science

Godbillon, C.: Dynamical Systems on Surfaces

Godement, R.: Analysis I, and II

Goldblatt, R.: Orthogonality and Spacetime Geometry

Gouvêa, F. Q.: p-Adic Numbers

Gross, M. et al.: Calabi-Yau Manifolds and Related Geometries

Gustafson, K. E.; Rao, D. K. M.: Numerical Range. The Field of Values of Linear Operators and Matrices

Gustafson, S. J.; Sigal, I. M.: Mathematical Concepts of Quantum Mechanics

Hahn, A. J.: Quadratic Algebras, Clifford Algebras, and Arithmetic Witt Groups

Hájek, P.; Havránek, T.: Mechanizing Hypothesis Formation

Heinonen, J.: Lectures on Analysis on Metric Spaces

Hlawka, E.; Schoißengeier, J.; Taschner, R.: Geometric and Analytic Number Theory

Holmgren, R. A.: A First Course in Discrete Dynamical Systems

Howe, R., Tan, E. Ch.: Non-Abelian Harmonic Analysis

Howes, N. R.: Modern Analysis and Topology

Hsieh, P.-F.; Sibuya, Y. (Eds.): Basic Theory of Ordinary Differential Equations

Humi, M., Miller, W.: Second Course in Ordinary Differential Equations for Scientists and Engineers

Hurwitz, A.; Kritikos, N.: Lectures on Number Theory

Huybrechts, D.: Complex Geometry: An Introduction

Isaev, A.: Introduction to Mathematical Methods in Bioinformatics

Istas, J.: Mathematical Modeling for the Life Sciences

Iversen, B.: Cohomology of Sheaves

Jacod, J.; Protter, P.: Probability Essentials

Jennings, G. A.: Modern Geometry with Applications

Jones, A.; Morris, S. A.; Pearson, K. R.: Abstract Algebra and Famous Inpossibilities

Jost, J.: Compact Riemann Surfaces

Jost, J.: Dynamical Systems. Examples of Complex Behaviour

Jost, J.: Postmodern Analysis

Jost, J.: Riemannian Geometry and Geometric Analysis

Kac, V.; Cheung, P.: Quantum Calculus

Kannan, R.; Krueger, C. K.: Advanced Analysis on the Real Line

Kelly, P.; Matthews, G.: The Non-Euclidean Hyperbolic Plane

Kempf, G.: Complex Abelian Varieties and Theta Functions

Kitchens, B. P.: Symbolic Dynamics

Kloeden, P.; Ombach, J.; Cyganowski, S.: From Elementary Probability to Stochastic Differential Equations with MAPLE

Kloeden, P. E.; Platen; E.; Schurz, H.: Numerical Solution of SDE Through Computer Experiments

Kostrikin, A. I.: Introduction to Algebra

Krasnoselskii, M. A.; Pokrovskii, A. V.: Systems with Hysteresis

Kurzweil, H.; Stellmacher, B.: The Theory of Finite Groups. An Introduction

Lang, S.: Introduction to Differentiable Manifolds

Luecking, D. H., Rubel, L. A.: Complex Analysis. A Functional Analysis Approach

Ma, Zhi-Ming; Roeckner, M.: Introduction to the Theory of (non-symmetric) Dirichlet Forms

Mac Lane, S.; Moerdijk, I.: Sheaves in Geometry and Logic

Marcus, D. A.: Number Fields

Martinez, A.: An Introduction to Semiclassical and Microlocal Analysis

Matoušek, J.: Using the Borsuk-Ulam Theorem

Matsuki, K.: Introduction to the Mori Program

Mazzola, G.; Milmeister G.; Weissman J.: Comprehensive Mathematics for Computer Scientists 1

Mazzola, G.; Milmeister G.; Weissman J.: Comprehensive Mathematics for Computer Scientists 2

Mc Carthy, P. J.: Introduction to Arithmetical Functions

McCrimmon, K.: A Taste of Jordan Algebras

Meyer, R. M.: Essential Mathematics for Applied Field

Meyer-Nieberg, P.: Banach Lattices

Mikosch, T.: Non-Life Insurance Mathematics

Mines, R.; Richman, F.; Ruitenburg, W.: A Course in Constructive Algebra

Moise, E. E.: Introductory Problem Courses in Analysis and Topology

Montesinos-Amilibia, J. M.: Classical Tessellations and Three Manifolds

Morris, P.: Introduction to Game Theory

Nikulin, V. V.; Shafarevich, I. R.: Geometries and Groups

Oden, J. J.; Reddy, J. N.: Variational Methods in Theoretical Mechanics

Øksendal, B.: Stochastic Differential Equations

Øksendal, B.; Sulem, A.: Applied Stochastic Control of Jump Diffusions

Poizat, B.: A Course in Model Theory

Polster, B.: A Geometrical Picture Book

Porter, J. R.; Woods, R. G.: Extensions and Absolutes of Hausdorff Spaces

Radjavi, H.; Rosenthal, P.: Simultaneous Triangularization

Ramsay, A.; Richtmeyer, R. D.: Introduction to Hyperbolic Geometry

Rautenberg, W.: A concise Introduction to Mathematical Logic

Rees, E. G.: Notes on Geometry

Reisel, R. B.: Elementary Theory of Metric Spaces

Rey, W. J. J.: Introduction to Robust and Quasi-Robust Statistical Methods

Ribenboim, P.: Classical Theory of Algebraic Numbers

Rickart, C. E.: Natural Function Algebras

Rotman, J. J.: Galois Theory

Rubel, L. A.: Entire and Meromorphic Functions

Ruiz-Tolosa, J. R.; Castillo E.: From Vectors to Tensors

Runde, V.: A Taste of Topology

Rybakowski, K. P.: The Homotopy Index and Partial Differential Equations

Sagan, H.: Space-Filling Curves

Samelson, H.: Notes on Lie Algebras

Sauvigny, F.: Partial Differential Equations I

Sauvigny, F.: Partial Differential Equations II

Schiff, J. L.: Normal Families

Sengupta, J. K.: Optimal Decisions under Uncertainty

Séroul, R.: Programming for Mathematicians

Seydel, R.: Tools for Computational Finance

Shafarevich, I. R.: Discourses on Algebra

Shapiro, J. H.: Composition Operators and Classical Function Theory

Simonnet, M.: Measures and Probabilities

Smith, K. E.; Kahanpää, L.; Kekäläinen, P.; Traves, W.: An Invitation to Algebraic Geometry

Smith, K. T.: Power Series from a Computational Point of View

Smoryński, C.: Logical Number Theory I. An Introduction

Stichtenoth, H.: Algebraic Function Fields and Codes

Stillwell, J.: Geometry of Surfaces

Stroock, D. W.: An Introduction to the Theory of Large Deviations

Sunder, V. S.: An Invitation to von Neumann Algebras

Tamme, G.: Introduction to Étale Cohomology

Tondeur, P.: Foliations on Riemannian Manifolds

Toth, G.: Finite Mbius Groups, Minimal Immersions of Spheres, and Moduli

Verhulst, F.: Nonlinear Differential Equations and Dynamical Systems

Wong, M. W.: Weyl Transforms

Xambó-Descamps, S.: Block Error-Correcting Codes

Zaanen, A. C.: Continuity, Integration and Fourier Theory

Zhang, F.: Matrix Theory

Zong, C.: Sphere Packings

Zong, C.: Strange Phenomena in Convex and Discrete Geometry

Zorich, V. A.: Mathematical Analysis I

Zorich, V. A.: Mathematical Analysis II